Micro Mechanical Systems

Principles and Technology

HANDBOOK OF SENSORS AND ACTUATORS

Series Editor: S. Middelhoek, Delft University of Technology, The Netherlands

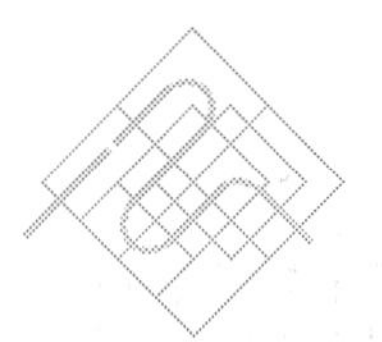

Volume 1	Thick Film Sensors (edited by M. Prudenziati)
Volume 2	Solid State Magnetic Sensors (by C.S. Roumenin)
Volume 3	Intelligent Sensors (edited by H. Yamasaki)
Volume 4	Semiconductor Sensors in Physico-Chemical Studies (edited by L. Yu. Kupriyanov)
Volume 5	Mercury Cadmium Telluride Imagers (A. Onshage)
Volume 6	Micro Mechanical Systems (edited by T. Fukuda and W. Menz)

HANDBOOK OF SENSORS AND ACTUATORS 6

Micro Mechanical Systems

Principles and Technology

Edited by
T. Fukuda
Department of Microsystem Engineering,
Mechano-Informatics and Systems
Nagoya University
Nagoya, Japan

and

W. Menz
Karlsruhe Research Centre
Institute for Microstructure Technology
Karlsruhe, Germany

1998
ELSEVIER
Amsterdam - Lausanne - New York - Oxford - Shannon - Singapore - Tokyo

ELSEVIER SCIENCE B.V.
Sara Burgerhartstraat 25
P.O. Box 211, 1000 AE Amsterdam,
The Netherlands

Library of Congress Cataloging-in-Publication Data

Micro mechanical systems : principles and technology / edited by T.
Fukuda and W. Menz.
p. cm. -- (Handbook of sensors and actuators ; v. 6)
Includes bibliographical references.
ISBN 0-444-82363-8
1. Microelectromechanical systems--Handbooks, manuals, etc.
I. Fukuda, T. (Toshio), 1948- . II. Menz, W. (Wolfgang)
III. Series.
TJ163.M48 1998
621--dc21 98-20529
CIP

ISBN: 0 444 82363 8

♾ The paper used in this publication meets the requirements of ANSI/NISO Z39.48-1992 (Permanence of Paper).

Printed in The Netherlands.

Introduction to the Series

The arrival of integrated circuits with very good performance/price ratios and relatively low-cost microprocessors and memories has had a profound influence on many areas of technical endeavour. Also in the measurement and control field, modern electronic circuits were introduced on a large scale leading to very sophisticated systems and novel solutions. However, in these measurement and control systems, quite often sensors and actuators were applied that were conceived many decades ago. Consequently, it became necessary to improve these devices in such a way that their performance/price ratios would approach that of modern electronic circuits.

This demand for new devices initiated worldwide research and development programs in the field of "sensors and actuators". Many generic sensor technologies were examined, from which the thin- and thick-film, glass fiber, metal oxides, polymers, quartz and silicon technologies are the most prominent.

A growing number of publications on this topic started to appear in a wide variety of scientific journals until, in 1981, the scientific journal Sensors and Actuators was initiated. Since then, it has become the main journal in this field.

When the development of a scientific field expands, the need for handbooks arises, wherein the information that appeared earlier in journals and conference proceedings is systematically and selectively presented. The sensor and actuator field is now in this position. For this reason, Elsevier Science took the initiative to develop a series of handbooks with the name "Handbook of Sensors and Actuators" which will contain the most meaningful background material that is important for the sensor and actuator field. Titles like Fundamentals of Transducers, Thick Film Sensors, Magnetic Sensors, Micromachining, Piezoelectric Crystal Sensors, Robot Sensors and Intelligent Sensors will be part of this series.

The series will contain handbooks compiled by only one author, and handbooks written by many authors, where one or more editors bear the responsibility for bringing together topics and authors. Great care was given to the selection of these authors and editors. They are all well known scientists in the field of sensors and actuators and all have impressive international reputations.

Elsevier Science and I, as Editor of the series, hope that these handbooks will receive a positive response from the sensor and actuator community and we expect that the series will be of great use to the many scientists and engineers working in this exciting field.

Simon Middelhoek

Contents

1. Introduction

W. Menz

Research Center Karlsruhe
Institute for Microstructure Technology
and
University of Freiburg
Faculty of Applied Science
P.O. Box 3640, 76021 Karlsruhe
Germany

Man thinks he is God knows who, when he calls a mite an elephant and sun a spark.

Georg Christoph Lichtenberg, 1742-1799.

1.1 Historical Background and Parallels to Microelectronics

Microsystems technology brings man away from the accustomed dimensions he can "grasp" and makes him enter a field which no longer corresponds to his natural sensory perceptions. He must learn to cope with these new possibilities, actually input his experiences, but not impose them inconsiderately on the novel technology. This development already started with microelectronics, but electronics itself is already an "abstract" thing to the normal person, and the conflict with personal experience emerged only in the process of dealing with mechanical microstructures.

When 50 years ago the transistor was invented, this marked the onset of a technological change which has had a permanent influence on the lives of all of us and will continue to do so. Microelectronics and the silicon material closely linked to it shapes our civilization in a significant way. Thus, it is justified to call the present era the "Silicon Age" as earlier epochs are called the "Bronze Age" or the "Iron Age" which had been paralleled by great cultural changes.

What makes microelectronics stand out from the many other technological developments so that one can speak rather of a technological leap than of an evolutionary development?

Conventional technologies such as extractive metallurgy or building construction were developed in the course of the centuries on the basis of practical experience and handed over to the next generation. Only later, when the possibilities were offered in the engineering sector, empirical experiences were supported by theory and science. The situation is quite different in microelectronics where, first, the theoretical fundamentals were elaborated before the very first step could be made towards implementation in engineering of a circuit element. The findings of quantum physics had not been the supplement but the prerequisite of microelectronics; empirical experience in the sense used before does not play a part in that technology. Also further development and optimization of microelectronics rely on quantum-theoretical knowledge. This fact is expressed most impressively by the term "band-gap engineering" which demonstrates how theoretical physics and industrial products get in direct touch here.

However, the direct relationship to theoretical-physical knowledge is not only apparent in the design of a microelectronic circuit but also in manufacturing technology. A decisive element in microelectronics technology is photolithography by which the patterns designed on the computer and optimized are transferred to the work piece (in this case the silicon monocrystal wafer). Photolithography is, on the one hand, the most expensive and costly investment in a semiconductor manufacturing plant; but, on the other hand, it is the process step accounting for the greatest successes in microelectronics.

It is noteworthy that by optical imaging only two-dimensional patterns can be transferred. At first sight, this seems to be a major drawback of the technology because we are accustomed to think, design and manufacture in three-dimensional categories. However, as optical imaging provides us with the means of, first, transferring patterns whose details are limited mainly by the wavelength of light only and, second, on account of freedom of wear of optical imaging, of working with extremely high accuracy of reproduction and, third, due to parallelity of optical imaging, of attaining very high flows of information, this drawback is more than outweighed by the technological advantages.

Extreme miniaturization and a high packing density have been achieved by optical transfer of the patterns onto the work piece so that despite the rise in total expenditure of manufacture a drastic reduction in costs of the single element has been possible. Also the reduction in costs has been accompanied by an improvement of quality. In this context, one should consider that at switching rates on the order of nanoseconds of the electrically active components the lengths of electric connections between these components must not be in excess of a few micrometers.

Another advantage of microelectronics consists in the consistent application of batchwise manufacture. Several hundreds or even thousands of ICs or storage elements are manufactured in parallel on a silicon wafer of 6, 8 or 12 inches diameter, and they, in principle, undergo the ever recurring process steps of

- coating,
- patterning,
- modifying (oxidizing, implanting, diffusing),
- etching.

As the processes attack simultaneously at all structures of a wafer and at all wafers of a batch, the scatter in manufacture is extremely small, the yield correspondingly high which, likewise, contributes to reducing the costs of a single element or an IC, although the costs in absolute terms of a semiconductor manufacturing process have risen by several orders of magnitude in the course of the past decades due to continuously growing requirements.

Analyzing the principles of the development philosophy of microelectronics, one can state that, unlike with conventional solutions, one has to do with hundreds, thousands or even millions of completely identical components which can be manufactured on a substrate at reasonable costs and with high packing densities by the methods described before. Intelligent linkage and superordinate management make of this "army" of "dull" single components a highly complex, high-performance system, the microprocessor.

Summarizing, the "recipe for success" of microelectronic circuit manufacture can be condensed to comprise four central steps:

(1) **Computer-aided design, optimization and simulation of the microelectronic circuit.** In microelectronics the design tools are highly developed so that in new development of a circuit expensive and time consuming process runs can be dispensed with in most cases.

(2) **Transfer of the computed patterns to the work piece by optical imaging.** The patterns are first transferred in series onto a set of masks. Then, the lateral structures are transferred in parallel onto the wafer using these masks.

(3) **Batchwise manufacture.** Processes are applied which simultaneously cope with the surfaces of all wafers of a batch so that the scatter in manufacture can be reduced. The expenditure in terms of technology and metrology of a process step is distributed among thousands of single components.

(4) **Linking many identical components with high packing densities** to become an intelligent system characterized by high performance. This is probably the decisive step in the development of microelectronics. It is a potential of this technology which is inaccessible to conventional technologies and which is the prerequisite and key for a success not experienced before.

1.2 The Motivation for Microsystem Technology

It seems appropriate here to take up once more the question why actually microelectronics, with all due respect of the technological details, means a leap in technology. Let us explain this by the example of a little arithmetical operation. During the past three decades the quality of the microelectric products, e.g. the switching rate of a transistor or the packing density of a storage cell, could be increased by more than three powers of ten. At the same time was it possible to reduce by more than three powers of ten the costs of manufacture of a circuit element or a storage cell. Now, if one defines a "quality factor," expressed as the product of improvement in quality and reduction in costs (per element), one obtains for microelectronics a factor of approximately 10^7. Such a value is not even approximated by any other technology. The quality factor for a technology such as steel production should hardly exceed the value of 100. This fact alone justifies the rating "technology leap" to be given to microelectronics.

Now the question evidently emerged whether the development "philosophy," the processes, the materials developed and optimized at high costs within the framework of implementation of microelectronics on an industrial scale, should not be applied to non-electronic problems as well, in other words that e.g. mechanical, optical structures or fluidic structures such as microelectronic circuits could be manufactured. Consequently, would it not be possible to develop the analog of the microprocessor, i.e. the "microsystem," and would it not be possible to have this system attain the maximum level of performance?

The answer to this question marked the start of development related to microsystems technology. Especially in US and European development laboratories, microsystems engineering is always approached from microelectronics as the point of departure.

In silicon based micromechanics one follows closely the technologies related to microelectronics. Anisotropic etching has been added as essential supplementing step by which three-dimensional structures with precisely defined surfaces and edges can be etched out of the silicon monocrystal. This is not in contradiction with the statement that microsystems engineering basically also features two-dimensional it only because the etched surfaces, i.e. the external shape of the microstructure body, cannot be arbitrarily chosen, but are necessarily determined by the crystal morphology. Therefore, also the variety of shapes of the microbodies produced by that technique will be limited. Neither any desired - angle nor circular structures can be produced.

On the other hand, resemblance to microelectronics obviously provides advantages which are not contestable. The possibility of integrating on a single chip microstructures and microelectronics monolithically is of high value. Evidently, also in that case it must be weighed up to which degree of integration such a procedure is still justifiable economically. The wafer, after all, undergoes some additional process steps and the question of the yield plays a role of growing importance.

As a matter of fact, microsystems technology is fed from other technological sources too. This is true above all in Japan where competence has been acquired in mechanical manufacture in the sub-micrometer range applied in mass in manufacture of high precision consumer goods such as video recorders or video cameras. One departs, so to speak, from precision mechanics and proceeds from the macroscopic range towards microsystems engineeringy. Just bear in mind that the very mechanism driving and amateur video camera is a highly complex mechanical microsystem manufactured in quantities of millions. Considering that development, the different names employed can be understood, namely "micromachine technology," the term used in Japan, whereas in Europe the current term is "microsystems technology" or MST and in the Anglo-Saxon countries "micro electro mechanical systems" or MEMS.

1.3 Microphysics and Design Considerations

So, even if conventional precision mechanics makes important contributions to microsystems technology, it should not be overlooked that the basic concepts of microsystems technology originate actually in microelectronics. Therefore, the consequences will be discussed below which result if mechanical structures are to be manufactured using the tools of microelectronics.

A crucial fact in microelectronics, namely two-dimensionality, will be addressed once more here and discussed in some detail. Even with a circuit penetrating several 10 µm down into the silicon wafer, its lateral extension of several centimeters is still greater by orders of magnitude compared to extension in z-direction. Mechanical structures which we use to manufacture in three-dimensional sizes, have to be converted into two-dimensionality in order to be able to transfer them to the work piece (substrate) using the means of photolithography, i.e. a method of projection with a defined focal plane. It is readily evident from many publications that the conventional machine element is projected too carelessly into the micrometer range by mere linear reduction without taking into account the difficulties arising from it. This will be explained more extensively by the following example. If you take a cube and reduce its size linearly by one order of magnitude, the surface will shrink by two orders of magnitude and the volume will shrink by three orders of magnitude.

Looking now at the surface to volume ratio, we can readily see that this ratio increases linearly with the reduction in scale. Now, if we do not consider a mathematical surface but a physical one, with the surface properties extending into the body down to a given layer thickness or where the influence of surface diminishes exponentially, this linear dependence undergoes variations, as represented schematically in Fig. 1.1. Surface properties such as hardness or corrosion resistance extend into the depth of the solid by at least some layers of atoms so that the surface effects play an overproportional part in microstructure bodies. Stiction is therefore an effect dominant in microsystems technology and difficult to control.

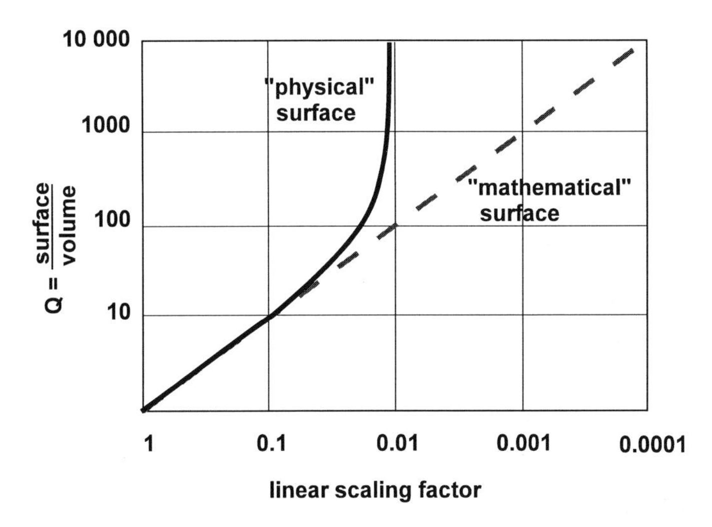

Fig.1.1 Influence of surface properties with increasing miniaturization of microstructures.

Other physical variables can neither be scaled down linearly to microstructures in an easy way. Looking e.g. at the Young's modulus of a material, one cannot expect that this parameter is independent of the linear dimension of a structure as soon as the latter attains the range of the grain sizes of the material. Therefore, material data on microstructures cannot always be extrapolated from tabulated values obtained in marcroscopic experiments. However, this dominating influence of the surface properties can also be seen under a positive aspect. It is known that by the methods of thin film technology material parameters such as the hardness, friction coefficients or corrosion resistance can be achieved which are unknown for solid material. Application of these methods allows structures to be provided with properties which make them appear advantageous compared to conventional structures.

Another property of mechanical microcomponents should not be left out in the listing here, namely the accuracy of fit of components. Taking the example of a conventional axle mounting, one can see that a 10 μm precision of the bearing clearance is not extraordinary with 100 mm diameter of the axle. The relative - allowance of fit in this case is consequently 10^{-4}. Compared with a microengineered design, e.g. the 100 µm diameter rotor of an electrostatic motor, the tolerance of the axial clearance will normally not be amenable to a reduction to less than 1 µm. The relative tolerance of fit in this case is only 10^{-2}, i.e. smaller by two powers of ten than in a conventional mechanical structure. This fact must be taken into account in designing mechanical microsystems. Rotary leadthroughs which, at the same time, are intended to seal the interior of a microsystem against an external contacting medium, cannot be made by that technology. This exerts decisive influences on the use of particular actuators in microsystems moving in a "hostile" macroscopic environment.

Mechanical, optical and fluidic structures manufactured with the tools of microelectronics are generally two-dimensional due to the use of photolithography and the optical transfer of the structures to the work piece. The fact that by multiple exposure and other technological tricks the image plane can be distributed among several structural planes cannot alter the principle of two-dimensionality, not even in those cases where the microstructure is three-dimensional after it has run through the process. The third dimension normally results automatically from the respective process steps and depends on the properties of the technology applied to manufacture the part. The limiting surfaces in silicon micromechanics are the crystallographic (111) surfaces; in the LIGA technique the third dimension is determined by beam guidance in synchrotron radiation. This fact cannot be repeated often enough in the educational process of a reorientation in the design of microsystems compared to conventional systems. It is important and decisive for the success of microsystems technology that in design the means offered by the technology used in manufacture are applied in an optimum manner and that this technology is not "overworked" by special tricks. It would not be reasonable to manufacture in microsystems engineering a micrometer sized four-speed gear unit just by linear reduction because this approach would not adequately take into account the potentials of the technology.

Reorientation in the design of mechanical components is not necessarily a disadvantage; microelectronics actually tells us which potential might be inherent in limitation to a few parameters if, by concentration on some focal points, a high degree of perfection can be achieved by this technology. The methods of microelectronics were not solely used at the time to produce conventional components on a reduced scale, but rather to enter completely new paths allowing the advantages offered by the new technology to be utilized in an optimum way. Also the designer in mechanical engineering must learn to reorientate and to fully exploit the potentials of microsystems technology.

1.4 From the microcomponent to the Microsystem

So far, only microstructure technology has been dealt with, i.e. the fundamental processes, allowing mechanical, fluidic, optical individual structures with very small dimensions to be implemented on a substrate. Although microstructure technology is a necessary prerequisite, it is not a sufficient one for the microsystem. It should be recalled that also in microelectronics the technological and economic breakthrough was not brought about by the invention of the transistor which, obviously, had been a necessary condition for the future development, but rather by the development of the microprocessor. But the microprocessor is the combination of many single elements which can be produced on a substrate at reasonable cost and with a high packing density in order to achieve in a further step, by intelligent linkage of these single elements, an overall performance which is many times higher than the sum of performances of the single elements. This fact must be considered also in the design of microsystems. Also in manufacture of microsystems it should be possible to make single components at reasonable costs and with small dimensions so that many elements can be combined into a system by intelligent linkage. In this way, applications in all sectors of life are conceivable which, by conventional methods, are not feasible or only insufficiently feasible and which go far beyond what seems to be achievable with the tools available in microelectronics.

The opinions regarding the definition of a microsystem differ widely; the relevant discussions are very vivid with differing motivations. But still, it can be stated that, basically, a microsystem consists of several blocks. It is visible from Fig. 1.2 that a complete microsystem comprises a subgroup of sensors, a further subgroup of actuators, a signal processing system, maybe additional mechanical structures such as alignment stops, holding devices, tools and the like, and a zone accommodating points of coupling to the macroscopic environment.

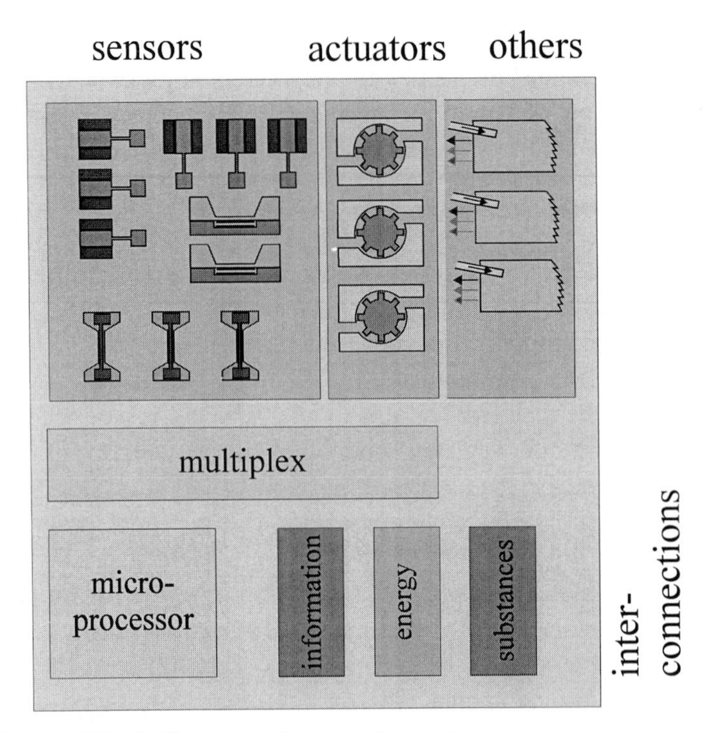

Fig.1.2 Block diagram of a complete microsystem.

A decisive advantage offered by microsystems technology is the possibility of implementing sensor arrays. By microstructure technology many sensor elements can be placed on a substrate at reasonable costs and with a high packing density. These sensors may be identical so that, by averaging, a better statistic overall statement can be made and, by redundancy, a higher reliability of the entire system can be achieved. On the other hand, the sensor elements can be gradated accordingly in their range of sensitivities so that tthe working range can be extended. Likewise, sensor elements with different characteristics or sensor elements with special geometric configurations are capable of simultaneously receiving a wealth of information from the environment.

The advantage of actuator elements lies in the possibility of building up redundant subsystems which can be activated upon failure of a component so that the reliability of the system is greatly improved. With these possibilities in hand, systems are conceivable which permanently monitor themselves and, in case of need, are also capable of self-repair.

Compared to microelectronics, the problems to be tackled in microsystems technology relate to the interface. Unlike in microelectronics, a great number of possible forms of energy and information transmissions have to be coped with.

Whereas in microelectronics the packaging and connection technology is restricted to mainly the provision of electric connections by wire bonding, die bonding and other techniques, a great number of novel techniques have to be developed in microsystems technology in order to be able to handle e.g. fluidic energy. But also acoustic or optical energy could be required for the system. The situation is similar in case of information transmission: Depending on the practical case, the information must be transmitted by electric, acoustic, optical, thermal, fluidic or other means into the system and out of the system. No standard technique is as yet available. When the system is applied in medical engineering, e.g. as a minimally invasive therapeutic system, drugs or biological substances must be handled by the system. This requirement means a great challenge to the packaging and connection technique in microsystems technology. A decisive step from microstructural engineering, i.e. from the single element, to an intelligent, adaptive microsystem having many uses consists in the provision of an efficient packaging and connection technique. It will decide in future whether or not microsystems technology can gain ground on the market over conventional solutions.

2. Photolithographic Microfabrication

Michael L. Reed
Department of Electrical and Computer Engineering

and

Gary K. Fedder
Department of Electrical and Computer Engineering and Robotics Institute

Carnegie Mellon University
Pittsburgh, Pennsylvania 15213
USA

Among the most developed of all microsystem technologies are those related to photolithographic microfabrication on planar substrates. In particular, technologies which are compatible with commercial integrated circuit (IC) processing offer a wealth of design choices and economies of scale. Integration of optical, mechanical, chemical, magnetic, piezoelectric, and other sensitive elements with electronic microcircuitry is a key goal of many microsystem technology programs.

In this chapter, we present an overview of planar microsystem technology. We will concentrate on systems based on single crystal silicon substrates, as these will remain the dominant choice for the foreseeable future. We describe both the various materials used in silicon fabrication and the unit processes used to deposit, pattern, and shape these materials. Appropriate combinations of unit processes result in integrated processes, of which there exist a multitude for fabricating micromechanical devices. We will describe in detail four integrated processes and resulting microstructures that are representative of the wide variety of micromechanical fabrication alternatives available: single-crystal silicon diaphragms fabricated using bulk micromachining, polysilicon thin-film microstructures built in a surface-micromachined process, a pressure-based flow sensor constructed from silicon and glass wafers bonded together, and a thermal-based flow sensor fabricated in a standard CMOS process.

2.1 Basic Concepts of Planar Processing

Planar processing is the most highly developed of all microsystem fabrication technologies. Based on lithographic pattern transfer, planar processing is well suited to high-volume, low-cost manufacturing. In this section we describe four key components of planar processing: the pattern transfer process; silicon micromachining; integration with electronics; and parallelism.

2.1.1 Pattern Transfer

The fundamental idea in all planar processing is lithographic pattern transfer from a master plate containing the pattern, known as a **mask**. By repeated application of photolithographic transfer steps, using different masks, many kinds of devices can be made using the same sequence of steps.

Figure 1 illustrates the general idea. A substrate (usually silicon, but other materials are also used) is coated with a thin film of another material, such as silicon dioxide (SiO_2) or a metal. The objective is to pattern the film into geometrically defined regions. This is done by removing (*i.e.,* **etching**) those areas of the film which do not contribute to the pattern. The key is to use a photosensitive material, called **photoresist**, which prevents the etching process from attacking the areas of the film we wish to leave intact. (While Figure 1 illustrates the general idea of lithographic pattern transfer, it by no means should be taken as more than representative of a generic process. There are many alternative ways to accomplish each of the steps, which are discussed in more detail in Section 2.3.4.)

The patterns of thin films built up on the substrate have many roles. For example, a patterned thin film of SiO_2 can act as a mask for a bulk silicon etch which will remove large areas of the substrate (bulk micromachining). Patterned areas of polycrystalline silicon may become structural elements in an electrostatic actuators (which are made using surface micromachining). Traces of continuous metal lines patterned lithographically act as electrical conductors which carry power and signals to the system. All of these different elements are constructed using the same basic set of lithographic patterning steps.

The key feature of photolithographic pattern transfer is parallelism. Silicon substrates used in commercial foundries are typically 15 or 20 cm (6 or 8 inches) in diameter, while the "chip" is only the order of 1 mm to 1 cm square. Therefore, each pattern transfer step contributes to the manufacture of not one device, but hundreds of devices in parallel. This is to be contrasted with serial manufacturing techniques, such as an assembly line, where each operation acts only on a single item. Tremendous economies of scale are possible with planar processing because there is little difference in cost to process one device as an entire wafer full of devices.

2.1.2 Bulk and Surface Micromachining

Silicon micromachining is the general term applied to techniques, originally based on planar microelectronic processing methods, which produce mechanical elements at the micron-to-millimeter length scale. Although many of these techniques have been borrowed from the field of microelectronics, there are a considerable number of processes which have no analog in the fabrication of integrated circuits. Silicon micromachining involves repeated application of a number of unit processes: oxidation, thin film deposition (including chemical vapor deposition, evaporation, sputtering, plating) photolithography, isotropic and anisotropic etching.

Microelectromechanical systems (MEMS) incorporate one or more kinds of mechanical elements, such as a membranes, cantilevered and doubly-clamped beams, wheels and axles, and the like. The construction of these functional elements from the basic building blocks of solid silicon and other materials is the purview of micromachining. Over the years, a number of

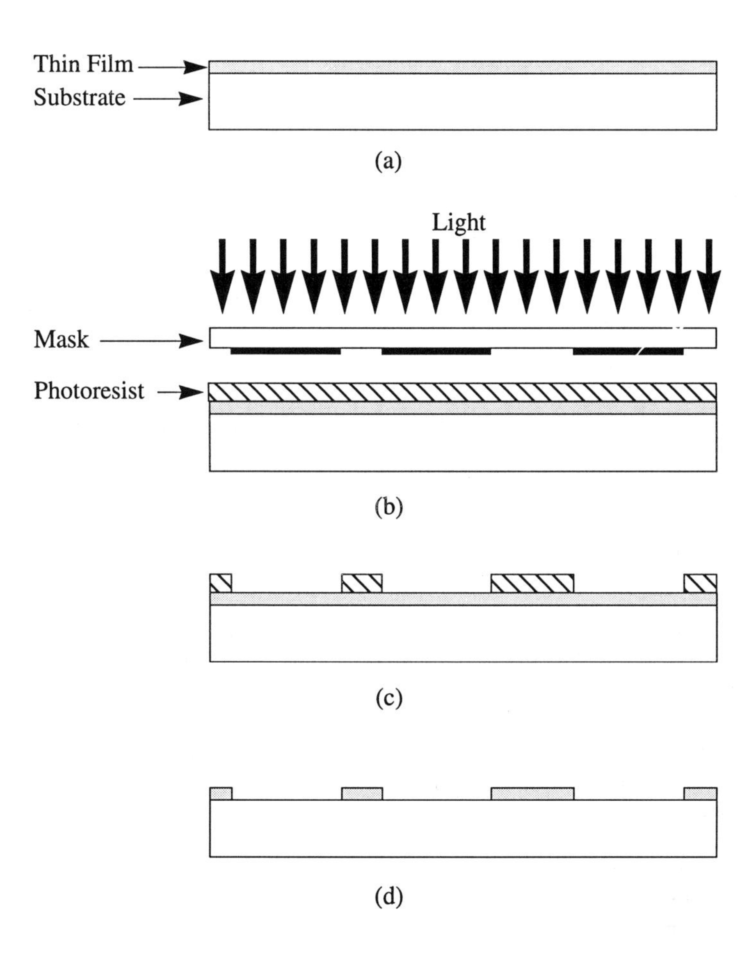

Figure 1. Fundamental steps in pattern transfer. Many variations are possible, but most processes make use of these basic steps in some form. (a) A flat substrate is coated with a thin film of material. The objective is to pattern this material into regions. (b) The film is coated with photoresist and aligned to a mask containing the patterns. Light shining through the mask exposes some areas of the resist, but is blocked by the dark features on the mask. (c) Exposure has made the photoresist more soluble in a developer. During the development process, unwanted areas are washed away. (d) The thin film is etched. The photoresist protects wanted areas on the film, but allows unwanted areas to be removed. After etching, the photoresist is removed, and the wafer is subjected to further processing. (In this and other similar diagrams, the horizontal and vertical scales are not the same.)

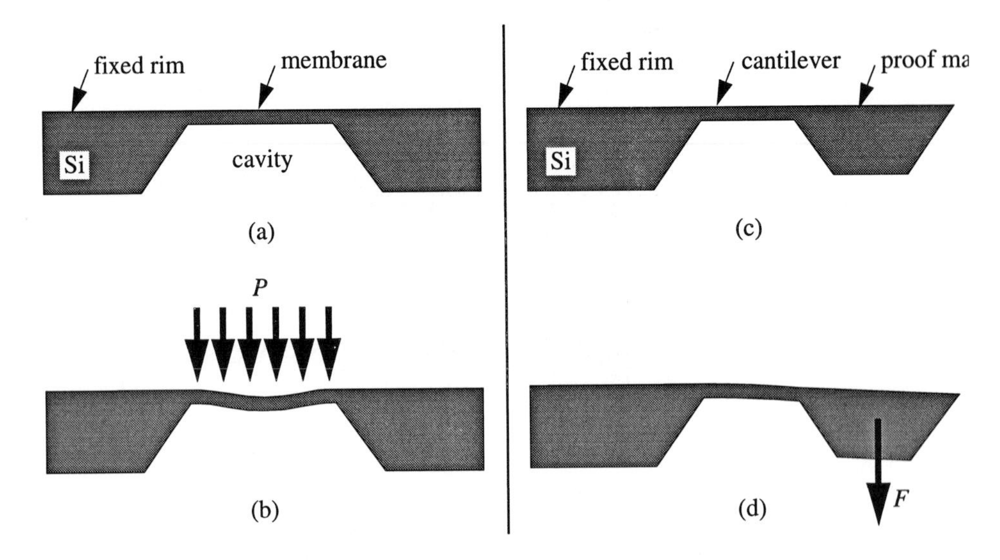

Figure 2. Examples of bulk micromachined devices, shown in cross section. (a) Pressure transducer made from a silicon membrane; (b) Pressure transducer with non-zero deflection due to a pressure differential, *P;* (c) Accelerometer made from a cantilevered silicon proof mass; (d) Accelerometer shown with non-zero deflection due to an external inertial force, *F*.

processes have been developed, some of them quite ingenious. These microsystem fabrication processes are conventionally divided into two classes: bulk and surface.

Bulk micromachining is the term applied to devices in which part of the silicon substrate is dissolved away to produce mechanical elements, such as beams, membranes, grooves, and other structures. Examples of two bulk-micromachined devices, a pressure transducer and an acceleration transducer (accelerometer), are illustrated in Figure 2. Both transducers are formed from the silicon in the substrate (*i.e.,* the "bulk" silicon). In Figure 2(a)-(b), bulk micromachining has produced a cavity with an overlying membrane out of the silicon substrate. A pressure difference between the two sides of the membrane induces a membrane deflection that can be detected. The bulk micromachined device shown in Figure 2(c) is an accelerometer consisting of a silicon proof mass suspended by a cantilever beam. The cantilever bends to resist the external inertial force, *F*, which is shown in the reference frame of the device in Figure 2(d).

Surface micromachining involves the use of sacrificial layers to produce elements largely confined to the vicinity of the silicon surface. In a typical surface micromachining process, alternating layers of SiO_2 and polysilicon are deposited and patterned into various shapes. The final process step is a **release** etch which removes all of the sacrificial SiO_2 but does not attack the polysilicon. (Figure 3 shows a simple process for producing a released cantilever.) The result is mechanical elements which are released from the substrate and are thus free to move.

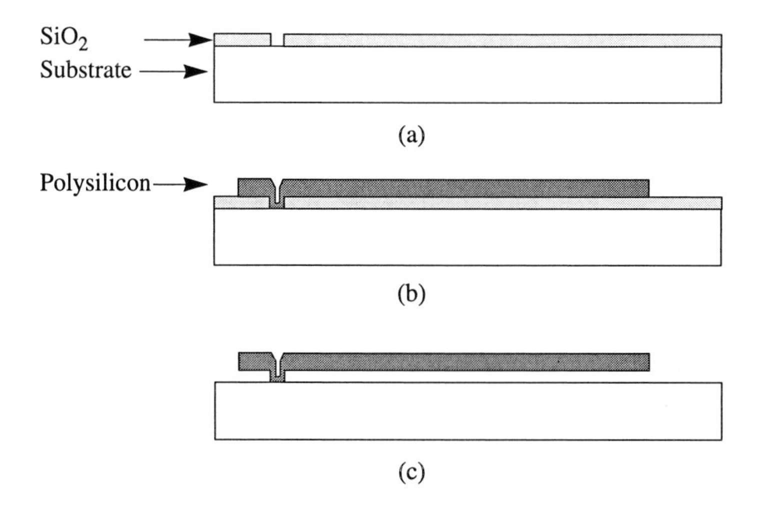

Figure 3. Example of a series of steps to produce a released cantilever using surface micromachining. (a) A thin SiO_2 layer is patterned to produce an anchor region. (b) Polysilicon is deposited and patterned into a cantilever. (c) Sacrificial etch of the SiO_2 results in a freestanding, cantilevered beam of structural polysilicon.

Of the two types of processes, bulk micromachining is the more established. Many commercial products such as accelerometers, pressure sensors, and velocity sensors use bulk micromachining. Surface micromachined devices with integrated electronics are a comparatively recent development. An example of a successful commercial surface micromachined device is the ADXL50 accelerometer produced by Analog Devices, pictured in Figure 4.

The distinction between bulk and surface micromachining exists as a convenient tool for classifying various kinds of processes, but should not be taken as an inflexible rule. For example, bulk processes generally make use of anisotropic silicon etchants (*e.g.,* potassium hydroxide, ethylenediamine/pyrocatechol, or hydrazine) to produce structures bounded by crystal planes in the silicon substrate. Surface processes, on the other hand, generally employ only lithographic patterning to define lateral geometries. There is no fundamental reason why the two kinds of processes cannot be combined. (For example, surface structures can be released by a bulk silicon etch [1].)

2.1.3 Integration with Electronic Circuitry

It is a cliche that devices are not useful, only systems are useful. For example, an acceleration sensing system consists of several parts:

- A seismic mass which resists acceleration due to its own inertia;
- A flexible suspension which supports the seismic mass and allows it respond to changes in external acceleration;

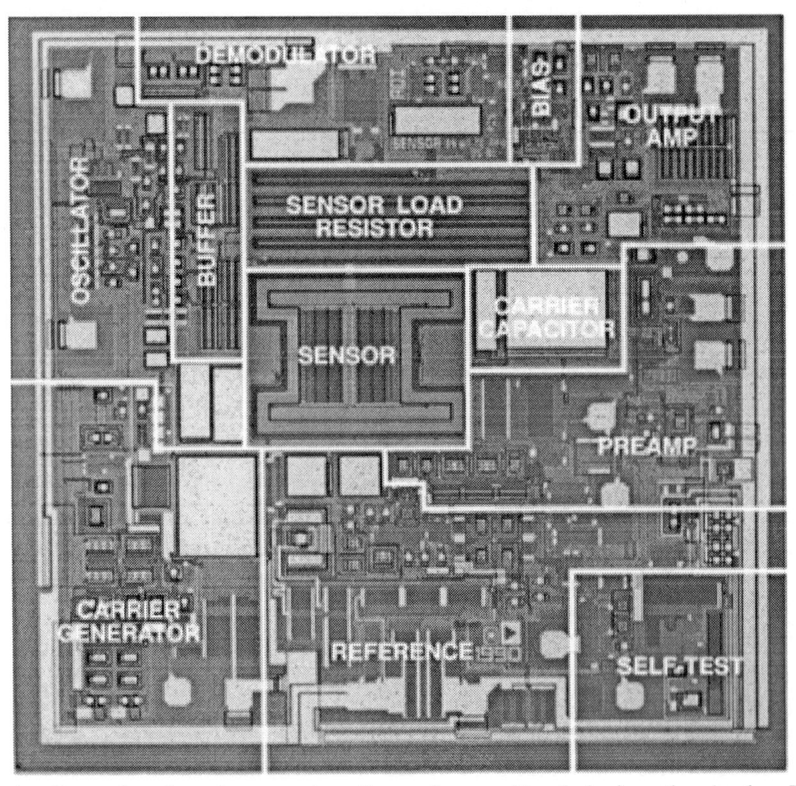

Figure 4. Example of an integrated surface micromachined device, the Analog Devices ADXL50 accelerometer, includes signal conditioning circuitry on chip. Courtesy of Analog Devices, Inc.

- A means to detect the motion of the seismic mass relative to the supporting frame. Typical means include capacitive (capacitance changes with deflection), piezoresistive (resistance changes with stress), and piezoelectric (dipole charge changes with stress) detection.
- An amplifier to strengthen the signal produced by the transducer;
- Electronic circuitry to process the amplified signal into a useful form.

With all of these elements in place, one has a minimum system capable of producing a voltage proportional to acceleration over some range.

An important question in designing this set of elements is, where should the system be partitioned? For example, it is possible to put the first three elements (seismic mass, suspension, and detection element) all on one chip, and the other two (amplifier and signal processing) on another chip. Such an approach allows for a clear delineation between the mechanical elements and a strictly microelectronic subsystem. This, in turn will lower the costs associated with each of the individual chips. However, there will probably be increased costs associated with hybrid packaging of the two chips, and decreased performance due to the physical separation of the sensing element and the first-stage amplifier.

Another system partition is to integrate some, or even most, of the electronic circuitry directly with the mechanical elements. This has many advantages, including (1) improved performance, especially signal to noise ratio; (2) reduced size and weight; (3) possible simplification in packaging; (4) reduced overall system costs. All of these come at a price: an increase in process complexity, since the chip must now contain both electronic and mechanical elements.

Where the system should be partitioned is a complex question with no single answer. The most economic partitioning will of course depend on the specific application, the number of parts to be manufactured, and the complexity of the given task. The point here is that MEMS technology has afforded us the **flexibility** to move the system partition line into another, previously unattainable region. So, instead of using a hybrid package with two chips because no other technology is available, one can now consider integration of electronic and mechanical elements into a monolithic structure.

While integrated microsystems of this nature have an increasingly important role, it is by no means the only useful application of micromachining technology. For example, ink-jet printer heads are a natural application for silicon bulk micromachining, requiring a nozzle, flow channels, and a localized heater to vaporize and eject an ink drop. However, there is little to be gained by putting thousands of transistors near the printer head. In this case, the most logical system partition does not include the signal processing with the mechanical elements.

2.1.4 Parallelism

In a previous section, we used a common, albeit misleading, characterization of microsystems, based on the size of the smallest elements. It is tempting to say that MEMS is synonymous with devices made up of micron-sized elements. But MEMS can be useful for other, different reasons. We can think of MEMS that are:

- useful because they are **small**;
- useful because they are **manufacturable**;
- useful because they can be **integrated** into larger systems.

This categorization, while useful to emphasize the various aspects of what makes MEMS useful, is of course somewhat arbitrary; the categories contain a great deal of overlap.

The first category, MEMS that are useful solely because they are **small**, would include, for example, a device designed to manipulate DNA inside a cell. Such a device must be small to function since the task cannot be performed with scissors and forceps. In this class, a single MEMS consisting of a small number of actuators can, by itself, do something useful.

The second class is microsystems that are useful because many are **manufactured** at once. In this class, the fact that the MEMS are small is important, but secondary to the economies of scale created by multiple and parallel fabrication. An example of a microsystem in this class is an ink-jet printer mechanism. The size of the printer head definitely leads to advantages, but it is the ability to fabricate thousands of individual devices at once that really makes a difference from the manufacturing standpoint. Acceleration and pressure sensors employing silicon planar processing technology are also in this class. Whether an accelerometer occupies 1 mm^3 or 10 mm^3 makes little difference in many applications, but the economies of scale resulting from manufacturing the former rather than the latter can be significant.

A third category are MEMS that are useful because many devices can be **integrated** into larger systems. In this class of distributed microsystems, the usefulness of a small-scale device is multiplied by arraying thousands or more identical elements. An analogy is an integrated circuit memory chip: each individual storage element, capable of storing or recalling one bit of information, is not as useful as a large array of storage elements properly configured. For example, integrated force arrays can be fabricated which multiply the minuscule force generated by a single electrostatic actuator simply by operating many devices in parallel. Other example systems that fall into this category include displays based on micro-mirror pixel elements and surface transport systems based on ciliary actuators.

Clearly, these three classes overlap, but it is convenient to distinguish ordinary micron scale MEMS, consisting of a few independent actuators, from distributed MEMS, which are inherently parallel and must work in a coordinated manner to be useful.

It is safe to assert that many future applications of MEMS will use distributed systems, rather than MEMS consisting of individual or few actuators. One reason for this is the fact that the forces, torques, and mechanical throws (*i.e.,* spatial range) of individual micron-scale actuators are simply too small for many applications [2]. This is certainly true for many near-term applications, where manipulation at the millimeter scale has more utility than micron-scale actuation. One way to create forces and throws at this scale is to design systems where many actuation, sensing and transmission elements are integrated.

It is equally true that the economic advantages conferred by exploiting the manufacturability of microsystem technology are an important driving force, more so than the novelty of using a tiny machine. It is those processes which allow one to build many devices at once, parallel processes, that are the manufacturable microsystem processes.

Therefore, photolithographic microsystem fabrication process, especially those based on planar silicon technology, encompass both of the classes of MEMS useful in the near term. Silicon microfabrication is inherently a parallel fabrication method, making many devices at once, so MEMS constructed in silicon are inherently manufacturable. Also, the methods of photolithographic pattern transfer are well-suited to the replication of individual elements, so distributed microsystems employing arrays of many identical devices or subsystems can also be introduced.

2.2 Materials

Photolithographically fabricated MEMS probably constitute 99% of the commercial MEMS market. Of these, an elemental analysis would reveal they are composed of mostly silicon. In this section, we will discuss this most important material, along with other materials commonly used to fabricate microsystems.

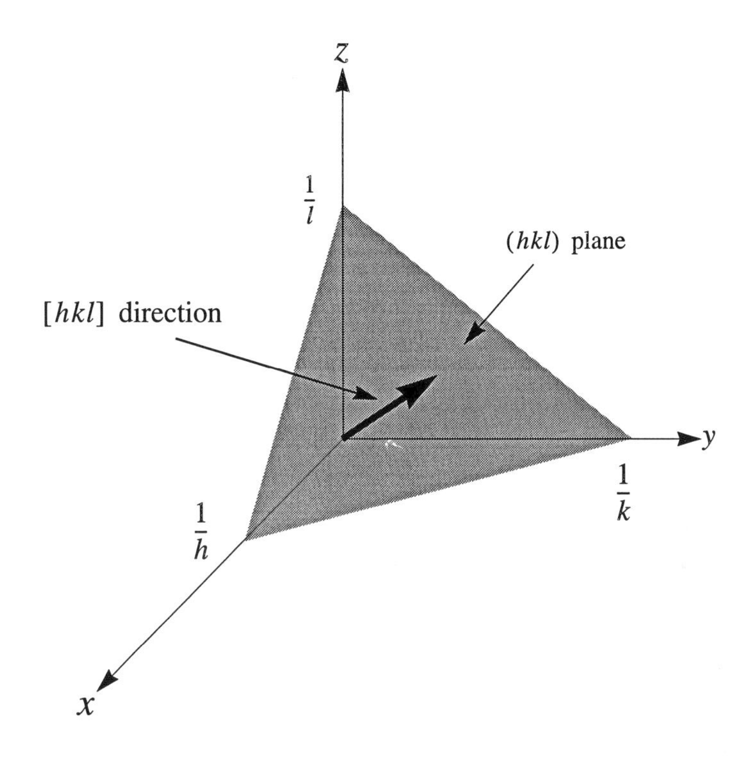

Figure 5. Miller index system. The shaded plane, with Miller indices (*hkl*), intercepts the axes at $1/h$, $1/k$, and $1/l$. The direction normal to the plane (*hkl*), directed away from the origin, is denoted [*hkl*].

2.2.1 Miller Indices

Central to most bulk micromachining processes is the use of anisotropic etching. In this process, different planes of atoms within the silicon bulk are exposed at different rates, resulting in clearly defined angles between the exposed planes. In order to unambiguously specify these planes, we use a system of coordinates known as Miller indices.

Miller indices are based on either a cubic cartesian or a hexagonal coordinate system. Because silicon has a crystal lattice easily described with cartesian coordinates, we use the former.[1] The Miller index of a plane, specified as a sequence (*hkl*) of three numbers enclosed in parentheses, identifies the plane with axis intercepts of

[1]Other materials, most notably aluminum oxide, have hexagonal symmetry and thus use a different Miller index system. A hexagonal Miller index can be easily recognized since it has four numbers, rather than three.

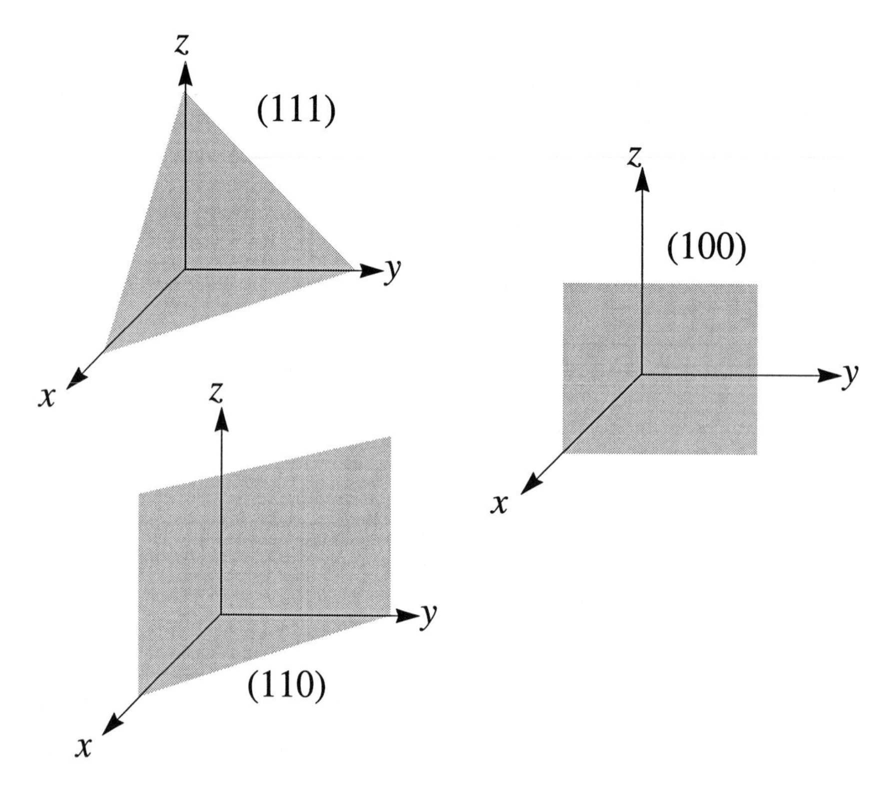

Figure 6. Important crystal planes in silicon include the (100), (111), and (110).

$$x = \frac{1}{h}$$

$$y = \frac{1}{k}$$

$$z = \frac{1}{l}$$

where h, k, and l are integers. This requirement can be met by considering parallel planes as equivalent. Thus, a plane intersecting at 1,1,1 is parallel to a plane intersecting at 2,2,2; both would be specified with the index (111). If a plane is parallel to one of the axes, then it is considered to intersect it at infinity, so the corresponding index is $1/\infty$, or zero. Also, if a plane has a negative intercept, this is denoted in the index with a bar: *i.e.,* a plane with intercepts of -1, ∞, 1/2 would have a Miller index of $(\bar{1}02)$.

A direction in space is specified using square brackets: [*hkl*]. This corresponds to a direction vector normal to the plane (*hkl*), where the tail of the vector is at the origin.

Because the silicon lattice looks the same along each of the three cubic axes, many of the planes are equivalent. For example, the (100), (010), and (001) planes, which are orthogonal to the *x*, *y*, and *z* coordinate axes, are equivalent in a physical sense. We group these planes, together with the $(\bar{1}00)$, $(0\bar{1}0)$, and $(00\bar{1})$ planes, into a **family** of planes specified by curly brackets: {100}. Important planes in micromachining are the {100}, {111}[(2)], and {110}[(3)] families of planes; see Figure 6. Similarly, the crystal directions are grouped into families of directions, specified by angle brackets (*e.g.*, <100>).

Silicon wafers are usually cut so that the planar face is {100}. Many anisotropic etches expose {111} planes, which is a slow-etching face. As a result, microstructures fabricated using these etches will have well-defined faces which meet at predictable angles. The intersection angle θ between planes (hkl) and $(h_1k_1l_1)$ is found from:

$$\theta = \operatorname{acos}\frac{hh_1 + kk_1 + ll_1}{\sqrt{(hh + kk + ll)\,(h_1h_1 + k_1k_1 + l_1l_1)}}$$

For (100) and (111) planes, this angle is $\operatorname{atan}(\sqrt{2}) \cong 54.7°$.

Visualization of the various crystal planes is made easier with a solid model detailing the Miller indices of the principal surfaces. Indeed, anyone contemplating serious bulk micromachining will find a model indispensable[(4)].

Commercial wafers are supplied with part of the circular perimeter ground into a straight line. This area is called the wafer flat, and is oriented normal to a <110> direction. This provides a useful, if somewhat crude, reference for aligning the mask patterns to the silicon crystal planes[(5)]. Often, a second, smaller flat is present. The position of this secondary, or minor, flat indicates the wafer orientation and doping type (see Figure 7).

2.2.2 Silicon

Silicon is the material of choice for many microsystems [3]. Most silicon micromachining leverages the enormous investment in IC processing. Because the electronics industry has created a strong demand, Si substrates with low-defect density, optically flat surfaces, and other desirable properties can be obtained at low cost. Also, many people have studied silicon in great detail, so information about many properties, both important and obscure, are readily available [4].

Silicon is mechanically strong and rigid, and has a high thermal conductivity. It is similar in strength to steel. Single-crystal silicon exhibits no creep or hysteresis and has low structural (internal) damping. It is unparalleled as a semiconductor material, so integration of mechanical elements with electronic devices is natural. It is also useful as a sensing element due to its large

---

[(2)]The {111} family includes eight planes: (111), $(\bar{1}11)$, $(1\bar{1}1)$, $(11\bar{1})$, $(\bar{1}\bar{1}1)$, $(1\bar{1}\bar{1})$, $(\bar{1}1\bar{1})$, and $(\bar{1}\bar{1}\bar{1})$.

[(3)]The {110} family has twelve planes: (110), $(\bar{1}10)$, $(1\bar{1}0)$, $(\bar{1}\bar{1}0)$, (011), $(0\bar{1}1)$, $(01\bar{1})$, $(0\bar{1}\bar{1})$, (101), $(\bar{1}01)$, $(10\bar{1})$, $(\bar{1}0\bar{1})$.

[(4)]A model template can be found at http://www.ece.cmu.edu/afs/ece/usr/mems/www/graphics/silicon-cube.html.

[(5)]The major flat is typically within 3° of the {110} plane.

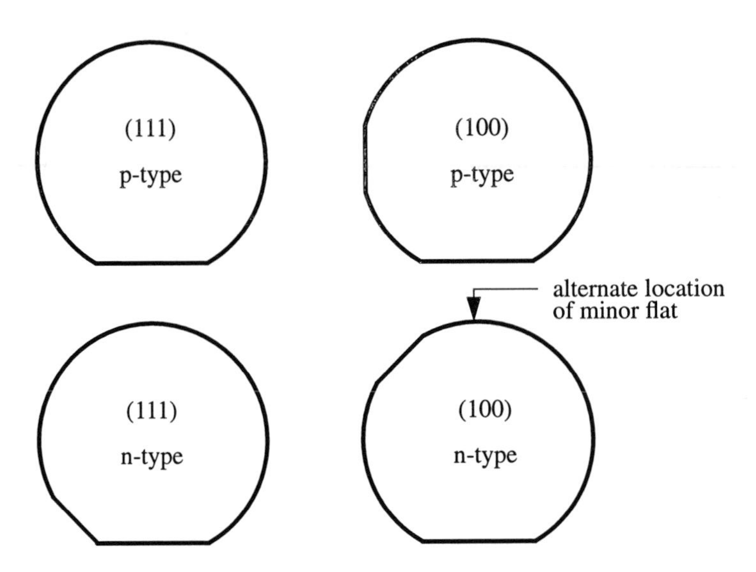

Figure 7. Standard location of major and minor flats on silicon wafers. In each case the major flat (at the bottom in these diagrams) corresponds to a {110} plane. Some manufacturers place the minor flat of the (100) n-type wafers directly opposite of the primary flat.

piezoresistive coefficient. The mechanical and electrical properties of silicon substrates are well controlled and reproducible due to the monocrystalline structure and the high quality and purity of commercially available material. A few important properties are summarized in Table 1.

One of the few undesirable features of silicon is its brittleness. Although it has a high yield stress, silicon does not strain very far before it shatters. Design of mechanical elements using silicon must take this into consideration. The natural cleavage planes in silicon are the {111} family.

When deposited in thin-film form, silicon is generally amorphous or polycrystalline. Amorphous silicon is not widely used in microsystems because it converts to polycrystalline form at fairly low temperatures (above 580°C). Polycrystalline silicon (polysilicon), on the other hand, is widely used both as a mechanical material (in surface micromachining) and as electronic elements such as MOS transistor gates and bipolar transistor emitter contacts.

2.2.3 SiO_2

Silicon dioxide, SiO_2, is an amorphous material used in microsystems as a dielectric in capacitors and transistors; as an insulator to isolate various electronic elements; and as a structural or sacrificial layer in many micromachining processes. Thin films of oxide are easily grown or deposited on silicon wafers using a variety of techniques that are reviewed in

Table 1: Relevant Properties of Silicon (300 K)

Property	Value
Density	2.33 g/cm^3
Atomic Weight	28.09
Atomic Concentration	5.0×10^{22} cm^{-3}
Lattice Constant	0.543 nm
Young's Modulus	190 GPa
Poisson's Ratio	0.28 (for <100>)
Coefficient of Thermal Expansion	4.2×10^{-6} K^{-1}
Specific Heat	0.7 J/g -°C
Thermal Conductivity	1.5 W/cm -°C
Dielectric Constant	11.9
Heat Capacity	20.1 J/mole-K

Section 2.3. High-quality oxide films provide excellent electrical insulation with resistivity values as high as 10^{10} Ω-m. Oxide films are also good thermal insulators, with a low thermal conductivity of around 1.4 W/m-K. Phosphorous-doped and boron-doped oxides, known as phosphosilicate glass (PSG) and borosilicate glass (BSG), respectively, will flow at temperatures above approximately 900°C. Reflow of PSG and BSG films is commonly employed for planarization purposes. Heavily doped PSG films are used as sacrificial material in many surface-micromachining processes because of the rapid etch rate in hydrofluoric acid.

The residual stress in oxide films, as grown or deposited on a substrate, is always compressive (on the order of 1 GPa). Therefore, micromechanical structures made from oxide are constrained in design and size to avoid mechanical buckling upon release from the substrate.

One of the crystalline forms of SiO_2, quartz, is useful as a microsystem substrate. Quartz sensors are generally bulk micromachined devices, and exploit the piezoelectric properties to convert mechanical displacement into electrical signals. Various ultrasonic transducers make use of quartz. Quartz has extremely low structural damping, making it an ideal material for mechanically resonant structures, such as low-noise oscillators.

Borosilicate glass substrates are used in many bulk micromachined processes. Specifically, electrostatic (or anodic) bonding of silicon to glass is almost always accomplished with Corning #7740 glass, because of its close match of thermal coefficient of expansion to that of silicon. Channels, depressions, and holes in glass substrates can be etched chemically or machined conventionally.

2.2.4 Others Materials

Many other materials find application in planar microsystem technology. A few of these include:

- Silicon nitride. A good insulator, nitride is mechanically very strong and has high resistance to many etchants used in microstructure fabrication. Consequently, it finds application as an intermediate masking material in many processes. Silicon nitride is stable in oxidizing atmospheres, making it useful as a mask for diffusion and for silicon oxidation (see the "LOCOS" process, discussed in Section 2.3.1). Like oxide films, thin-film stoichiometric silicon nitride, Si_3N_4, is amorphous and has a high resistivity ($>10^{13}$ W-m) and low thermal conductivity. Stoichiometric silicon nitride films have a large tensile residual stress on the order of 1 GPa, making films over 1000 Å-thick prone to cracking. Low-stress nitride films are made by increasing the silicon content of the material, forming $Si_{1x}N_{1.1x}$. Such films can produce be used to make thin free-standing membranes; it is thus also useful as the support layer in x-ray masks. However, silicon-rich nitride films are mildly conductive. Oxynitride films, $Si_xO_yN_z$, are commonly used as inter-metal dielectrics in standard CMOS processes. The composition of oxynitride films is adjusted to provide near zero stress.
- Metals. Metal films, especially aluminum and aluminum alloys, are used as electrical conductors. More recently, they have been successfully employed as structural elements in surface micromachined electrostatic actuators [1]. Other metals commonly used in standard microelectronic processing include such refractory metals as tungsten and titanium. Metals will deform for large deflections; however, robust metallic microstructures for relatively small deflections have been successfully produced. Nickel films are both highly conductive and have a high magnetic permeability; thus, nickel is useful in applications related to magnetic field sensing. Relatively thick (>1 mm thick) electroplated nickel films have been used to make high-aspect-ratio microstructures. Bimetallic chrome/gold films are generally used as highly conductive interconnect on glass substrates, because of the superior adhesion of chrome to glass. Chrome and gold are not attacked by silicon anisotropic wet etchants.
- Piezoelectric Materials. In addition to quartz, thin films of zinc oxide, ZnO, have been employed as actuation elements in applications such as microphones and hydrophones. Films of $PbZr_xTi_{(1-x)}O_3$, PZT, show promise as actuators due to their high piezoelectric coefficient, but remain difficult to deposit and integrate with electronic devices. Polyvinylidenefluoride is a piezoelectric and pyroelectric polymer widely used in tactile and thermal sensors.
- Polyimides. Polyimide films are useful as support structures, electroplating molds, and as sensor materials. In this latter application, swelling due to humidity is most often exploited. Polyimides are sometimes used as a sacrificial material for release of aluminum microstructures in an oxygen plasma etch.

2.3 Unit Processes

Photolithographic microfabrication processes make use of the order of a dozen or so fundamental unit processes, which are performed sequentially. In this section we describe some of the more common unit processes. Several texts [5] [6] [7] [8] cover microfabrication unit pro-

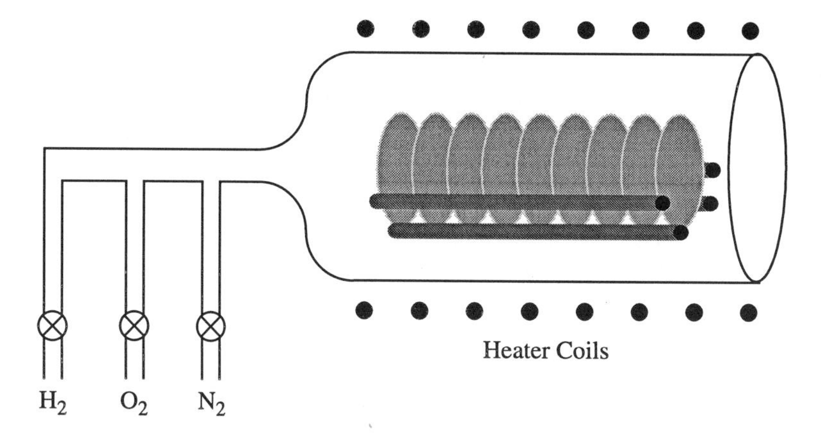

Figure 8. Arrangement of thermal oxidation system. The silicon wafers are stacked vertically inside a quartz boat, which is inserted into a resistively heated quartz furnace tube. A gas manifold supplies nitrogen (for flushing), oxygen, and hydrogen. Wet oxidation can be performed by bubbling oxygen through a source of water; alternatively, hydrogen burning provides a clean source of water.

cesses in much greater depth than we have room to provide in this overview.

2.3.1 Thermal Oxidation

Thermal oxidation is the process by which a film of silicon dioxide, SiO_2, is grown on the surface of a silicon wafer. By *grown* we mean that the silicon in the film originates in the wafer bulk; in contrast, the silicon in *deposited* films originates from outside of the wafer. The distinction is important because the properties of Si-SiO_2 interfaces are highly dependent on how the oxide film is formed. In particular, grown oxides usually have a much lower density of electronic defects which adversely affect the device performance.

Silicon can be oxidized with either dry oxygen, or water. The oxidation rate is much more rapid with water as the reactant; hence it is preferred for films more than about 0.2 μm thick. Dry oxygen results in slightly denser films with better interface properties.

Thermal oxidation is usually performed in a furnace, Figure 8, at temperatures from 900°C to 1100°C.[6] Typical oxidation furnaces are hot wall, resistively heated, open quartz tubes. Wafers are stacked vertically in a slotted quartz wafer boat so that thirty or more wafers can be oxidized at once. Provision is made for supplying oxidants, and nitrogen for flushing the fur-

---

[6]Higher temperatures are possible, and are sometimes employed when very thick oxides (> 2 μm) are desired, but the furnace tube tends to deform over time if it is held at too high a temperature.

nace. The oxidation process is rate controlled by processes in the oxide or at the interface, not transport of reactant inside the furnace. Therefore, precise flow control of the reactant gases is not necessary, and there is little variation in oxide thickness from the edge to the center of a wafer. To reduce contamination by unwanted impurities, the source gases are supplied from condensed liquid tanks. Water for oxidation is conveniently formed by the reaction of pure oxygen and hydrogen, or can be supplied by routing oxygen through a container of liquid water.

The oxidation kinetics of silicon are adequately described by the Deal-Grove model [5]. The oxide thickness is given by

$$t = \frac{x_{ox}}{B/A} + \frac{x_{ox}^2}{B}$$

where t is the oxidation time, x_{ox} is the oxide thickness, and B/A and B are known as the linear and parabolic rate constants, respectively. According to the Deal-Grove model, the reaction is rate limited at the beginning of the oxidation step by the availability of surface reaction sites. This is a straightforward first-order surface reaction with linear kinetics:

$$x_{ox} \approx \frac{B}{A} t$$

As the reaction proceeds, the oxide layer forming on the surface presents a barrier to oxidant. The reacting species must therefore diffuse through the oxide layer to reach the interface where there are available silicon atoms. However, since the oxide layer is monotonically increasing in thickness, the transport time is also increasing. The result is that in this "diffusion-limited" regime the oxide thickness is given approximately by

$$x_{ox} \approx \sqrt{Bt}$$

Both the linear and parabolic rate constants are thermally activated, therefore precise temperature control is necessary in the oxidation system in order to obtain uniform and reproducible oxide thicknesses.

From the standpoint of microsystems technology, the fact that oxidant diffuses through the oxide layer and reacts at the Si/SiO_2 interface is of immense consequence. It means that the interface is formed inside what was a crystalline silicon lattice and is therefore completely free of surface impurities. Furthermore, the intersection of an p-n junction with an interface results in controlled, reproducible electronic properties, unlike the results which occur when a junction intersects an unoxidized surface. Electronic states inside the bandgap, which originate at excess dangling bonds at the interface and lead to undesirable electronic properties such as reduced channel mobility in MOS transistors, are easily neutralized by a low temperature hydrogen anneal at the end of the fabrication process.

An important use of thermal oxidation of silicon is the formation of *field oxide*, which is used to isolate transistors from each other. Islands of field oxide are formed using a mask of Si_3N_4 which oxidizes at a much slower rate than silicon. This process is known as the local oxidation of silicon (LOCOS) and results in the structure shown in Figure 9. A thin oxide is grown

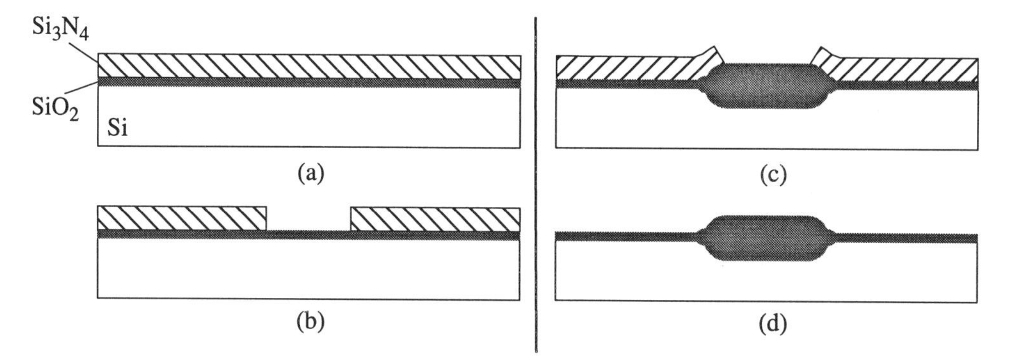

Figure 9. Local oxidation of silicon (LOCOS process). (a) Thin films of SiO_2 and Si_3N_4 are grown and deposited on the silicon surface. (b) The nitride layer is patterned using photolithography. (c) Oxidation of the wafer produces silicon dioxide growth only where the nitride has been removed. (d) After stripping the nitride, a local area of silicon dioxide remains. The transition region between the thin and thick oxide regions is known as the "birds beak" after its approximate shape.

prior to the nitride deposition to compensate for tensile stress in the nitride film.

2.3.2 Chemical Vapor Deposition (CVD)

Chemical Vapor Deposition, CVD, is a method for depositing thin films. CVD is generally used for polysilicon, insulators like SiO_2 and Si_3N_4, and some metals (particularly tungsten). In CVD, the components of the film are transported via reactants in the form of gases. The reaction is driven in conventional CVD by the elevated temperature. Ideally, the reaction forming the film occurs only on the wafer surface; if the reaction occurs in the gas stream ("gas-phase nucleation") then particles will form instead of a robust film.

CVD is most often carried out at a low pressure, typically 0.15 to 2 torr. The equipment is similar to oxidation furnaces, except that the tube is sealed and evacuated. Because many of the gases used in CVD are toxic and/or pyrophoric, special care must be taken in handling the exhaust stream. Generally, CVD equipment is designed so that the deposition process is rate limited by reactant transport. Therefore, precision mass flow controllers are used to meter the flow of reactant gases into the chamber. Temperature need only be controlled to a few degrees, much less than is required with a reaction rate limited process like silicon oxidation.

Polysilicon deposition can be accomplished by the pyrolytic decomposition of silane, SiH_4, into elemental silicon and hydrogen gas. Deposition can be done at temperatures as low as 400°C, although better films are produced at higher temperatures (600 - 700°C). Of particular importance in microsystems incorporating mechanical elements of polysilicon is the residual stress in the film. This has been found to be a strong function of the deposition temperature and later thermal treatments [9][10].

SiO_2 films are formed from the reaction of silane and an oxidant, typically oxygen or N_2O. Suitable oxide films can be grown at 450°C or lower, and can thus be deposited over films of aluminum. SiO_2 films are often heavily doped with phosphorus to form phosphosilicate glass, PSG. Doping is accomplished by introducing PH_3, phosphine gas, into the reactant stream during deposition. PSG has the desirable property of "flowing" at elevated temperatures (above 900°C) and is thus used as a planarization layer.

Films of Si_3N_4 are formed similarly, except that ammonia gas, NH_3, is reacted with the silane. Nitride deposition takes place at higher temperatures, typically 800 - 900°C. Again, stress in the films is of concern and can be controlled by varying the proportions of the reactant gases. Unlike grown nitride films, deposited films may not be stoichiometric and may contain appreciable amounts of hydrogen.

A special case of CVD is *epitaxy*, the growth of single crystal films as an extension of the underlying substrate. Silicon epitaxy is widely used to produce thin films with different doping densities than the starting wafer. Epitaxy uses either pyrolytic decomposition of silane, or reactions involving one of the chlorosilanes ($SiHCl_3$, SiH_2Cl_2, SiH_3Cl), or silicon tetrachloride, $SiCl_4$. The silane reaction takes place at lower temperatures, but must be run at high flow rates to prevent gas-phase nucleation; thus the efficiency is low. Reactions with chlorine containing gases are easier to control but need higher temperatures. Epitaxy is generally restricted to the initial stages of microsystem fabrication while a bare silicon surface is still present.

Instead of a hot-wall furnace holding a boatload of wafers, some equipment processes only a single wafer at a time. The generic name for such processes is *rapid thermal processing* (RTP), which derives from the fact that the wafers heat to processing temperature at extremely high rates, above 200 K/s. RTP equipment generally employs banks of high intensity quartz-halogen lamps surrounding a cold-wall quartz environmental chamber. Heating of the silicon wafer is by direct free carrier absorption of the intense light. RTP can be used for CVD with the introduction of reactive gases; it is also used for thermal oxidation and annealing of deposited films.

An alternative to conventional CVD processing is plasma-enhanced CVD, or PECVD. In this technology, the reaction is driven not with thermal energy but by collisions with energetic electrons in a glow discharge. The plasma is generally set up with an rf generator of a few hundred watts, which is sufficient to create a discharge with a fractional ionization of approximately 10^{-5}. The electrons in the tail of the distribution, with energies exceeding a few electron volts, can both ionize and break atomic bonds when they collide with neutral gas molecules. As a result, PECVD processing can take place at very low ambient temperatures. (Even though the electron temperature of the plasma may exceed 20,000 K, the low ionization fraction coupled with the low background pressure results in only a modest amount of generated heat.)

PECVD is commonly used to deposit silicon nitride films using silane and ammonia. Hydrogen incorporation is even higher than for conventional low-pressure CVD. The films produced by CVD thus have higher etch rates than comparable films deposited by other techniques. Although room temperature deposition is possible, substrates are generally heated to 300°C or more to improve adhesion. A particular advantage of PECVD is that very high growth rates, on the order of microns per minute, can be obtained. Also, while stress in the films can be considerable, it can be controlled by varying the frequency of the rf generator used to sustain the plasma.

2.3.3 Evaporation and Sputtering

Evaporation and sputtering are largely physical deposition processes, in contrast to CVD, which relies on chemical reactions.[7] In both types of processes the material to be deposited starts out as a solid and is transported to the substrate surface where a film is slowly built up. In evaporation, the transport takes place by thermally converting the solid into a vapor. In sputtering, atoms or molecules of the desired material are removed (from the "target") by energetic ions created in a glow discharge.

The configuration of an evaporation system is shown in Figure 10. The system consists of a large bell jar evacuated to a low base pressure, generally less than 10^{-7} torr. (The lower the base pressure, the fewer the impurities incorporated into the growing film.) The substrates are placed near the top of the vacuum chamber, with the side receiving the film face down. The material to be deposited in converted into a vapor, which condenses onto the substrates.

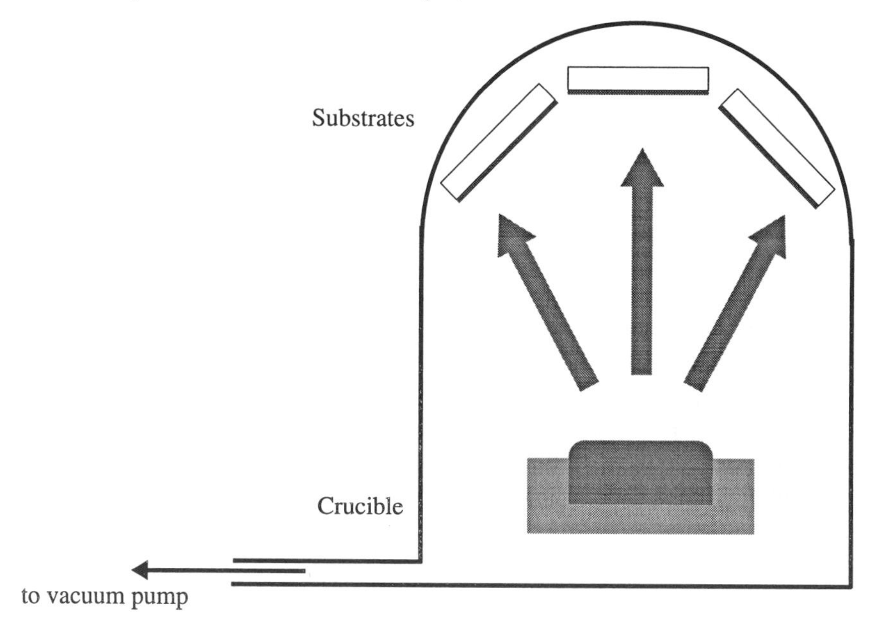

Figure 10. Evaporation system. Inside a vacuum chamber, the material to be deposited is evaporated and transported to the substrates in the vapor phase. The material condenses on the substrates, building up the film. Evaporation systems differ depending on the method used to melt the source material.

Because the base pressure in the system is low, the mean free path of the evaporated atoms is high. This means that collisions between source atoms and background gas molecules are

---

[7]Chemical reactions take place in some processes, such as reactive sputtering.

rare; thus there is little scattering of the vapor as it travels from the crucible to the wafers. The result is that evaporated films are not particularly conformal. The cloud of atoms striking the substrate can be shadowed by features on the surface of the substrate. To prevent this, the substrates are mounted on a planetary system which rotates them about their own axis and rotates the hemispherical planetary assembly about its own axis. Also, since the source emits vapor over a well-defined solid angle, the substrates must be placed at a considerable distance from the source to obtain reasonable wafer-to-wafer uniformity.

Evaporators use one of three basic methods to melt the source material and form a vapor:

- Thermal evaporators pass a current through a refractory metal filament coated with the desired source metal.
- "Flash" evaporators utilize a resistively heated platen and a spool of the source material drawn into a wire. As the spool is unwound, the end of the wire contacts the platen, where it is immediately vaporized.
- Electron-beam ("e-beam") evaporators use an intense beam of electrons to heat a plug of source material inside a crucible.

E-beam systems are the most common, as they are capable of high deposition rates and, since only the source metal itself is heated, provide extremely pure films. Thermal and flash systems have the advantage of not illuminating the sample with soft x-rays which are produced when the electrons from an e-beam system strike the source. This is generally not a problem except where electronic defects arising from radiation damage in SiO_2 layers is of concern.

Evaporation is most suited to the deposition of elemental metals. Compounds are problematic since the individual components have their own individual vapor pressures. Alloys can be deposited using "co-evaporation"; this involves multiple sources, each of which heats a single element.

Sputtering systems are more widely used in microfabrication. In sputtering, energetic ions are created in a glow discharge plasma by electron impact ionization of the neutral gas molecules introduced into the system (generally argon). These ions are attracted to an electrically biased *target* containing the source material (Figure 11). As the ions impinge on the target, they dislodge atoms (elemental targets) or molecules (compound targets) which then redeposit on the substrates. dc power sources are generally used for metal (*i.e.,* conducting) targets, while rf sources are necessary for insulators. A further refinement is the *rf magnetron* source, which includes an annular magnet to increase the path length of the plasma electrons (which increases the ion density, and hence the deposition rate) and also confines the plasma somewhat away from the target (which allows the target to operate at a lower temperature, increasing reliability).

Sputtering is the preferred deposition method for several reasons:

- The substrates remain much cooler than in evaporation for comparable deposition rates. In fact, it is possible to deposit films on room temperature substrates, although improved adhesion is generally found when the substrates are slightly heated (≈200°C).
- Sputtering does not require the mechanically complicated planetary systems needed in evaporators to ensure uniform step coverage. Sputtered films are naturally more confor-

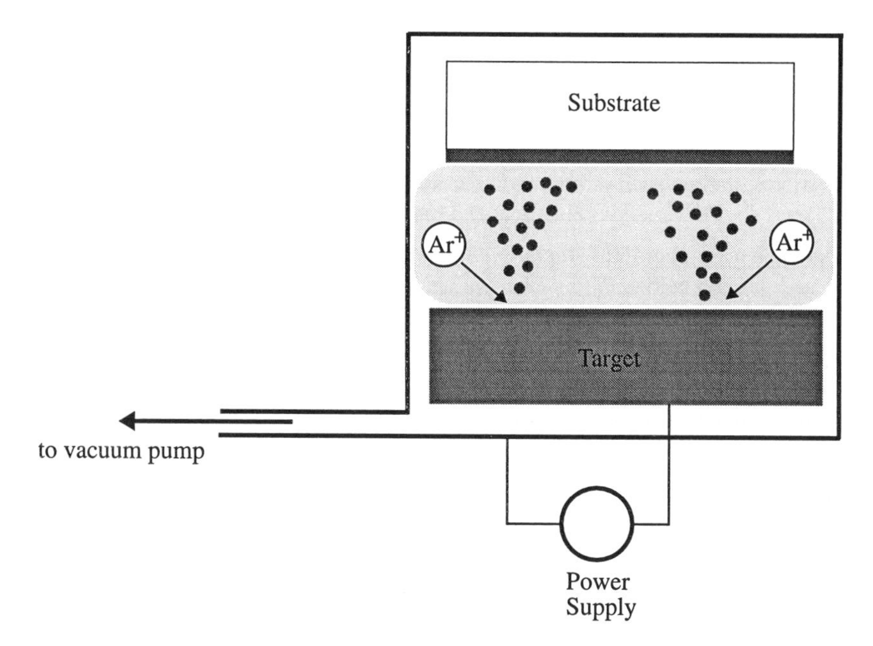

Figure 11. Sputtering principle. A glow-discharge plasma created by a DC or RF power supply produces ionized gas molecules (Ar^+ in this diagram). These ions are attracted to the target, where they knock off ("sputter") atoms of the desired material. These atoms subsequently find their way to the substrate, building up a thin film.

mal than evaporated films since the deposited atoms undergo scattering in the plasma, which causes them to strike the substrate over a larger range of angles.

- The vacuum chamber of a sputtering system can be made much more compact than an evaporator. The smaller chamber leads to a reduction in pumpdown time, increasing the throughput of the system. Also, sputtering lends itself to load locking, which further improves the system throughput.
- The sputtering process is more flexible. Deposition of compound films can easily be accomplished by using an rf source and the appropriate targets.

Instead of using an inert gas like argon, it is possible to strike a plasma using a reactive gas like oxygen or even nitrogen. Doing so results in chemical reactions between the sputtered atoms and the background gas. For example, a pure zinc target sputtered with oxygen results in a deposited film of zinc oxide; or a film of silicon nitride can be formed by sputtering a silicon target with nitrogen. The latter technique is known as *reactive sputtering*.

In both evaporation and sputtering it is common to monitor the growth rate by mounting a piezoelectric quartz crystal of known size and shape so that it also has material deposited on it. Because the resonant frequency of the crystal is proportional to its mass, monitoring this parameter as material is being deposited provides a measure of the growth rate. The film thickness can be determined by integrating the growth rate over the deposition time.

2.3.4 Photolithography

Photolithography is a key step in microfabrication. It is ideally suited to the patterning of planar substrates, and as such is ubiquitous in micromachining. By repeated application of a few basic steps, a wide variety of electrical and mechanical structures are created.

The main steps in photolithography are illustrated in Figure 1. An example of a complete procedure is described below. In this example we will assume a silicon substrate coated with a film of SiO_2:

1. The wafer undergoes a *prebake* (also called a *singe*) for a few minutes at 200°C or higher. The purpose of this step is to drive off loosely bonded water on the oxide surface which would otherwise prevent good adhesion of photoresist.

2. The wafer is treated with an adhesion promoter. A common material for oxide surfaces is hexamethyldisilizane (HMDS). The adhesion promoter is generally applied with a spin technique (described below).

3. Photoresist is applied to the wafer surface by spin coating, or *spinning*. In this operation, the substrate is placed on a vacuum chuck which holds the wafer in place. Next, a quantity of resist is dispensed from a nozzle while the chuck is rotated at a slow speed. Finally, the speed is ramped up to several thousand rpm which causes all but a thin layer of resist (generally about 1 to 2 μm thick) to be thrown off. By carefully controlling the rotation rate and the resist viscosity, uniform layers with reproducible thickness can be obtained.

   Difficulties arise with spinning when the surface to be covered has severe topography. Reasonably robust resist layers can be produced over features with less than 0.5 μm or so step heights. However, many micromechanical devices have features exceeding this constraint. For example, layers of structural polysilicon are generally more than 1.0 μm thick; multiple layers of structural polysilicon and intervening sacrificial oxides can result in surface features many microns high. Spinning resist over such features results in severe thinning over large steps and a nonuniform thickness across the wafer. Variations in resist thickness result in nonuniform exposure and dimensional variations after subsequent processing.

   Recently, Linder *et al.* [11] have used an electroplated resist which provides a conformal coverage over extremely large steps. In this work they were successful in coating steps over 600 μm high - the entire thickness of a silicon substrate.

4. Next, most of the solvents in the photoresist are driven out in the *softbake* step. This can done either in a convection oven (typically 90°C for 30 minutes), in an infrared oven (~3-4 minutes), on a hotplate (~1 minute), or with a microwave source, which takes only a few seconds. The softbake also improves the adhesion of the photoresist to the substrate.

5. The wafers are then placed in an exposure tool and *aligned* so that the patterns projected onto the substrate during the exposure step are registered properly with previously defined patterns. If this is the first mask, the patterns are aligned to the wafer flat. In some bulk-microma-

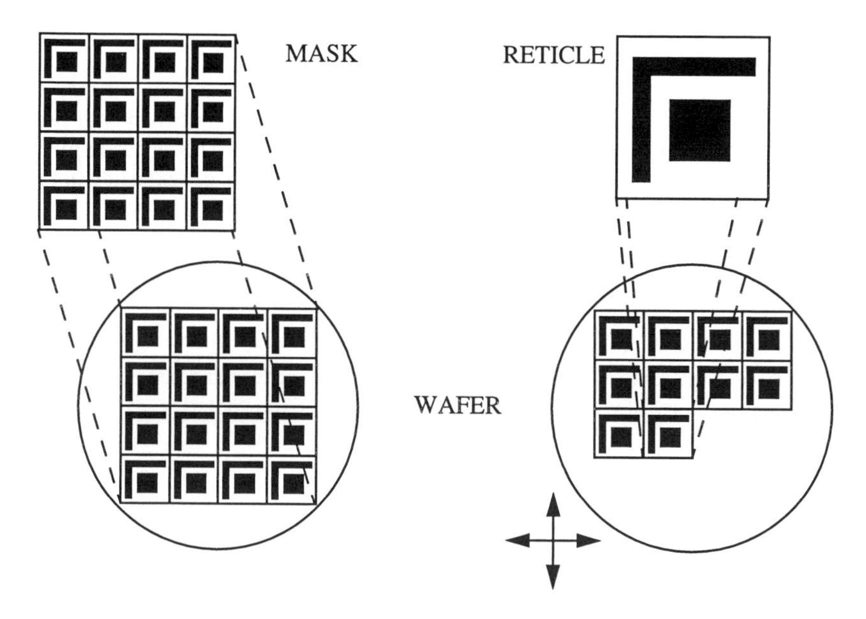

Figure 12. Whole-wafer (left) and step-and-repeat (right) exposure systems. In whole-wafer systems, features on a 1:1 mask are projected onto the wafer. In a step-and-repeat system, a single subfield of the patterns (carried on the reticle) is repeatedly reduced and exposed on the wafer.

chined processes, where precise alignment to the crystal lattice is critical, the wafer flat may not provide an accurate enough reference. In this case, alignment marks in the form of pits can be etched into the wafer using anisotropic etching which will be automatically aligned to the lattice. Subsequent masks are then aligned to these pits.

Exposure tools take one of two forms: a whole-wafer exposure system, and step-and-repeat systems. The idea behind these two systems is illustrated in Figure 12. In a whole-wafer exposure system, the lithographic patterns are contained in a mask at the same dimensional scale as the desired result on the wafer. In some systems the mask is brought into contact with the substrate, but more commonly reflection optics are employed which obviate the need for direct contact; such systems are known as proximity aligners. Whole-wafer systems expose the entire substrate in one pass. Step-and-repeat systems, known as steppers, repeatedly expose a smaller area, generally a single die, by moving the wafer in an x-y pattern with a precision stage. The patterns are contained on a glass plate called a reticle and are reduced during the exposure process. (The amount of reduction depends on the stepper design; common ratios are 10:1, 5:1, and 4:1.) Stepping systems are capable of higher precision for volume manufacturing since dimensional integrity need be maintained across a much smaller area (*i.e.*, the die area rather than the entire substrate area). The price paid for this is the need for an extremely accurate system for controlling the stage movements. Laser interferometry is used to position the stage, while the entire stepper is placed in an environmental chamber to lessen the effects of ambient temperature changes.

In many bulk-micromachined processes there is a need to print patterns on both sides of the substrate. Clearly, these patterns must be aligned to each other. This is accomplished with special backside alignment tools which use either two masks at once (exposing both sides of the wafer), or through-wafer infrared illumination which allows for front-to-back alignment.

6. In both kinds of systems, the resist is *exposed* to light from an intense source as it passes through the clear areas of the mask or reticle. The light and dark patterns on the mask are thus transferred to the resist. The effect of the light is to change the solubility of the resist in a solution called the developer. Resists that increase in solubility upon exposure to light are called positive resists and are nearly universally used.(8) Because dark areas on the mask correspond to areas of resist which remain after development, the "polarity" of the mask pattern is duplicated in the positive resist.
7. The exposed wafer is then *developed* by either spraying or dipping the substrate in an appropriate solvent. Most positive resists are developed in proprietary alkaline solutions (*e.g.*, dilute NaOH or KOH). The exposure process increases the solubility of the resist by a factor of about one thousand, so unexposed areas are largely untouched while the exposed portions are quickly washed away. However, since the solubility of unexposed resist is finite, careful control of the development process is essential in controlling the pattern dimensions.
8. After development, the wafers are rinsed, dried, and inspected under a microscope. (Up to this point, lithography is performed in "yellow rooms" with special illumination to prevent undesired resist exposure. Once the resist is developed, there is no problem with direct white light illumination like that found in a microscope.) The purpose of this "*develop inspect*" step is to check for lithography defects such as incomplete development, resist lifting, under/overexposure, and the like. Another important task during the develop inspect is to measure "critical dimensions": the linewidth of special test patterns which are used to gauge the control of the lithography process (Figure 13). Pattern widths are measured with a calibrated microscope. Control of feature linewidth depends critically on exposure energy (which in turn depends on factors like illumination uniformity and resist thickness) and developer conditions (concentration, time, temperature).

If all is well at this point, then processing proceeds to the next step. If, however, there are problems such as incorrect exposure or development, resist scumming, lifting, or particles, then the resist can be *reworked*. This is accomplished by stripping the resist and returning to the first step. Because the photolithographic processes (up to the etch step) are all low temperature and do not affect the underlying layers, multiple reworks are possible without harm. Photolithography is very sensitive to environmental conditions such as relative humidity, and can even be affected by the level of air pollution. Thus day-to-day variations in exposure and development time, along with a certain percentage of rework, are a normal, expected part of the lithography process.

---

(8)The exposed portions of negative resists tend to swell in the developer, which closes off spaces less than about 2 μm. Thus negative resists are not as suitable for fine line lithography.

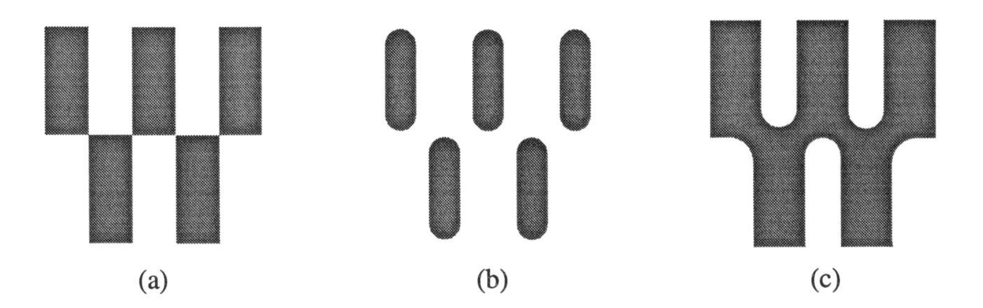

Figure 13. Example of a "tuning fork" critical dimension pattern used to gauge lithographic process control. (a) Correctly exposed pattern. The lines and spaces ("critical dimensions) are of equal width, and the corners of the pattern just meet. (b) Appearance of an overexposed or overdeveloped pattern: spaces are wider than the lines. (c) Appearance of an underexposed or underdeveloped pattern: lines are wider than the spaces.

9. The wafers now undergo a *hardbake*; typical conditions are 150 to 180°C for 1 hour in a convection oven. The purpose of the hardbake is to remove any remaining solvents and rinse water from the photoresist, and to provide even more adhesion of the photoresist patterns to the substrate.The hardbake step is particularly important prior to vacuum processing, such as dry etching or ion implantation, to prevent residual solvent from bursting out of the resist.

10. The wafers are now ready to be *etched.* Ideally, the areas where the photoresist remains on the wafer is impervious to the etchant, therefore the film will be etched only in the spaces between the resist features. Section 2.3.5 will cover etching processes in more detail. The photoresist patterns can also be used as a mask for ion implantation.

11. The wafers are inspected again under a microscope - the *etch inspect* step. It is important to gauge whether or not the etch process has progressed enough. Usually a visual inspection is sufficient; however, critical dimensions may again be measured in some processes. If insufficient etching has taken place, the wafers can generally be returned to the etch bath or plasma system for further processing. Of course, if too much etching has taken place there is no way to "unetch" the film.

12. Once the etching has been deemed complete, the photoresist is *stripped.* Stripping can be performed with a wet process, typically with solvents such as trichloroethane, acetone, and alcohol, but excessively hardened resist is more often stripped with an oxygen plasma. Residual resist is then removed with a mixture of sulfuric acid and hydrogen peroxide.

13. The last lithographic step is a final inspection under a microscope. The primary purpose of the *final inspect* step is to verify that all of the photoresist has been removed. (Because photoresist is an organic material, there must not be any residues remaining which could contaminate equipment used in subsequent steps, especially furnaces.) Often, critical dimensions are again measured as a process control measure. If critical dimensions are out of the specified range, then the wafer lot may have to be scrapped. While this results in a loss of product, it

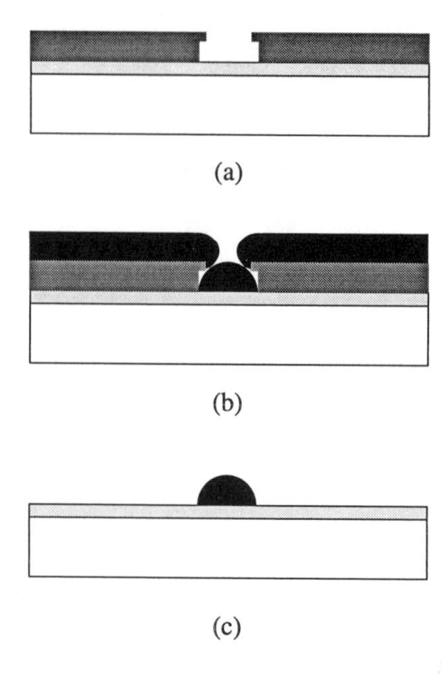

Figure 14. Liftoff processing. The resist is soaked in chlorobenzene before exposure. The result is that a thin layer at the top of the resist becomes less soluble in the developer. (a) After development, the resist profile is reentrant. (b) Evaporation of metal on top of the resist layer results in cracks near the overhanging portion of the space. (c) After removing the resist, a metal line remains in the space. Note that no metal etching is required with this process.

is preferable to continuing the fabrication process on doomed material.

Liftoff Processing - There are many possible permutations of this example photolithographic process. We will describe one: "liftoff processing," which is used for metals like gold which are difficult to etch. In the liftoff process, a reverse polarity mask is used. Clear areas on the mask correspond to places where metal will be left on the wafer. The process proceeds similarly to conventional processing, except that the photoresist is soaked in a chemical (*e.g.*, chlorobenzene) which reduces its solubility in the developer. The key to liftoff is that the chemical is only absorbed into a thin layer on the top of the resist.

After exposure and development, the resist profile looks as depicted in Figure 14(a). Because the top layer of resist is less soluble, the development rate is slower; this results in an a reentrant profile with an overhanging ledge. Next, the metal is deposited, preferably with evaporation, on *top* of the resist. Because the overhanging ledge in the resist partially shadows the opening, the metal is not conformally coated along the hole in the resist, but develops microcracks between the film on top of the resist and the metal deposited on the substrate. (Sputter deposition, being more conformal than evaporation, can be problematic. It is possible to perform liftoff processing using sputtered films, but the process window is more narrow.) At this point the cross-section is as shown in Figure 14(b). The final step is to "lift off" the resist layer, and the excess metal on top of the resist, by immersing the substrate in solvent. The solvent enters the resist layer through the microcracks, dissolving it. After rinsing, the desired structures are left: isolated areas of metal corresponding to clear areas on the mask.

The primary advantage of liftoff processing is that no metal etching need be performed. This is valuable in cases where the metal etch is dangerous or inconvenient,[9] or the etchant is incompatible with the layer underlying the metal.

Mask Generation - In any lithographic process, the information is contained in the mask (or reticle). Mask generation is a specialized process which itself uses a photolithographic technique to expose a resist layer on the mask plate. Exposure tools for mask generation are generally one of three types:

- Direct reduction camera. In this method, a magnified version of the artwork containing the mask patterns is photographed directly onto a reticle through a special camera which reduces the image. The reticle is then stepped onto a mask, using a step-and-repeat camera similar to the steppers used in wafer photolithography.
- Flashing tools. Using a computer, the mask pattern is decomposed into rectangles. "Flashers" use mechanical jaws to define a rectangle through which the mask plate (or reticle) is exposed to light. The plate is positioned using a precision x-y stage. In this way, the mask pattern is built up over time with thousands of individual flashes of light.
- Electron-beam exposure tools. "E-beam" tools are essentially a scanning electron microscopes with a precision x-y beam scanner and a high speed beam blanker. They also have a precision x-y stage because the area which can be exposed to the beam is only a few cm^2. Because the electron beam spot size can be made very small (*i.e.,* less than 10 nm) e-beam exposure systems are capable of producing exceptionally fine lines. It is also possible to expose wafers directly with an e-beam system ("e-beam direct write") but this is not practical for manufacturing as the exposure time for a single layer can be many hours, too long for mass production.

In each case a mask plate, consisting of low-thermal-expansion glass coated with an opaque material such as chrome, and a layer of photoresist, is exposed and developed. The pattern is transferred from the mask resist to the chrome layer using wet chemical etching or plasma etching.

2.3.5 Etching

Most planar processing steps are subtractive, in that a film of material is deposited everywhere, and the parts not desired are removed. Processes used to remove the unwanted areas are called etches, and include isotropic and anisotropic etches, wet chemical and plasma etches, and various physical removal methods. The choice of an etching process is largely a matter of finding a chemical or reactive species that selectively attacks the film one wants to remove, while leaving the other areas of the wafer undamaged. *Selectivity* of an etchant is defined as the ratio of etch rates of the two materials. A good etchant for a specific material must have a high

---

[9]For example, etches for gold films are commonly compounds of cyanide.

selectivity with respect to all other materials present.

Thin-Film Etching - Common wet chemical etches for some example films include the following:

- *Si:* Isotropic wet etches of silicon and polysilicon thin films can be accomplished with mixtures of hydrofluoric and nitric acids ($HF:HNO_3:H_2O$). Acetic acid ($CH_3COOH:H_2O$) often replaces part of the water as it provides better surface wetting and smoother surfaces after etching. The mixture of hydrofluoric, nitric, and acetic acids is called HNA. The mechanism of this nondirectional silicon etch is that the silicon is first converted to an oxide by reaction with the nitric acid. Then, the hydrofluoric acid attacks the oxide. For this reason, this etchant is not particularly selective against films of SiO_2. Silicon isotropic etching is usually performed at room temperature.
- *SiO_2:* Hydrofluoric acid ($HF:H_2O$), generally buffered with ammonium fluoride (NH_4F) to slow the depletion of fluorine atoms. HF etching is performed at room temperature. Buffered HF (BHF) is extremely selective against photoresist and silicon, etching these at minuscule rates. (Unbuffered HF tends to attack resist.) With careful control of etchant concentration, the etch rate is highly reproducible, making timed etching a useful method of endpoint control. Another way to determine endpoint when etching films of SiO_2 on Si is to make use of the fact that SiO_2 is hydrophilic, while Si is hydrophobic. When etching is complete, water will no longer wet the wafer surface but will instead bead up (or "sheet off" when the wafer is held at an angle). Etch rate is dependent on the density of the oxide and the dopant concentration. Thermally grown dry oxide will etch the slowest, while PSG and BSG etch much more rapidly. Densification of PSG and BSG by annealing at high temperatures will reduce the etch rate and, more importantly, produce a more uniform and reproducible etch rate.
- *Si_3N_4:* One of the few chemicals which will attack silicon nitride fast enough to use as an etchant is phosphoric acid, $H_3PO_4:H_2O$, at 170 - 180°C, near or at the boiling point. Concentrated HF can also be used if no oxide films are present. Silicon-rich nitride has a smaller etch rate in HF; this material can be used as a masking film for all but the longest HF etches. Masking the hot phosphoric acid etch requires an oxide mask; resist is attacked by the etchant. Therefore, nitride is more commonly etched with plasma processes, described below.
- *Al:* Although metallization layers of aluminum can be etched in heated mixtures of phosphoric, acetic, and nitric acids ('PAN etch'), the results are often less than satisfactory. Ragged edges can result because of the presence of various grain orientations (which etch at different rates) and grain boundaries. The substrate must be agitated during etching to avoid gas bubble formation on the surface. Also, aluminum layers used in microsystems are usually alloyed with silicon (to prevent the growth of metal spikes into the substrate, which can short out shallow p-n junctions) and copper (to prevent electromigration failure), which complicate the etching process. For these reasons, plasma or reactive ion etches are more commonly performed.

Anisotropic Silicon Etching - Several wet chemical etchants are *directional*, in that they show a

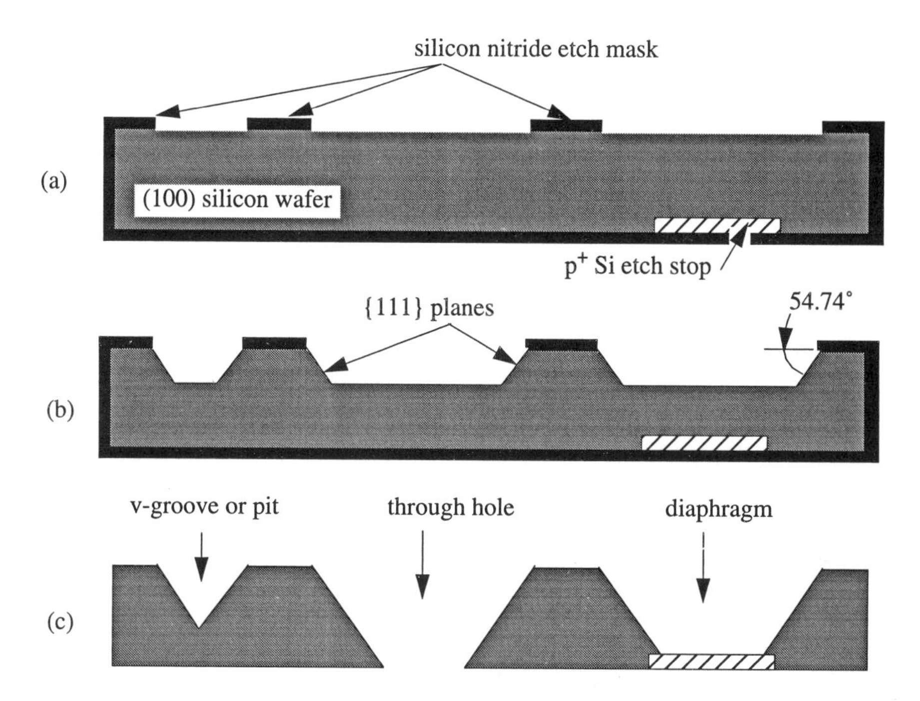

Figure 15. V-groove etching in silicon. (a) A dark-field mask pattern is replicated in silicon nitride on one side of a (100) silicon wafer. A heavily boron-doped silicon etch stop is diffused on the other side. (b) Immersion in an anisotropic etchant, such as KOH, preferentially attacks {100} planes but has a slow etch rate against {111} planes. (c) Continued etching results in the limiting {111} planes meeting to form a v-groove (left side). If the clear area of the mask is sufficiently wide, the cavity will penetrate through the entire wafer to form a through hole (center). Etching stops on the p^+ layer to form a silicon diaphragm (right side).

strong selectivity to different crystal planes in the silicon. Such etchants are *anisotropic*, and are widely used in bulk micromachining to form mechanical elements from the silicon substrate. Generally, the {111} planes are the slowest etching.

Figure 15 illustrates the anisotropic etching to form pits, grooves, through-holes, and diaphragms in a (100) silicon wafer. A silicon nitride layer protects the wafer from the etchant except in areas where it has been patterned. Partial etching exposes the {111} crystal planes. Anisotropic etchants, such as KOH and EDP, preferentially etch the {100} and {110} planes and stop on {111} planes. Complete etching forms a through hole, a diaphragm, and a v-shaped groove or pyramidal pit. Anisotropic etchants are useful to make thin diaphragms of silicon. Alkaline etchants show a strong selectivity for n-type silicon. Conversely, heavy boron doping of a thin layer (p^+ Si) will cause the etch to nearly stop, resulting in a freestanding membrane.

Common anisotropic etches for silicon include:

- Alkali hydroxide solutions, especially KOH, but also NaOH, CsOH, and tetramethyl ammonium hydroxide (TMAH). Aqueous potassium hydroxide, KOH, is quite useful because it is very selective; the ratio of {100} to {111} etch rates is over 200:1 at certain concentrations. The addition of isopropyl alcohol improves the selectivity at the expense of etch rate.
- EDP, a mixture of ethylenediamine ($NH_2(CH_2)_2NH_2$), pyrocatechol ($C_6H_4(OH)_2$), and water. While not as anisotropic as KOH, EDP has the advantage of etching silicon dioxide, and also some metals, very slowly. There is thus considerably more process flexibility when using EDP; however, the handling of the etchant is more complicated since it is toxic and must be refluxed.
- Hydrazine, N_2H_4, mixed with water, is useful when a large {100} etch rate is needed; *i.e.*, when etching through-holes in the wafer. The {100}/{111} etch rate selectivity is small (about 10:1). Hydrazine also demands careful handling; it is more commonly known as a rocket fuel.

A common difficulty encountered with anisotropic etching is hillock defects. Hillocks are small, pyramidal formations, generally bounded by {111} or near-{111} planes, which materialize on the otherwise smooth etched surfaces. The origin and mechanism of formation of hillocks is not completely understood. In KOH, hillocks can be eliminated by etching at a sufficiently high concentration (above 35 weight percent).

Electrochemical silicon etching can also be used to selectively protect regions. The silicon

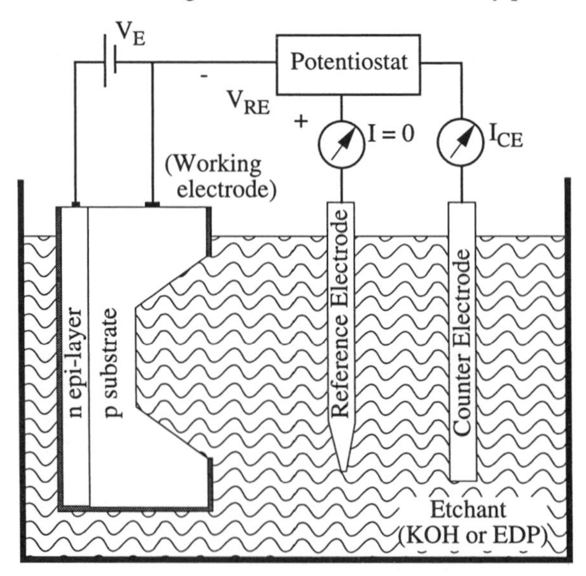

Figure 16. Standard configuration for electrochemical etch-stop.

etch rate is strongly dependent on the potential difference between the substrate and the etchant. Application of a small potential difference (under 1.0 V) between one side of a p-n junction and a metal counterelectrode in the solution results in an increase in etch selectivity against the side of the silicon connected to the anode. A four-electrode configuration, shown in Figure 16, is a common electrochemical etch-stop system. The potentiostat controls the potential of the etchant with respect to the substrate and the epitaxial layer. The potentiostat keeps the voltage, V_{RE}, between the working electrode (substrate) and reference electrode at a fixed value such that the substrate will etch in the solution. The reference electrode has a high-impedance (I=0) input to the potentiostat which provides a reproducible measurement of the bulk solution potential. The error between the reference-electrode voltage and the desired voltage is fed back to drive the counter-electrode current, I_{CE}, and null the error. The n-type epitaxial layer is set at a higher potential, $V_{RE}+V_E$, above the "passivation potential" of n-type silicon. The etch will stop on the epitaxial layer and form a passivating oxide layer. Therefore, the electrochemical etch-stop technique provides reproducible dimensional control of the diaphragm thickness, which is set by the epitaxial layer thickness.

Anisotropic etching, by definition, exploits a differential in etch rate between various crystal planes. Edges where two planes (of the same family, which etch at the same rate) meet expose a range of planes to the etchant. If the corner formed by these planes is convex, the etch will "find" the fastest etching plane and expose it further. However, if the corner is concave, the fastest etching plane will be removed, exposing the *slowest* etching planes. For example, a square open area in an oxide mask will produce a pit in the silicon bounded by four {111} planes, the slowest etching. Overetching of this structure will only slightly change the dimensions of the pit. But the same pattern of opposite polarity will be undercut at the corners of the square mask, resulting in a pyramid bounded by the fastest etching planes. (Which planes are fastest depends critically on the etch composition; typical convex corner undercutting plane families are {212}, {311} and {411}.) Overetching of this structure will drastically change the height and volume of the resultant pyramid.

One drawback of through-wafer bulk micromachining is the loss of usable die area per device. For example, an 800 µm-wide square diaphragm etched out of a standard 500 µm-thick silicon wafer requires an 1.5 mm-wide hole in the back side of the wafer to accommodate the {111} sidewall slope. The area necessary to fabricate the device is almost four times greater than the diaphragm. This limitation may be eliminated with recent developments in deep silicon dry etching that can produce almost vertical sidewalls.

Stiction and Drying - Etches of sacrificial layers, such as the removal of temporary oxide layers in surface micromachining, present special problems. One of these is *stiction*, the unwanted adhesion of released microstructures to the substrate. Stiction can be initiated by bridges of rinse liquid remaining after release etch processing. Because the dimensions encountered in microsystem elements are small, surface tension forces are extremely large. The pressure difference across a meniscus with radius of curvature r is given by the Laplace-Young equation:

$$\Delta P = \frac{\gamma}{r}$$

where γ is the surface tension constant of the system. For example, a water droplet with a radius of curvature equal to half of a beam-substrate gap distance of 2 μm exerts a pressure of 70 kPa, nearly an atmosphere. Such large forces are sufficient to break micromechanical elements. If they survive the rinse process, they can adhere to the substrate by other mechanisms, such as van der Waal bonding or electrostatic charge transfer.

Stiction during device fabrication is minimized by several techniques. Critical point drying at high pressure, or sublimation drying, prevent a liquid/gas meniscus from forming altogether. Another method is to support the structures with rigid materials, like photoresists or polymers, able to withstand the surface tension forces during the wet release etch. After the microstructures are released from the substrate, the support materials are ashed in an oxygen plasma. Another alternative employs temporary breakaway support beams to counteract the surface tension forces.

Dry Etching Processes - Dry etching is the term applied to etching methods involving gaseous plasmas. These include "plasma etching," an isotropic process where the substrate is unbiased, and reactive ion etching (RIE) where physical bombardment by reactive ions provides a degree of anisotropy. (Ion milling, which is the physical removal of material by an intense low-energy beam of ions, usually argon or oxygen, can also be considered a dry process.) Dry etching avoids the problems of stiction and has other advantages, including fewer problems with chemical disposal compared to wet etching. Like plasma CVD, the energetic electrons in the plasma provide the driving force for the process. For example, molecules of a fluorine containing gas such as CF_4 or SF_6 release highly reactive fluorine radicals when impacted by an electron in the plasma; the fluorine reacts with silicon to form a volatile product (SiF_4).

Plasma etching is clean and compatible with many materials used in microsystems, but requires complicated equipment and considerable process development to obtain reproducible results. The etch rate and directionality of plasma-based etching is dependent on several variables including plasma chemistry, power, pressure, wafer temperature, and gas flow rates.

2.4 Integrated Processes

2.4.1 Bulk-micromachining process

A bulk-micromachined piezoresistive pressure sensor, shown in Figure 17, is similar to commercially available silicon pressure sensors. A rim of bulk p-type silicon supports a thin diaphragm made from an n-type silicon epitaxial layer (epi-layer) with built-in silicon piezoresistors. The epi-layer controls the thickness and uniformity of the diaphragm [12]. A differential pressure, P_1-P_2, across the two sides of the diaphragm causes a displacement. Stress is induced in the piezoresistors which, for small deflections, is proportional to the pressure difference. The first sensor with diffused piezoresistors integrated in a bulk-micromachined diaphragm was reported in 1962 by Tufte *et al* [13]. In 1994, this kind of pressure sensor was responsible for about 70% of the commercial micromechanical sensor market [14].

The process flow for the pressure sensor is shown in Figure 18. The process begins with a (100) p-type silicon substrate on top of which is grown a epi-layer of lightly doped n-type silicon, as shown in Figure 18a. The diaphragm is ultimately made from the epitaxial layer and generally has a thickness between 5 μm and 20 μm, depending on the desired pressure sensitivity of the device.

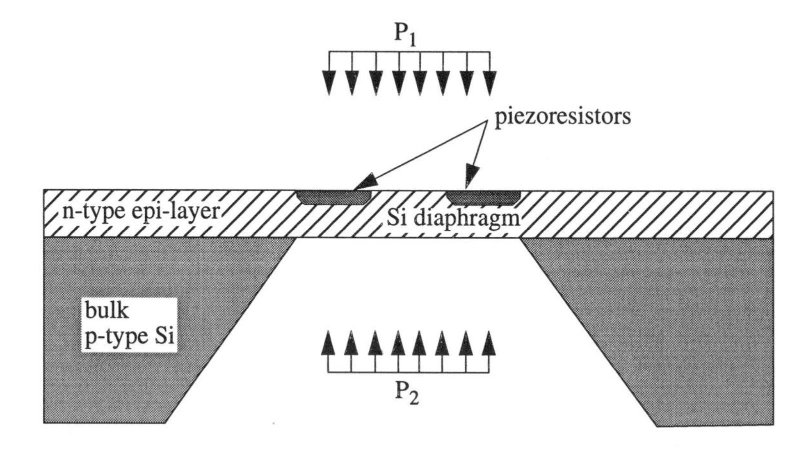

Figure 17. Schematic cross-section of a piezoresistive pressure sensor. The silicon diaphragm deflects upon application of a differential pressure, P_1-P_2. Piezoresistors embedded in the diaphragm measure the stress from the diaphragm deflection.

The initial step in making the pressure sensor is formation of the silicon piezoresistors by ion implantation. A silicon dioxide layer is grown on the wafer surface by thermal oxidation to act as a mask for the ion implantation (Figure 18b). The oxide must be thick enough to mask the implant. Alternatively, a mask material of LPCVD oxide or nitride layer could be selected. In the first photolithography step, the photoresist is patterned for subsequent plasma etching of the thermal oxide (Figure 18c). A set of heavily doped p-type piezoresistors are then created by ion implanting boron (Figure 18d). The boron penetrates the epi-layer in areas where the oxide has been etched away.

The oxide implant mask is removed in hydrofluoric acid (HF) and followed by deposition of a new layer of LPCVD oxide. The oxide is patterned and etched on the front side to provide contact cuts (vias) to the piezoresistors (Figure 18e). Metallization is sputtered and patterned to form interconnect and bonding pads (Figure 18f). The metallization is typically a bimetal layer of chrome and gold. Aluminum will etch in the subsequent wet silicon etch, so gold is chosen for metallization. A chrome layer is necessary to improve adhesion to the underlying silicon dioxide material. The metallization can be patterned with etching or a liftoff technique. A separate photolithography step patterns an opening in photoresist on the back side of the wafer, and fluorine-based plasma etching removes the exposed oxide (Figure 18g).

The bulk etching is left as the last step in the process (Figure 18h), since no further lithography is possible after the diaphragm is formed (at least not with a manufacturable yield). The substrate is etched from the back side in EDP, but a simple timed etch will not provide an adequate yield for the thin diaphragms. Process variations across the wafer and from batch to batch can cause the diaphragm thickness to vary greatly. Several etch-stop techniques exist to control the diaphragm thickness. However, the requirement of a lightly doped silicon layer for the

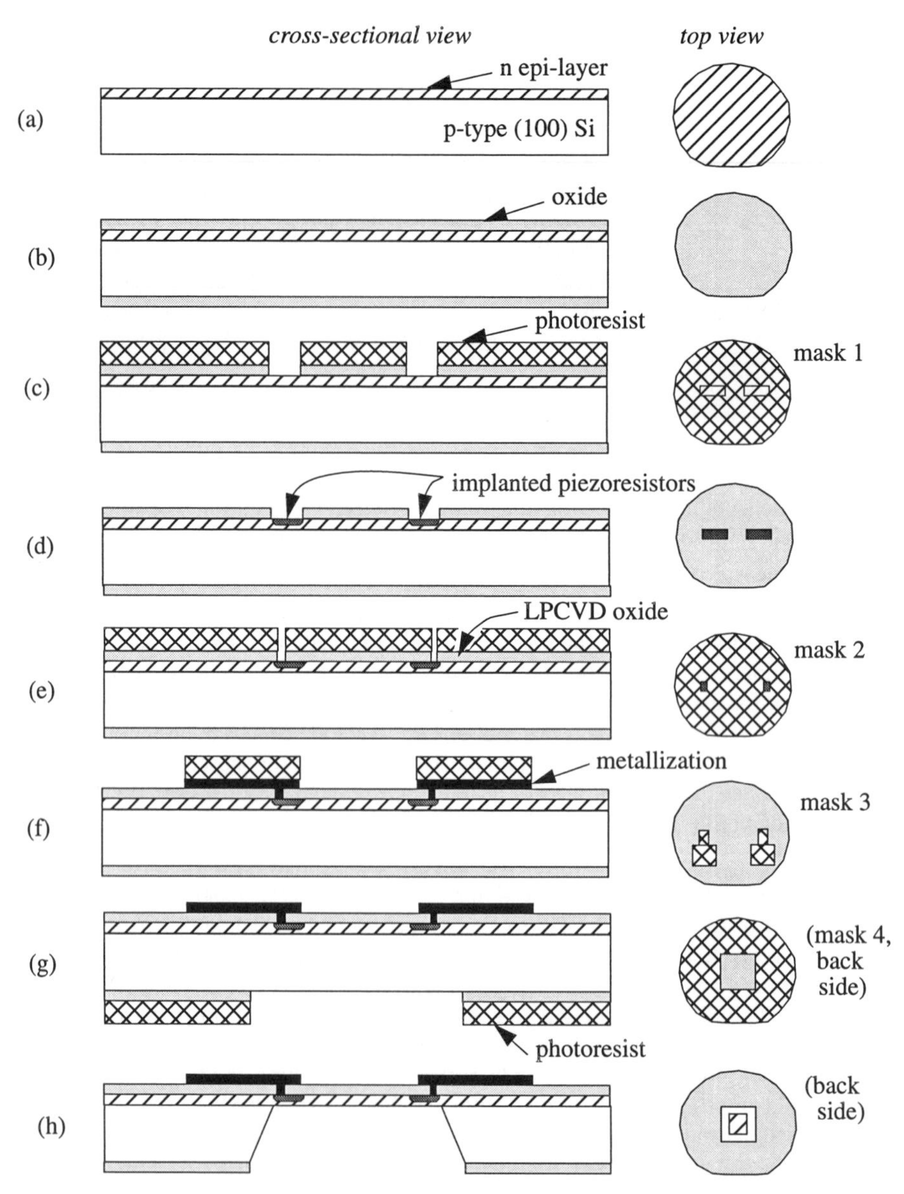

Figure 18. Process flow for a piezoresistive pressure sensor, showing parallel cross-sectional and top views of the wafer. (a) Si substrate with n-type epitaxial layer; (b) grow thermal SiO_2; (c) pattern photoresist and etch SiO_2; (d) implant boron for p-type piezoresistors; (e) etch SiO_2 layer and deposit LPCVD SiO_2; (f) sputter and pattern metallization; (g) pattern photoresist on back side of wafer and etch Si_3N_4; (h) etch silicon with electrochemical etch-stop on epi-layer.

piezoresistive pressure sensors rules out use of a p^+ etch stop. An electrochemical etch will attack the p-type substrate and stop upon reaching the lightly doped n-type epitaxial layer.

2.4.2 Surface-micromachining process

In contrast to bulk-micromachined structures, surface-micromachined structures are made from thin-film materials on the surface of the substrate. Most surface micromachined structures are composed of three primary kinds of materials that can be categorized according to their function: insulator, spacer, and structure, as illustrated in Figure 19. The particular structure shown in Figure 4 is a simple cantilever beam attached to the substrate at one end and free to move at the other end. An insulator material on top of the substrate is usually necessary to isolate the substrate from dc electrical signals on the beam, and to provide a chemically inert surface for subsequent processing steps. A structural material forms the micromechanical structures, and is chosen based on its mechanical and electrical qualities. A spacer material is used to set the vertical spacing between the substrate and the structure. The spacer layer will eventually be removed, and therefore it is often called a 'sacrificial layer'.

A common structural material for surface micromachining is polycrystalline silicon (polysilicon) thin films. Polysilicon surface micromachining is used to make a variety of devices, including accelerometers, gyroscopes, resonator oscillators, mirror arrays, rotary motors, and Fresnel lenses. Polysilicon has several desirable qualities. It a stiff material with a Young's modulus of around 165 GPa; it has low structural damping; and it can be deposited uniformly with a relatively low residual stress using LPCVD and thermal annealing. Fine-grained amorphous silicon is an alternative structural material that has been successfully used to make a variety of devices.

A detailed polysilicon surface-micromachining process flow is given in Figure 20, and is based on the original process developed at the University of California at Berkeley around 1983 [15]. We will describe each step in this process and highlight the reasoning and trade-offs in each step.

(a) Silicon is chosen as the starting substrate to be compatible with standard integrated-cir-

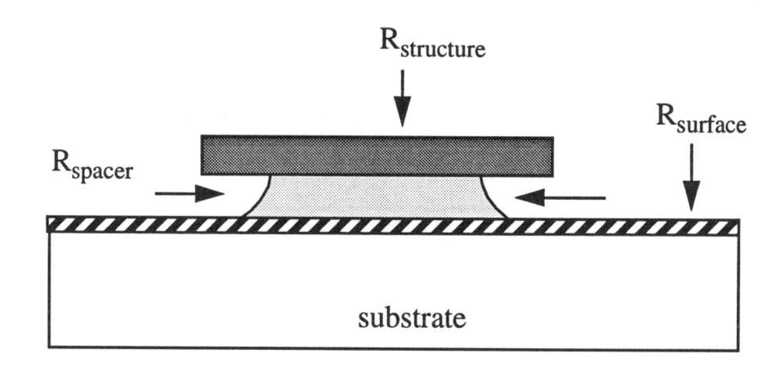

Figure 19. A cross-section of the generic surface-micromachining process showing the three functional layers - surface, structural, and spacer - and illustrating the three corresponding etch rates, $R_{surface}$, $R_{structure}$, R_{spacer}.

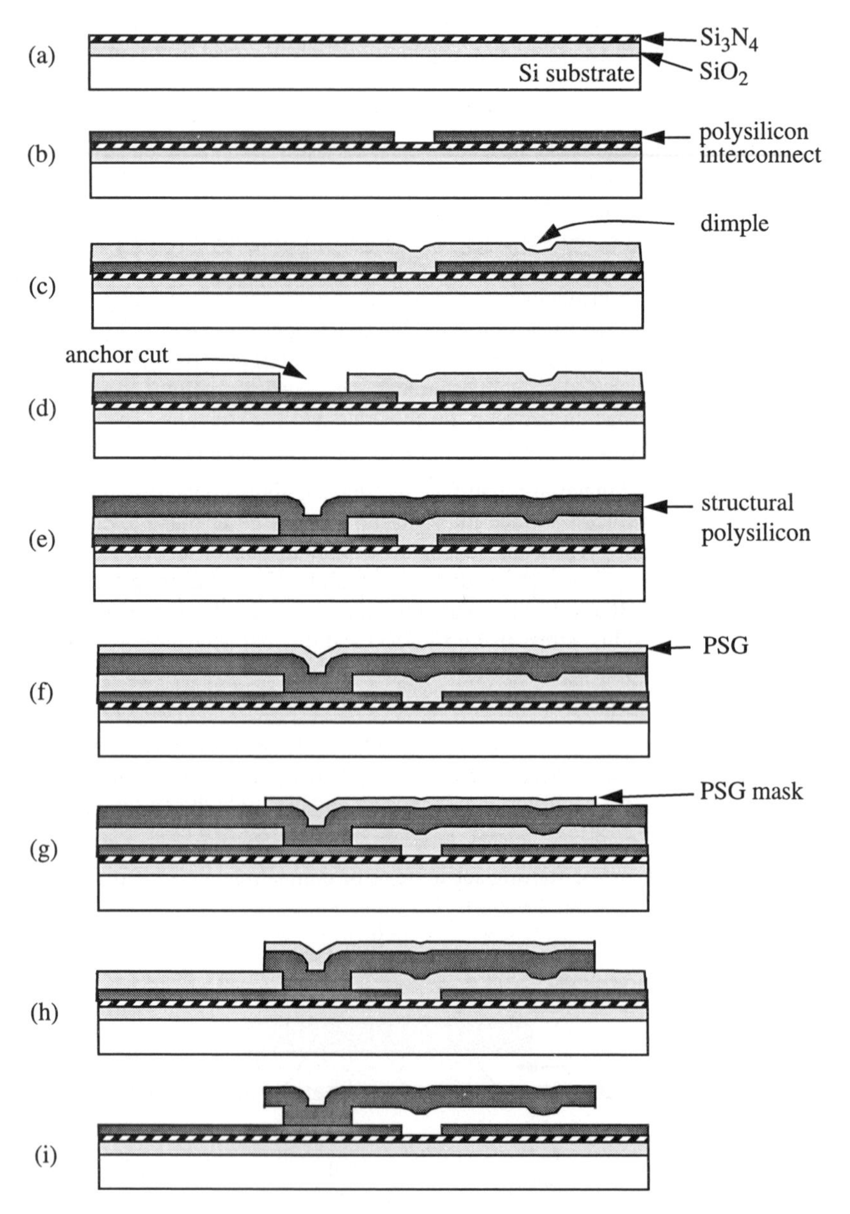

Figure 20. Polysilicon surface-micromachining process flow. Back-side films are not shown. (a) grow thermal oxide and deposit LPCVD silicon nitride; (b) deposit and pattern LPCVD polysilicon; (c) deposit 2 μm PSG and pattern dimples; (d) define anchor cuts in PSG; (e) deposit 2 μm undoped polysilicon; (f) deposit 5000 Å PSG and anneal at 1000°C, 1 hour.; (g) etch PSG mask; (h) etch polysilicon structure; (i) wet etch sacrificial PSG in HF; dry structure.

cuit microfabrication steps. The substrate doping does not affect the mechanical quality of the microstructures, and therefore can be chosen based on other factors, such as substrate conductivity and circuit compatibility. A thin layer of oxide, 1000 Å to 5000 Å thick, is grown by a thermal oxidation step to provide electrical insulation with a high breakdown voltage. A 5000 Å-thick layer of LPCVD silicon nitride is deposited on top of the oxide to protect the underlying oxide from attack by hydrofluoric acid (HF) during later steps in the process flow. Silicon-rich nitride ($Si_xN_{1.1x}$) is deposited because it has a much lower tensile stress than stoichiometric silicon nitride (Si_3N_4).

(b) Next, a 3000 Å-thick layer of *in situ* doped n-type LPCVD polysilicon is deposited. This layer is defined by the first mask in the process and will form electrical interconnect between and under the microstructures. The photolithography process module described earlier is used to pattern the wafer with photoresist. The exposed polysilicon is etched down to the underlying nitride layer in a CCl_4 plasma and the photoresist is stripped.

(c) The next step is deposition of the spacer material, 2 µm-thick LPCVD phosphosilicate glass (PSG). Thinner layers do not allow adequate spacing for many applications and can lead to structures permanently sticking to the substrate. Thicker spacer layers are difficult to etch in subsequent steps. Thicker spacer gaps between conducting structures lower the available electrostatic force and capacitance per unit area. These considerations may be important for specific micromechanical designs. The 2 µm PSG spacer thickness is a compromise that is commonly used.

(d) A second mask and photolithography step defines the photoresist for contact cuts (electrical vias and mechanical anchor cuts) into the PSG. The PSG is etched in a CF_4-based plasma until the first, interconnect polysilicon layer is reached.

(e) The polysilicon structural material is now deposited on top of the spacer layer. The polysilicon film can be deposited by LPCVD as *in situ* doped or undoped. In the latter case, a high-temperature furnace anneal is used to drive in phosphorous (n-type dopant) from the PSG in contact with the polysilicon. The deposition rate of undoped polysilicon in a conventional LPCVD furnace is about 1 µm/hr, which is three to four times faster than *in situ* doped polysilicon. The structural thickness ranges from less than 1 µm to about 10 µm, being limited at the high end by the etching technology and unreasonably long deposition times. A common thickness for polysilicon microstructures is 2 µm, which balances a need for vertical stiffness with the limitations of etching and deposition.

(f) A 5000 Å-thick PSG layer is deposited on top of the structural layer. Next, the PSG is patterned by the third mask and photolithography step and subsequently etched in a fluorine-based plasma. The PSG is used as an etch-resistant mask for the thick structural polysilicon etch. Photoresist alone does not hold up under long etch time required to etch through polysilicon thicknesses greater than 2 µm. This is especially true along the edges of the anchor cuts where the photoresist is thinned due to flow. The PSG is conformal and easily covers these edges. Furthermore, the PSG provides well-defined vertical sidewalls for the microstructures, which is often important for mechanical applications.

(g) The exposed structural polysilicon is etched in a chlorine-based plasma until the spacer PSG is encountered. At this point, the polysilicon has a large residual (built-in) stress relative to the underlying PSG. A high-temperature furnace anneal, commonly 1000 °C for 1 hour, is used to relieve stress in the structural material. The stress reduction is caused by recrystallization of the polysilicon at the high temperature. Rapid thermal annealing (RTA) has also been shown to relieve residual stress in polysilicon.

(h) The wafer is now immersed in HF until the sacrificial (spacer) PSG is etched away under the structures, thereby releasing the structures so they are free to move. The wafer is then rinsed and dried. A critical part of the surface-micromachining process is the final drying step. Surface tension forces during drying can force the microstructures down to the surface of the wafer and subsequently stick in place. Several techniques have been suggested to relieve the sticking problem during drying. Drying methods include high-temperature evaporation in methanol, freeze drying in t-methyl alcohol, supercritical-point drying in CO_2, and the use of temporary polymer pillars for support during drying followed by a dry etch removal of the polymer.

Two examples of suspended surface-micromachined structures are shown in Figure 21 [16]. In both examples, the structures are anchored to the substrate through slender beam springs. More complex structures can be formed by adding an extra structural polysilicon layer to the process. The third polysilicon layer can be patterned to form a hub for electrostatic rotary micromotors, as shown in Figure 21(a) [17][18]. Completion of the two-polysilicon-layer process forms the rotor and stator for the micromotor. Then, a sidewall spacer of LPCVD silicon nitride is created by an anisotropic RIE. An HF etch undercuts the PSG below the rotor, followed by deposition of a very thin (700 nm) PSG spacer layer. The third LPCVD polysilicon layer conformally fills in the anchor hole to form the hub. Release in HF removes the PSG and allows the rotor to rotate about the anchored hub. Figure 21(b) shows an example electrostatic synchronous micromotor built in this process [18]. Hinges and slider structures can also be constructed from the extra polysilicon layer [19][20].

One of the first products that has exploited polysilicon surface micromachining is the Analog Devices ADXL50 accelerometer, shown in Figure 4. The micromechanical structure is near

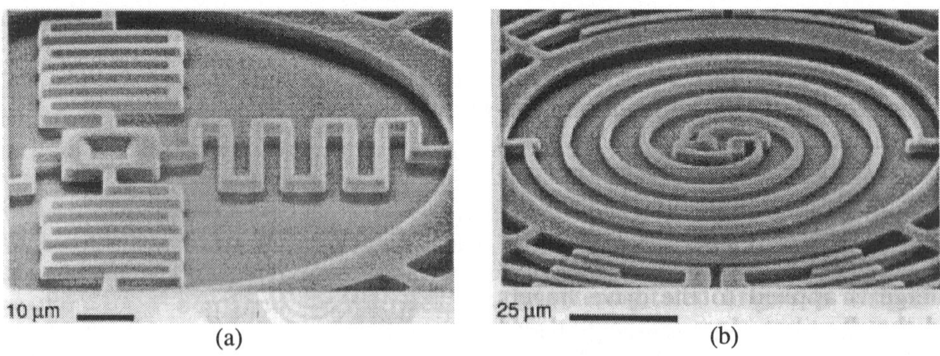

Figure 21. Examples of polysilicon surface-micromachined structures [16]. (a) A set of meander springs suspend a structure. (b) Two Archimedian spiral springs suspend a rotary interdigitated comb-finger structure.

the center of the die and is surrounded by sensing and signal conditioning circuitry. The microstructure, shown close up in Figure 21, is a polysilicon proof mass suspended by tethers. When the chip is accelerated in the direction indicated in the figure, the suspended proof mass displaces relative to the fixed cantilever fingers. This displacement is detected by measuring the capacitance between the fixed fingers and fingers attached to the proof mass. For small displacements, the signal is proportional to acceleration. The accelerometer has a full scale range of ± 50 g in a hermetic TO-100 (10-pin can) package.

2.4.3 Dissolved-wafer process

The dissolved wafer process was developed by researchers at the University of Michigan and has been used to create pressure sensors [21], tactile sensors, accelerometers, flow sensors [22], microvalves, and neural probes. Like conventional surface-micromachining, the dissolved-wafer process allows the fabrication of arbitrarily shaped structures with no constraints on spacing between components. However, structures are made from single-crystal heavily boron doped silicon bonded to a glass substrate.

An example device, an ultrasensitive microflow sensor fabricated in the dissolved-wafer process, is shown in Figure 24 [22]. A schematic of the flowmeter in Figure 24(a) shows the principle of operation. The flow channel and pressure-sensor diaphragm are made from a p+ silicon layer bonded to a glass substrate. Gas flows into a hole in the glass base of the device, through a micron-sized flow channel, and then out another hole in the glass. The pressure difference across the micro-channel gives a measure of flow. Figure 24(b) gives a different perspective of the flowmeter, identifying the fluidic interconnections in more detail. A scanning electron micrograph (SEM) of the device is shown in Figure 24(c).

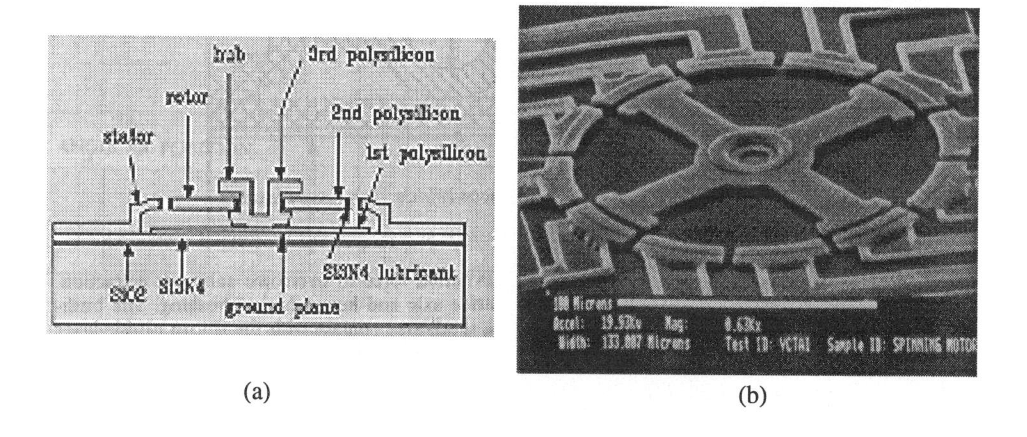

(a) (b)

Figure 22. A surface-micromachined electrostatic micromotor. (a) Micromotor cross-section, showing the role of the three polysilicon layers. After [17]. (b) Scanning electron micrograph of a synchronous micromotor with 4 rotor and 12 stator poles. After [18].

The fabrication process for the flowmeter begins with a moderately doped (100) p-type silicon substrate on top of which is thermally grown a 0.5 μm-thick oxide layer (a). The oxide is patterned by first photolithography step and etched. (b). Recesses about 3 μm deep in the silicon are then etched with KOH (c). The oxide acts as an adequate mask for the short exposure to KOH. The oxide is now removed in hydrofluoric acid (HF) and replaced with a new layer of 1.2 μm-thick thermal oxide. After another photolithographic step, the oxide layer is etched to define a diffusion mask (d). The subsequent deep boron diffusion, typically between 12 μm to 15 μm deep at 1175°C, sets the thickness and rigidity of the microstructural support posts (e). An additional photolithography and etch is used to re-pattern the oxide for a subsequent shallow boron diffusion. The shallow diffusion defines the diaphragm thickness, which sets the device sensitivity and is typically between 1 μm and 3 μm thick (f). Placing the deep diffusion before the shallow diffusion allows use of the same oxide mask for both steps. After the diffusions, the thermal oxide is stripped off in HF and a layer of LPCVD oxide is deposited. The LPCVD oxide is patterned and etched such that it only covers the diaphragm (g). The addition of the compressive oxide film to the diaphragm serves two purposes: to electrically insulate the diaphragm against electrical shorts if it touches on a counter electrode, and to compensate for tensile residual stress in the heavily boron-doped silicon material. The optimal oxide thickness will produce a composite diaphragm with near-zero stress and a subsequent increase in pressure sensitivity.

Next, the patterned side of the silicon wafer is electrostatically bonded to a glass substrate

Figure 23. Analog Devices ADXL50 accelerometer. The central polysilicon proof mass is suspended by four tethers. Interdigitated comb-fingers on both sides of the central spine capacitively detect lateral motion of the proof mass (The lateral motion is up and down in the photograph.). Courtesy of Analog Devices, Inc.

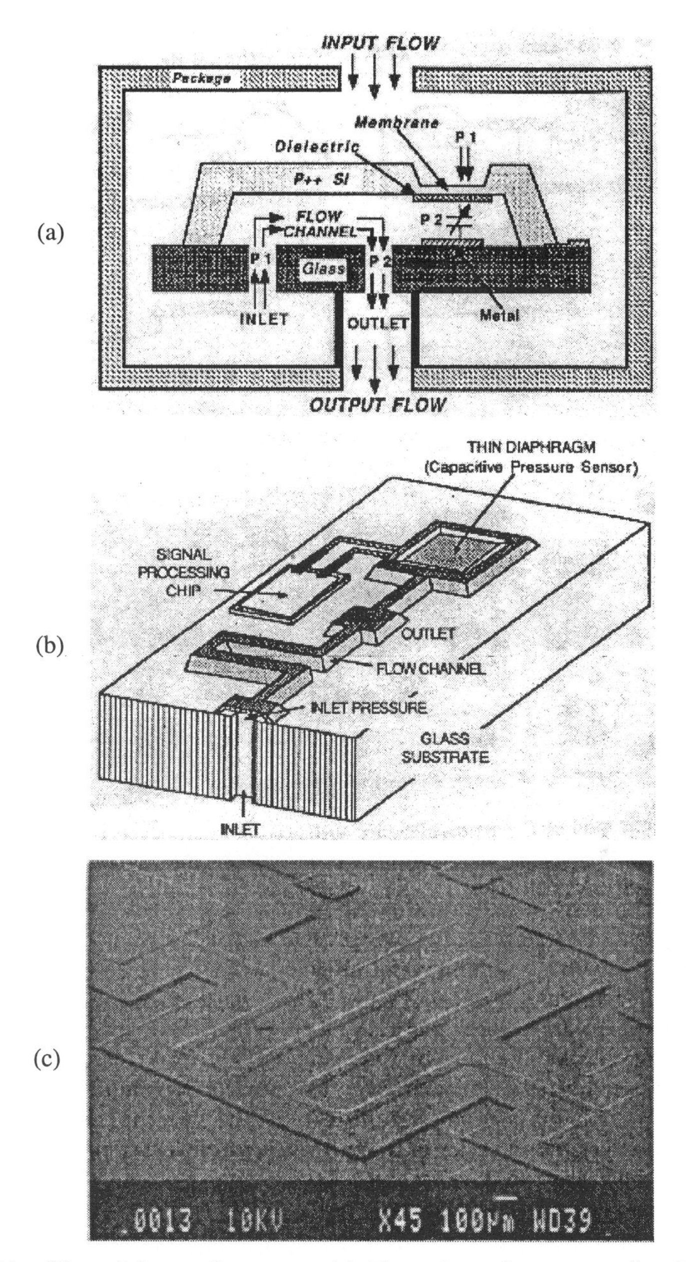

Figure 24. Ultraminiature flowmeter. (a) The schematic cross-section inside a package illustrates the device operation. (b) The perspective view gives a simplified device layout. (c) The scanning electron micrograph shows the inlet, outlet, and meandering flow channel. After [22].

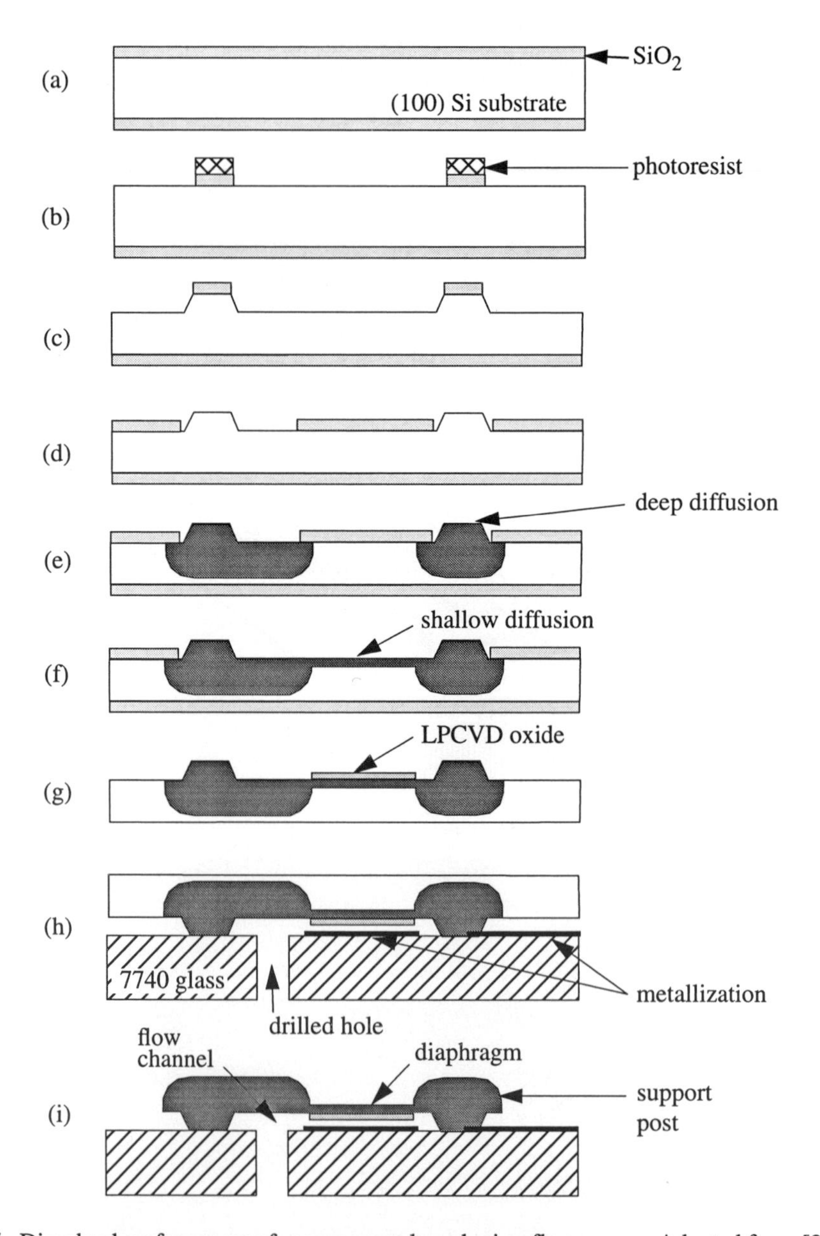

Figure 25. Dissolved-wafer process for a pressure-based microflow sensor. Adapted from [22]. (a) silicon substrate with thermal oxide coating; (b) oxide patterning; (c) etch recess in silicon; (d) deposit and pattern oxide diffusion mask; (e) deep boron diffusion; (f) shallow boron diffusion; (g) deposit LPCVD oxide for diaphragm stress compensation; (h) anodic bonding of wafer to glass substrate; (i) dissolve bulk Si wafer in EDP.

(h). Since the glass is transparent, alignment is simple and does not require a double-sided alignment tool. The bonding takes place at moderate temperatures, requiring a matching of temperature coefficients of expansion (TCE) for the silicon and glass wafers. The TCE of Corning #7740 borosilicate glass is very closely matched to that of silicon for temperatures below about 500°C and is used in most silicon-glass anodic bonding. Prior to bonding, an evaporated gold/chrome metallization layer is patterned on the glass substrate. The chrome acts as an underlying adhesion layer for the gold interconnect. Holes are drilled into the glass to form inlet and outlet ports for the flow channel.

The bonded silicon/glass wafer stack is immersed in ethylenediamine/pyrocatechol/water (EDP), which etches the bulk silicon wafer without attacking the heavily boron-doped silicon regions, or the glass, metallization, or dielectric layers (i). After etching, the wafer stack is rinsed in hot water and isopropyl alcohol to remove any residual EDP. The remaining boron-doped silicon features form the free-standing microstructures on the surface of the glass wafer.

Several enhancements to the dissolved-wafer process have been reported. Fine-pitch lateral microstructures are formed by RIE on the heavily doped diffused layer prior to anodic bonding (between steps (g) and (h) in Figure 25) [23]. An example of a laterally actuated microstructure fabricated in this process is shown in Fig. 12. Recently, microstructures up to

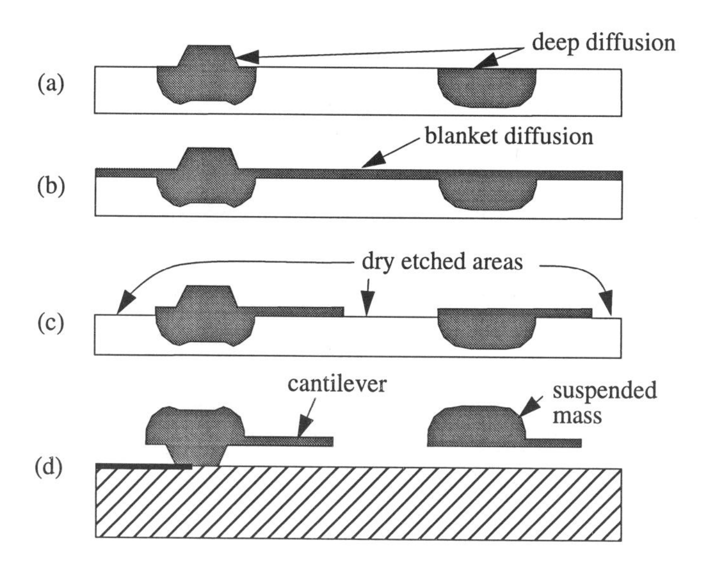

Figure 26. A modified dissolved wafer process which produces fine-pitch microstructures. Adapted from [23]. (a) Process after recess etch and deep boron diffusion; (b) Blanket boron diffusion; (c) Plasma etching of microstructures; (d) anodic bonding and wafer dissolution.

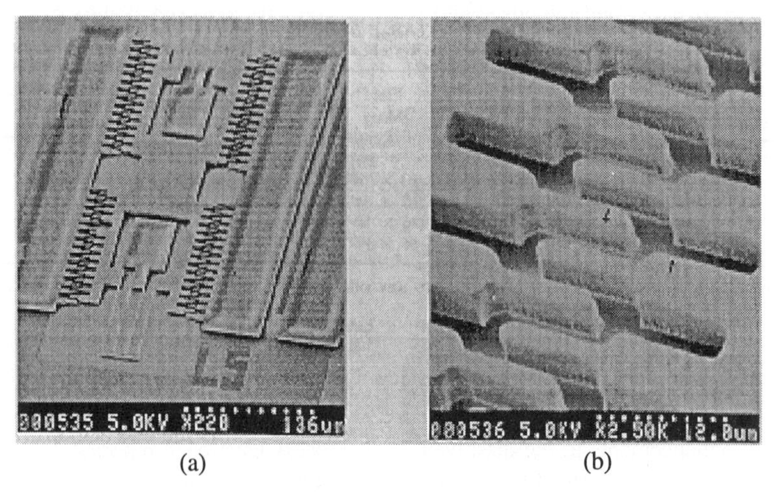

(a) (b)

Figure 27. Example microstructure fabricated with a dissolved wafer process [23]. (a) SEM of the laterally driven structure. (b) A magnified view of the comb fingers.

35 μm thick have been fabricated by using a chlorine-based plasma generated by an electron cyclotron resonance (ECR) source [24]. CMOS circuitry can be integrated with the heavily boron-doped structures by using electrochemical etching during the wafer dissolution. The electrochemical etch stops on an n-type epitaxial layer containing the CMOS devices.

2.4.4 Standard CMOS micromachining process

Several techniques based on silicon undercut of dielectric layers have been developed to create thin-film microstructures from standard CMOS processes [26][27]. Oxide cantilevers and bridges on bare silicon wafers can be fabricated by undercutting the silicon beneath the patterned oxide using a wet anisotropic etchant. The etchant selectively etches the silicon without appreciably attacking the oxide. This micromachining technique was first demonstrated in 1977 for a small voltage-addressable array of optical modulators [25].

Most of the basic microstructural unit processes are available in standard CMOS; however, the process steps are placed in a fixed order. Instead of modifying the process flow, design of microstructures is accomplished solely by manipulating the layout. In general, the microstructure layout violates CMOS design rules, but these design rule violations can be tolerated by standard CMOS foundry services because the process sequence is unaffected.

An abbreviated process description, shown in Figure 28, illustrates the fabrication of a composite oxide/aluminum bridge in a standard single-metal n-well CMOS process. Only a subset of steps from the full CMOS process are shown; process steps that do not affect the microstructure are omitted. Many texts are available to provide an overview of the entire standard CMOS technology [6][7].

Fabrication begins with a (100) lightly n-type silicon substrate which has been selectively oxidized to form a field oxide (a). An optional p-channel source/drain boron ion implantation

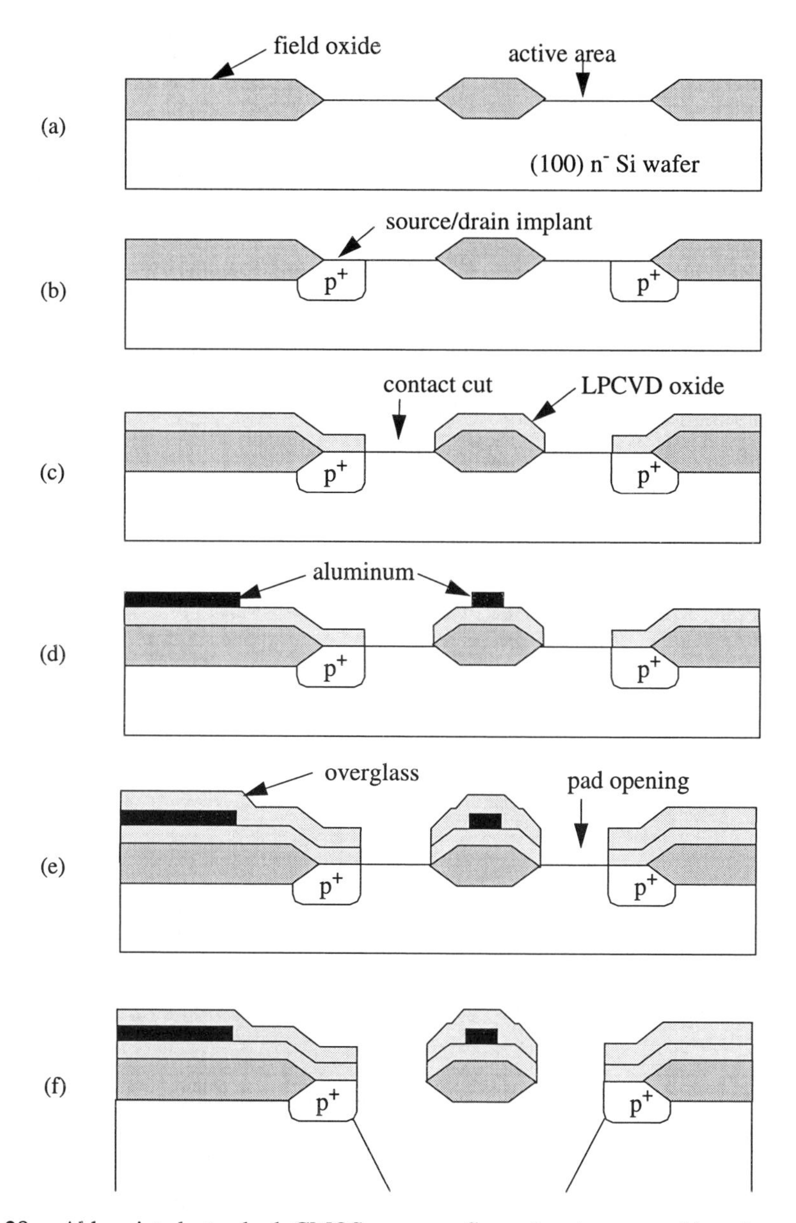

Figure 28. Abbreviated standard CMOS process flow showing an oxide microstructure formation using stacked vias. (a) After field oxidation; (b) After p-channel source/drain boron implant; (c) After oxide contact cut; (d) After aluminum etching; (e) After opening overglass for bond pads, showing via stacking; (f) after silicon wet-etch to release a cantilever beam with an embedded aluminum conductor.

acts as a lateral etch stop for the final silicon wet etching step (b). The etch stop ensures that the silicon etch does not encroach on circuitry that may be very close to the microstructures. Next, contact cuts are etched through the intermediate oxide layer (c). Aluminum is then sputter deposited, patterned, and etched (d). In a normal CMOS process, the design rules prohibit etching of aluminum over the contact cut. However, this design rule is ignored in the microstructure process. The aluminum deposited on the oxide sidewall is thicker than in planar areas above the oxide. Therefore, special attention to the etching may be necessary to avoid "stringer" formation of aluminum on the via sidewalls. Next, an oxide "overglass" layer is deposited using plasma-enhanced CVD. Contact cuts are etched in the overglass for bond pads. The overglass cut is stacked over the previous via cuts to define the oxide microstructure (e). Again, there may be difficulties with stringers in this step, so attention must be paid to the oxide etch. Finally, a silicon wet-chemical etch in EDP releases a cantilever beam with an embedded aluminum conductor (f). The p^+ layer provides protection against any silicon undercut on the substrate.

A similar process flow to that shown in Figure 28 can be formed for CMOS processes with multiple aluminum and oxide layers. Via cuts can be stacked together to form thicker microstructures with multiple embedded aluminum and polysilicon conductors. Polysilicon/aluminum thermocouples have been fabricated on thermally isolated fixed-fixed beam structures in this kind of process. These structures are used as infrared detectors and as thermal flow sensors. For example, the two-dimensional thermal flow sensor shown in Figure 29 consists of a polysilicon resistive heater located on the central portion of an oxide microbridge and surrounded by four polysilicon/aluminum thermopiles (a thermopile is a set of thermocouples) [28]. Thermal isolation of the thermopiles from the substrate is provided by the microbridge. The temperature difference between the thermopiles gives a measure of air flow in two directions.

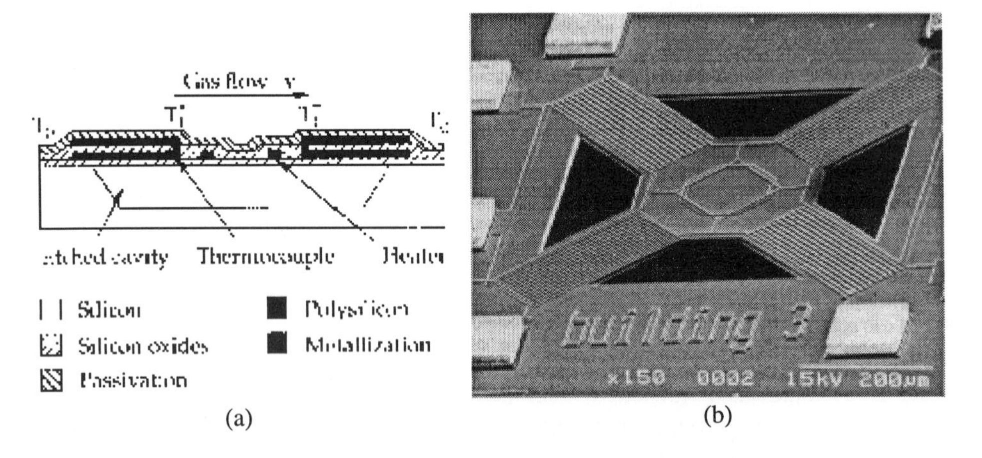

Figure 29. A two-dimensional gas flow sensor in CMOS IC technology. (a) Cross-section of the final microbridge structure. (b) SEM of the fabricated device. Adapted from [28].

2.5 Limitations of Planar Processes

The chief limitation of all processes based on photolithography is that one does not have the same control over out-of-plane dimensions as in-plane geometry. For example, on the same mask level there can be structures with widths of 1 μm, 10 μm, and 1000 μm, but the heights of all of the structures resulting from that mask level will be about the same. This limitation is imposed by the nature of exposure tools, which are most commonly optimized for microelectronics. In general, IC processes are tailored for lithographic tools with a short depth of focus (less than 1 μm is sufficient). For example, difficulties with step coverage of metal interconnects are prevented by planarizing inter-metal dielectrics.

Microsystems employing mechanical elements do not have the same set of constraints as conventional microelectronic devices. For example, scaling of MOS transistors to smaller dimensions results in decreased capacitance, with a corresponding decrease in stored charge and faster switching speeds. In the world of digital systems, this is the path of progress. However, an electrostatic actuator has a stored energy density proportional to the capacitance, so instead of minimizing this quantity, a designer may wish to maximize it. Increasing the dimensions of elements in the direction orthogonal to the wafer surface is an obvious way to achieve this goal.

Bulk micromachining technology represents one way to overcome the limitations of limited control in the z-direction. Other methods include electroplating of thick films, lithography on thick photoresist films using ultraviolet or x-ray exposure sources, and direct-write physical deposition and etch methods. All of these aim to produce high-aspect-ratio microstructures, where the vertical height of the patterns are not restricted by the conventional depth of focus limitation. A true three-dimensional microsystems fabrication technology, which maintains the manufacturing advantage of mass parallelism, is still in the future.

2.6 Acknowledgments

The authors wish to acknowledge the Defense Advanced Research Projects Agency for support of their research programs, and to thank the many individuals who have contributed their time and efforts in support of this project. We would especially like to acknowledge the support of Prof. Dr. Henry Baltes, and the entire staff of the Physical Electronics Laboratory of ETH Zürich, Switzerland, who hosted one of the authors (MLR) during the writing of this chapter.

References

[1] G. K. Fedder, S. Santhanam, M. L. Reed, S. C. Eagle, D. F. Guillou, M. S.-C. Lu, L. R. Carley, "Laminated High-Aspect-Ratio Microstructures in a Conventional CMOS Process," Proceedings of the Ninth IEEE Workshop on Micro Electro Mechanical Systems, San Diego, CA, February 11-15, 1996, pp. 13-18.

[2] K. D. Wise, "Integrated Microelectromechanical Systems: A Perspective on MEMS in the 90s," in Proceedings of the Fourth IEEE Workshop on Micro Electro Mechanical Systems (MEMS-91), Nara, Japan, January 1991.

[3] K.E. Petersen, "Silicon as a Mechanical Material," Proceedings IEEE 1982.

[4] S. M. Sze, *Semiconductor Sensors*, John Wiley and Sons, Inc., 1994.

[5] W. R. Runyan, K. E. Bean, *Semiconductor Integrated Circuit Process Technology*, Addison-Wesley Publishing Company, 1990.

[6] R. C. Jaeger, *Introduction to Microelectronic Fabrication*, (Modular series on solid-state devices, v. 5), Addison-Wesley, Reading, MA, (1988).

[7] S. M. Sze, *VLSI Technology*, McGraw-Hill, New York, (1983).

[8] J. L. Vossen, *Thin Film Processes*, Academic Press, New York, (1978).

[9] L.-S. Fan and R.S. Muller, "As-deposited low-strain LPCVD polysilicon," in Proceedings of the Workshop on Solid-State Sensors and Actuators, Hilton Head, SC, 1988, page 55.

[10] P. Krulevitch, R. T. Howe, G. Johnson, and J. Huang, "Stress in undoped LPCVD polycrystalline silicon," in Proceedings of the International Conference on Solid-State Sensors and Actuators, Montreaux, 1991, page 949.

[11] S. Linder, H. Baltes, F. Gnaedinger, E. Doering, "Photolithography in Anisotropically Etched Grooves," Proceedings of the Ninth IEEE Workshop on Micro Electro Mechanical Systems, San Diego, CA, February 11-15, 1996, page 38.

[12] S. K. Clark and K. D. Wise, "Pressure sensitivity in anisotropically etched thin-diaphragm pressure sensors," *IEEE Trans. Electron Devices*, **ED-26**, 1887, (1979).

[13] O. N. Tufte, P. W. Chapman, and D. Long, "Silicon diffused-element piezoresistive diaphragms," *J. Appl. Phys.*, **33**, 3322-7 (1962).

[14] *Microelectromechanical Systems Market Study*, System Planning Corporation, (1994).

[15] R. T. Howe and R. S. Muller, "Polycrystalline silicon micromechanical beams," *J. Electrochem. Soc.*, **130**, 1420-1423, (1983).

[16] W. C. Tang, T. C. H. Nguyen, and R. T. Howe, "Laterally driven polysilicon resonant structures," *Sensors and Actuators*, **20**, 25-32, (1989).

[17] L.-S. Fan, Y.-C. Tai and R. S. Muller, "IC-processed electrostatic micromotors," *Sensors and Actuators*, **20**, 41-47, (1989).

[18] Y.-C. Tai and R. S. Muller, "IC-processed electrostatic synchronous micromotors," *Sensors and Actuators*, **20**, 49-55, (1989).

[19] L.-S. Fan, Y.-C. Tai and R. S. Muller, 'Integrated movable micromechanical structures for sensors and actuators," *IEEE Trans. Electron Devices*, **35**, 724-730, (1988).

[20] R. Yeh, E. J. J. Kruglick, and K. S. J. Pister, Surface-micromachined components for articulated microrobots," *J. Microelectromech. Sys.*, **5**(1), 10-17, (1996).

[21] H. L. Chau and K. D. Wise, "An ultraminiature solid state pressure sensor for a cardiovascular catheter," *IEEE Trans. Electron Devices*, **35**(8), 2355-2362, (1988).

[22] S. T. Cho, K. Najafi, C. E. Lowman, K. D. Wise, "An ultrasensitive silicon pressure-based microflow sensor," *IEEE Trans. Electron Devices*, **39**(4), 825-835, (1992).

[23] Y. B. Gianchandani and K. Najafi, "A bulk silicon dissolved wafer process for microelectromechanical devices," *J. Microelectromech. Sys.*, **1**(2), 77-85, (1992).

[24] W.-H. Juan and S. W. Pang, "Released Si microstructures fabricated by deep etching and shallow diffusion," *J. Microelectromech. Sys.*, **5**(1), 18-23, (1996).

[25] K. E. Petersen, "Silicon as a mechanical material," *Proc. IEEE*, **70**, 420-457, (1982).

[26] M. Parameswaran, H. P. Baltes, Lj. Ristic, A. C. Dhaded, and A. M. Robinson, "A new approach for the fabrication of micromechanical structures," *Sensors and Actuators*, **19**, 289-307, (1989).

[27] J. C. Marshall, M. Parameswaran, M. E. Zaghloul, and M. Gaitan, "High-level CAD melds micromachined devices with foundries," *IEEE Circuits and Devices*, **8**(6), 10-17, (1992).

[28] J. Robadey, O. Paul, and H. Baltes, "Two-dimensional integrated gas flow sensors by CMOS IC technology," J. Micromech. Microeng., **5**, 243-250, (1995).

3. Micromachining by Machine Tools

T. Masuzawa

Institute of Industrial Science, University of Tokyo
7-22-1 Roppongi, Minato-ku, Tokyo 106, Japan

1. HISTORICAL ASPECTS

Micromachining as an application of machining with machine tools has apparently a different history from that based on photolithograph. It has a long history since the Industrial Revolution as the main means for parts production. Since the large part of those parts produced in 19th century was in a range from a millimeter to meters in its dimension, the machine tools had been developed towards the most suitable style for this dimension range.

The development of watch industry related the first miniaturization of machining dimension. This sub millimeter oriented production required various special machine tools. Swiss lathe is a typical model which satisfied this difficult requirement. Although those machine tools were more or less specified for watch production, the technology itself should form a valuable background for the micromachining in next generation.

When electronics became an important technology, the production of vacuum tube requested other types of micromachining. This was a sign of the beginning of explosive miniaturization in vast fields of industry. Vacuum tube changed its shape and size quickly towards MT (miniature tube), and finally to acorn tubes. During this development transistor was born. After transistor became a practical device, all kinds of electronic parts and relating mechanical parts and components rushed into the race of miniaturization. In this flow, machine tools have continuously supported the production of up-to-date miniaturized parts. This means that the miniaturization in machine tool technology has been rather continuous.

However, it must be mentioned that there was a kind of phase change during this development. It is introduced by so-called nontraditional machining techniques, including EDM, ECM, LBM and USM. The reason why these new techniques were developed and joined the group of machine tools is that the improvement of material and the request of miniaturization sometimes caused difficulties in the application of conventional machine tools. In recent stage of development, micromachining by machine tools consists of variety of machining methods based on different phenomena and styles of equipment.

2. BASICS OF MICRO MACHINE TOOLS.

2.1. Shape specification element (SSE)

In micromachining, the dimension of products is as small as sub millimeter. Consequently, the error acceptable in most cases is around a micrometer, or recently more often sub micrometer. It means that micromachining is in the category of precision or ultra precision machining. In this field, it is important to recognize the element which specifies the shape and dimension of the product. Hereafter in this chapter, such an element is defined as a shape specification element, or SSE. In machine tools, the SSE's are very often solid tools and their paths. Image tools such as in laser machining and electron beam machining are rather exceptional ones. On the other hand, in photolithograph, the SSE is a resist mask and it determines only two dimensions. To obtain the dimension in another coordinate, the processing time is usually used as an indirect SSE. The above fact suggests that the micromachining with machine tools, especially those with solid tools, is naturally suitable for machining 3D micro products. In this case, all three dimensions can be precisely determined by copying the shape and path of a real object, the tool.

2.2. Convex and concave shapes

For the micromachining using solid tools, there is a significant difference between the conditions in machining convex and concave shapes. In convex shapes, there is little limitation in the shape and size of the tool, because there is an open space around the workpiece, or the machining area. On the other hand, in machining a concave shape, the working space is limited by the shape of the product. Usually, this limitation is not severe at the beginning of machining and gradually becomes serious according to the increase of depth, and worst at the finishing stage.

According to the above geometrical problem, very tiny, and often slim, tools are required for micromachining concave shapes such as holes, slits and cavities. The larger the aspect ratio of the product is, the longer and slimmer tools are required. Therefore, the fabrication of such tiny tools is the first problem of micromachining. This is basically a subject of micromachining convex shapes, although sometimes combined with machining concave shapes.

In machining convex shapes, the limitation of tool dimension is not significant as mentioned above. However, another problem arises when a slim products such as the tools for micromachining holes or slits are targeted. It is the rigidity and strength of the workpiece. Such workpieces like microtools and other pin type microproducts are easily deformed by millinewtons of machining force. Therefore, the applied machining method must be carefully selected from the viewpoint of machining force.

2.3. Unit removal (UR)

One of the more important issues in precision machining is the unit removal or the UR. The UR is the volume, or the three dimensional bulk, which is removed from the workpiece according to a unit of machining phenomenon. Taking EDM for an example, the difference of the workpiece dimension between before and after a discharge, a crater, is the UR (cf. Fig. 1a). When the process is continuous like turning, two dimensional UR is more useful. The hatched area in Fig. 1b corresponds to the UR in such a case.

Since the UR eliminates the smallest controllable dimension, a UR as small as possible is desirable for precision machining and, in particular, micromachining. In this sense, photolithograph is very suitable for micromachining, because the UR is possibly close to the order of atoms. In the processes by machine tools, the UR is usually much larger. However, since in many cases in micromachining the required tolerance is well over tens of nanometers and sometimes even a micrometer, the UR's of many machine tool processes are on an acceptable level.

Small UR is also important in the viewpoint of machining force. In mechanical processes the machining force increases when the UR increases. Consequently, the choice of machining condition so as to realize a small UR is very important. Also, the development of technology for realizing smaller UR is essentially requested in these processes.

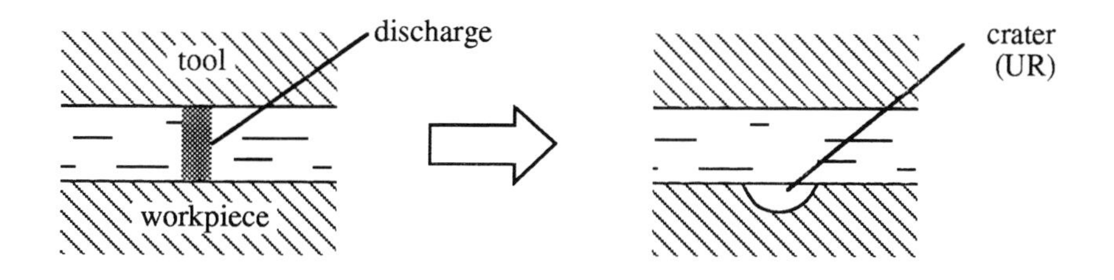

a. Unit removal in EDM

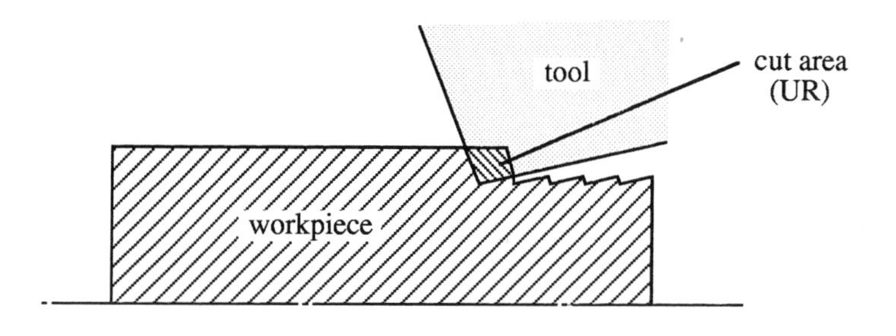

b. Unit removal in turning.

Fig. 1. Unit removal in machining process.

2.4. Miniaturization of machine tools

In principle a machine tool for micromachining can be as large as that for normal size products. However, a miniaturized machine tool is more suitable for micromachining.

The most important reason relates the deformation by heat. Since the deformation length of the machine by the temperature change is proportional to the size of the machine, smaller machines are less sensitive to the heat environment. Therefore, higher precision is more easily realized by a miniaturized machine tool.

Another reason is that strong force is not required for machining in most micromachining processes. This allows all structure be less rigid, and consequently all components of the machine can be smaller.

There is another advantage in the miniaturization of machine tool. If a machine is designed very small in dimension, a wide choice is allowed in the selection of the parts material. For instance, materials which are more expensive but higher quality can be used for composing the machine tool. Materials which are superior in characteristics but obtainable only in small sizes can also be included in the choice.

3. MICROTOOL PREPARATION

For micromachining of concave shapes, microtools must be prepared. The preparation of microtools is mainly a subject of micromachining convex shapes such as pins, spindles, needles and rods. As described previously, there is less limitation in the size and shape of the tools used in this preparation. The problems which should be solved are the realization of small UR and small machining force.

Among mechanical machining processes, grinding is one of the most suitable methods for this purpose. This is because the UR can be very small if very fine abrasives are used to compose the grinding wheel, and because the machining force can be small if the cut depth and feed speed are small. Ultra high precision cutting has similar possibility. But, grinding is more advantageous because the process is carried out by multi cutters in parallel operation, which recovers the disadvantage of small UR and slow feed and realizes reasonable productivity.

With the above advantages and the fact that it has a long history in the application for tool making, grinding has been the most popular method in microtool preparation. A typical application is the fabrication of microdrills. Most commercial microdrills are produced by applying grinding in all steps of the process, making outlines, cutting grooves and forming the cutting edges. Grinding can be applied for microtools which has the dimension (in diameter) of hundreds of micrometers (in some cases down to tens of micrometers), and is suitable for the medium hardness of object materials such as of tool steels.

For the smaller tools, i. e. tens of micrometers or less in diameter, and for harder materials such as carbide alloys and sintered diamond (SD), there are difficulties in the application of grinding. In such cases, EDM (electrical discharge machining, or electrodischarge machining) is a powerful technology. In EDM machining force is very small and negligible in most cases, because the machining phenomenon is based on the thermal melting of the material and the removal takes place on this molten material. Since the induced thermal power density can be well over 10^8W/cm^2, any kind of materials are melted or vaporized. Therefore, the choice of the workpiece material depends on neither mechanical nor thermal properties. As for another basic requirement, small UR, it is also satisfied by controlling the discharge energy by reducing the current and duration time of discharge pulses. It is also competitive in productivity with grinding, because the discharge frequency can be increased up to megahertz order. The detail of the tool making by EDM, particularly by wire electrodischarge grinding (WEDG) will be described in the next section.

Some microtools consist of both, convex and concave shapes. In the preparation of this kind of tools, micromachining technologies for concave shapes which will be described in the following sections should also be applied.

The groove formation that is, for example, required for twist drills can be achieved by grinding using a very thin grinding wheel, or by WEDG using a very thin wire electrode and a tilted wire guide.

4. WIRE ELECTRODISCHARGE GRINDING (WEDG)

The most powerful method for machining micropins and similar products, including various microtools is wire electrodischarge grinding (hereafter referred to as WEDG)[1]. This is a variation of EDM and innovated more than ten years ago. The background and the basic concept of this technology are as follows.

4.1. EDM

EDM is a technology for machining metals, alloys and other electroconductive materials. The material removal phenomenon is based on melting and evaporation by the heat of electrical sparks. When an electrical spark takes place in dielectric fluid, a crater is formed on each electrode. By adjusting the electrical condition and by selecting the material combination of electrodes, one of the craters can be extremely larger than the other. The material which bears this larger crater will be the workpiece, and the other will be the tool electrode (usually called 'electrode'). The integration of craters deforms the workpiece (Fig. 2). In original EDM technique, the shape of the electrode is copied into the workpiece. Later, the employment of a wire as the electrode introduced WEDM (wire EDM) technique. In WEDM, the electrode (wire) cuts the workpiece along a programmed path, similarly to the process of a jig saw. After WEDM was introduced, the original type of EDM is called 'sinking EDM', as well as simply 'EDM'. Both types of EDM are widely used in die/mold making and other metal processing.

4.2. Background of WEDG development

In case when a fine pin is requested to be machined by EDM, a rectangular metal block has been used as the electrode. Fig. 3a shows the basic setup of this conventional method. The workpiece is rotated in the range of 10 to 3000 rpm. The block electrode is fixed on a table, and the table is controlled to give the feed perpendicularly to the rotation axis. By copying the flat surface of the block, a straight cylindrical pin is produced.

This system has been practically used, but there are some problems. One of them is the wear of the block electrode. Though low-wear materials are selected, the electrode wears far more quickly compared with the tools for mechanical cutting or grinding. Therefore, the prediction of the machined diameter of the pin is difficult, and usually the diameter must be measured several times during the operation. It makes the operation much complicated. Moreover, as the measurement of very fine pin is difficult to be done on the machine, the pin or the holder of the pin must be removed from the machine several times for being brought to the measuring equipment (usually a measuring microscope). This procedure give nothing but the errors, eccentricity and tilting, which are introduced in resetting the pin.

Another problem is the unevenness of the gap distance along the axis of the pin. The gap distance is affected by the concentration of the debris in the gap. When the difference of the maximum and the minimum debris concentration within the machining area is large, the evenness of the gap distance will be lost because the debris concentration influences on the breakdown voltage and, consequently, the gap distance. In this conventional system the machining area includes the entire length which is going to be machined. It means that the condition is the worst from this viewpoint, because wider area normally allows more variation of the debris concentration. It will result in bad straightness and the limitation in the minimum machinable diameter.

4.3. Outline of WEDG

Concerning with the electrode wear, a wire electrode is a solution on the analogy of WEDM as shown in Fig. 3b. However, it is not easy to apply it to micropins. The main reason is that the machining accuracy is not high enough for aiming at micromachining, because it is difficult to suppress the wire vibration completely. The amplitude easily reaches 10 μm or more. Also the deviation of the wire caused by the original curl or by the machining force may

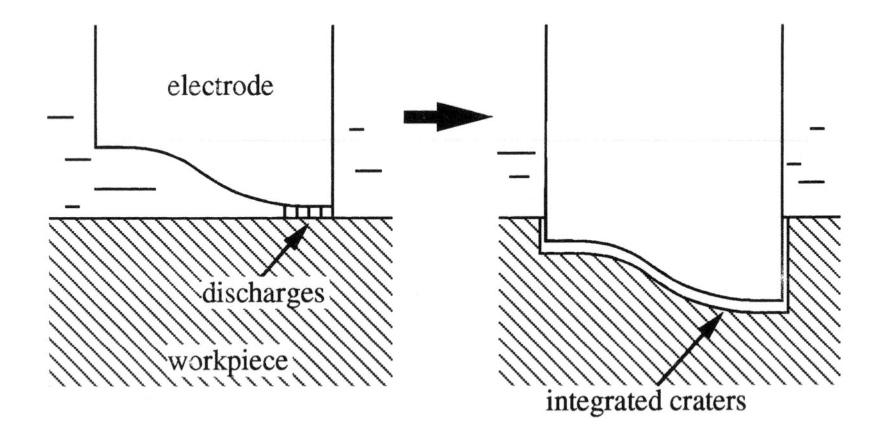

Fig. 2. EDM process.

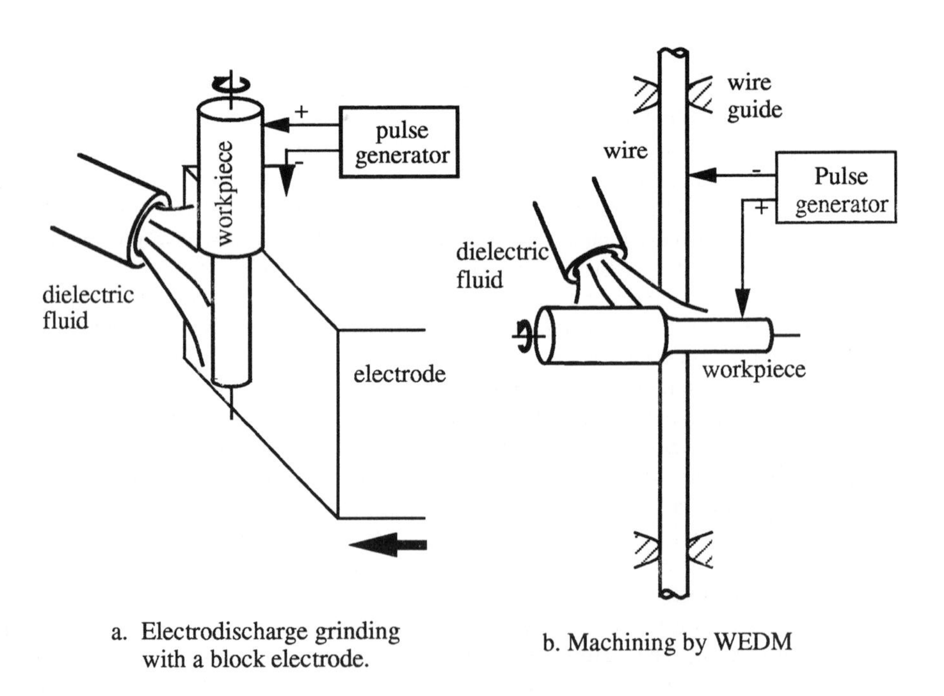

a. Electrodischarge grinding with a block electrode.

b. Machining by WEDM

Fig. 3. Machining of micropins by conventional methods.

The WEDG is based on the idea that the vibration and the deviation of the wire may be zero at the point where the wire contacts with the wire guide. Fig. 4 shows the fundamental

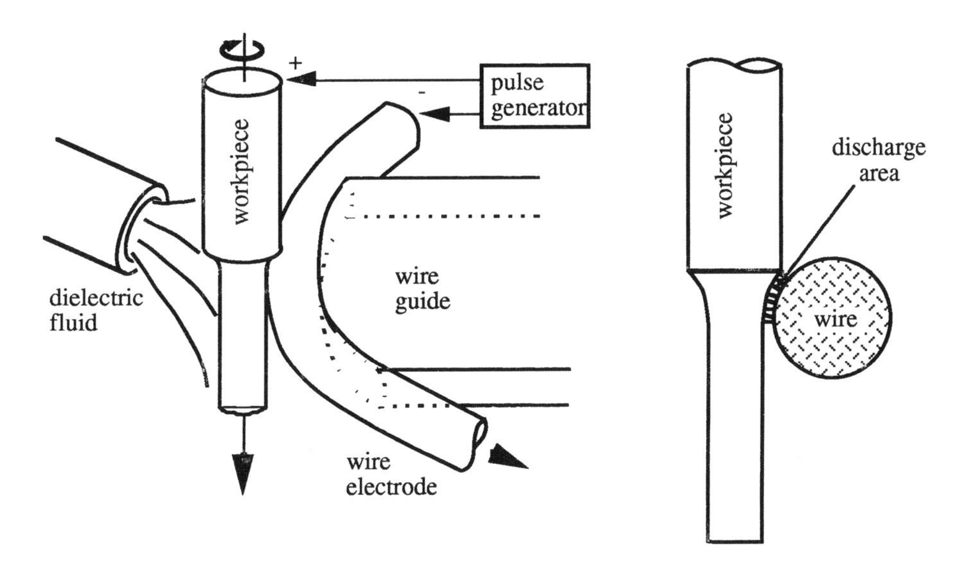

Fig. 4. Principle of WEDG.

Fig. 5. Discharge area in WEDG.

setup of the WEDG method. A wire electrode travels, sliding along a wire guide. Electrical discharge takes place between the wire and the pin. What is most important is that the discharge area must be within the section of the wire between A and B, where the wire is contacting with the wire guide. Some tension must be given to the wire to maintain the contact. However, it is unnecessary to give such high tension as what is applied in usual WEDM, where it is necessary to keep the middle part of the wire straight only by tension. Since the removal rate is very small for such micropins, the traveling speed of the wire is not necessarily large, typically between several millimeters and several centimeters per minute, while it is usually a thousand times faster in commercial WEDM machines.

This setup ensures the accurate position of the wire edge, consequently the high accuracy of the finished pin.

Concerning with the other problem, the unevenness of the gap distance, this method is also promising. Fig. 5 shows the close-up section view of Fig. 4. The workpiece is fed downwards, and the discharge area is limited at the front edge of the wire, as it is indicated in the figure. The straightness of the machined part depends only on the variation of the gap distance at point S. Although the gap distance at S will be affected by the debris concentration in the discharge area, the variation should be small, because the discharge area is very small comparing with the conventional method with a block electrode. Since the discharge area is very small in this system, the debris concentration is low and, consequently, its variation in time is also small. This leads to very small variation of the gap distance, and high accuracy.

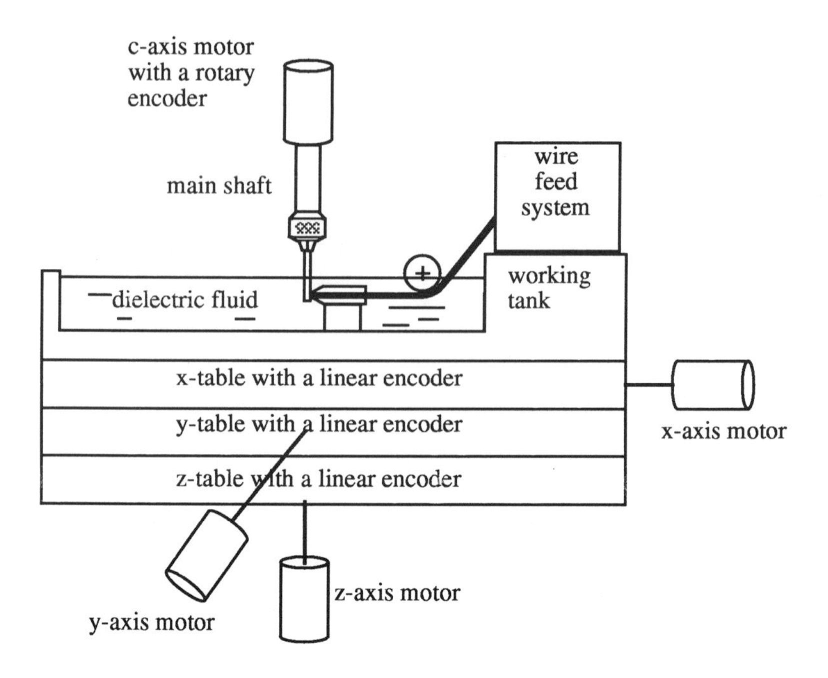

Fig. 6. Scheme of a WEDG machine.

4.4. Equipment

A typical scheme of the WEDG system is shown in Fig. 6. The system is composed of five main parts:

1. Wire feed system with a wire guide, a driving motor and a braking system.
2. A three-axis (x, y and z) slide table which is numerically controlled.
3. Main shaft of which the rotational angle is numerically controlled (c-axis).
4. A pulse generator for discharge.
5. A computer system which controls the 4 axes according to the designed shape of the workpiece and discharge status.

The workpiece blank is chucked on the main shaft and submerged into the working tank on the xyz table. The working tank is filled with the working fluid or the dielectric fluid. Instead of submerging in the fluid, the operation by pouring the fluid through a nozzle set near the working point is also possible. EDM oils or ion-exchanged water is usually used as the working fluid.

A metal wire for normal WEDM machines are usually used as the wire electrode because the diameter of this kind of wire is well guaranteed.

The wire is fed with a speed around 10mm/s with a tension around 2N (for the brass wire of ϕ0.2mm). These values are much smaller than in WEDM as mentioned before.

As for the pulse generator, relaxation type ones are often used because it is suitable for the very short pulses required in micro EDM. Switching type generators are used only when a high productivity is the main requirement. Positive polarity is applied to the workpiece because the removal rate of the anode is much larger than the cathode when a very short pulse discharge is applied.

The gap between the wire and the workpiece must be controlled so as to maintain a stable repetition of discharges. The feedback for this control is usually given by the average current through the gap. This gap control is usually given through the z-axis movement, while other axes are used for special applications. One of such applications is to cut off the workpiece.

The main shaft is rotated in a speed between 500 to 3000 rpm when the product has a round section. For other applications, the rotational angle is fixed or controlled according to the sequence for generating the required workpiece shapes.

By controlling the x, y, z and c (rotation) axes, micropins of various shapes can be machined. Some of the basic machinable shapes are shown in Fig. 7. The minimum radius at the corners is limited by the radius of the wire. A thinner wire is used when a smaller radius is requested. A wire guide with appropriate groove size should be selected for different wire diameters.

Some example pins which are produced by WEDG are shown in Fig. 8.

The WEDG is easily combined with various micromachining processes on the same machine. Combination with micro EDM is the simplest application because the same setup by replacing the wire and the wire guide with a workpiece is ready for micro EDMing. Only the

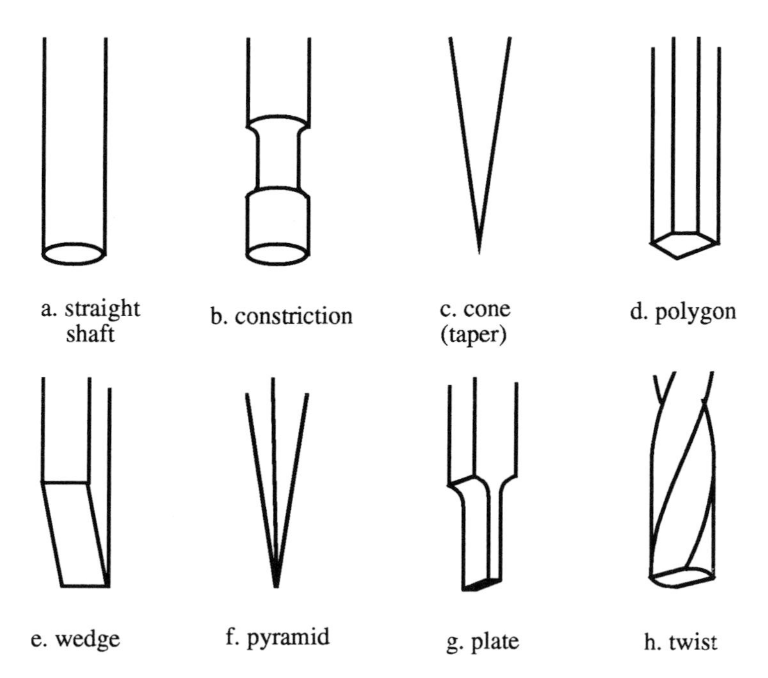

Fig. 7. Machinable shapes by WEDG.

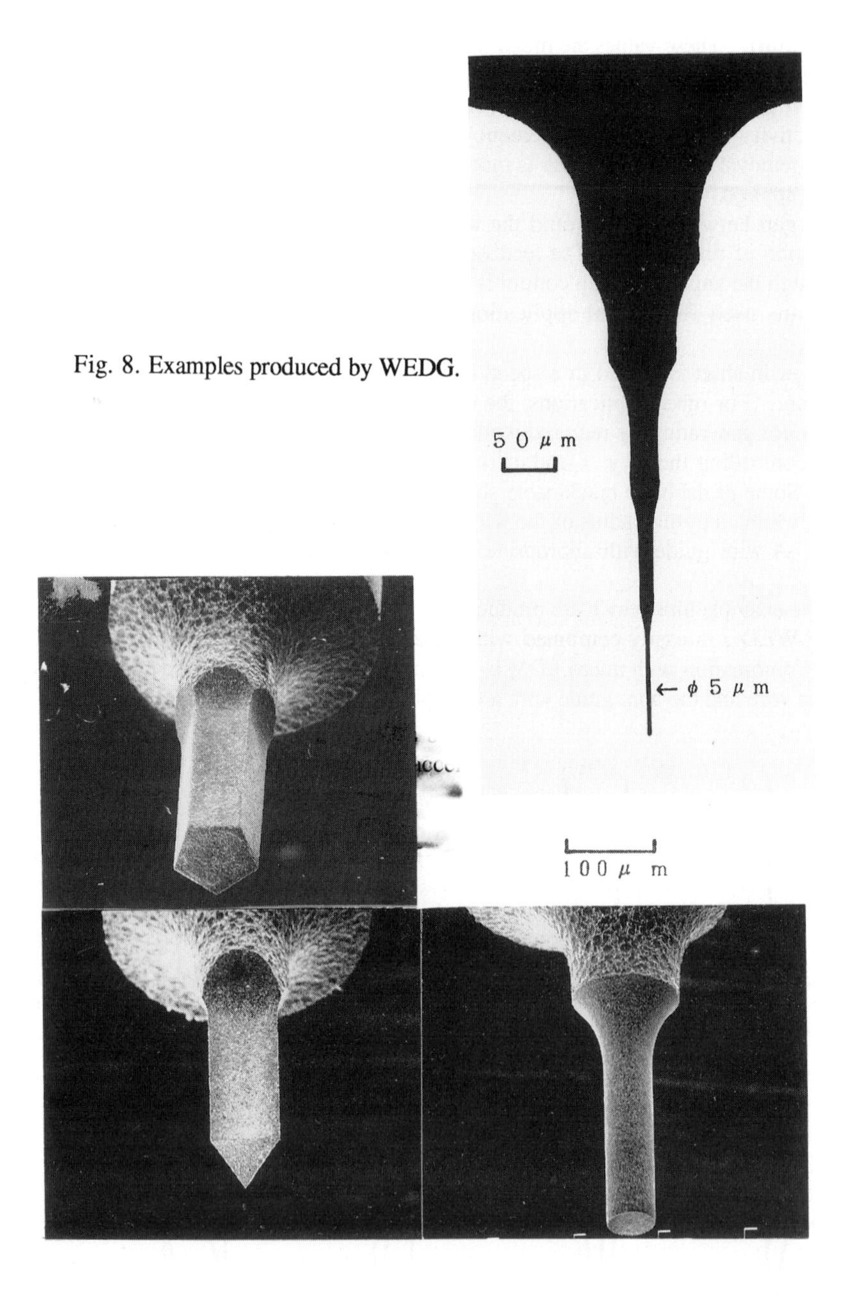

Fig. 8. Examples produced by WEDG.

electrical polarity for the pin should be changed to the negative. Also for mechanical micromachining processes, the WEDG can easily be applied for tool making on the same machine. Attachment of the wire feed system including the wire guide and the pulse generator enables to fabricate the microtools on the machine for such machining processes.

5 MICRO EDM

Those pins produced by WEDG can directly be the electrodes for micro EDM. Various types of holes and cavities can be machined[2,3] by using such electrodes as shown in Figs. 7 and 8.

5.1. Straight microholes

Straight holes are machined simply by feeding the electrode along its axis. The section shape of a hole is determined by that of the electrode with a slight enlargement according to the discharge gap. Any type of convex section is obtainable for the electrode by WEDG.

For holes with a round section, the electrode is often rotated during machining. The rotation improves the roundness of machined holes. An electrode with different section from a round one is sometimes useful to improve the machinability of deep microholes according to the better removability of debris in the gap.

When a concave section hole is machined, sequential operation by changing the electrode or moving it in x-y plane, is necessary. An example for machining a hole with the cross type section is shown in Fig. 9. This machining can be accomplished by turning the electrode 90 degrees after the first machining of a slit and machining another slit over the machined one. Fig. 10 is an example of applying two electrodes with different shapes. This kind of operation is easily realized when micro EDM is combined with the WEDG. The sequence of electrode modification for this example is shown in the figure.

5.2. Cavities

Cavities are obtained simply by stopping the electrode feed at an appropriate position on z-axis. When the precision near the bottom, particularly at the corner, is required, electrode is corrected by the WEDG and several times of finishing operation is given.

For a cavity with a complicated shape, a sequential operation similar to the machining of concave section holes can be applied. An example is shown in Fig. 11. Such sequential operation with simple electrodes has advantages in the machining accuracy to the operation by the copy machining with an entirely shaped electrode, because the correction is much easier for such simple electrodes than for a complicated electrode.

5.3. 3D machining by end-milling type MEDM

By controlling the position of the workpiece along a designed path, three dimensional shapes can be generated with a simple electrode. Fig. 12 shows some examples machined by this process using cylindrical electrodes. The slits in Fig. 12a was machined with vertical electrode feeds shifting the horizontal position 1μm for each stroke. The slit width is 12μm. 500 strokes produce a slit. The cavity in Fig. 12b was machined with a ϕ30μm electrode by moving it along a path which composes a pyramid sunk in a steel ball for miniature ball bearing.

6. MICRO MECHANICAL MACHINING (MMM)

Mechanical machining has a disadvantage that it essentially requires machining force.

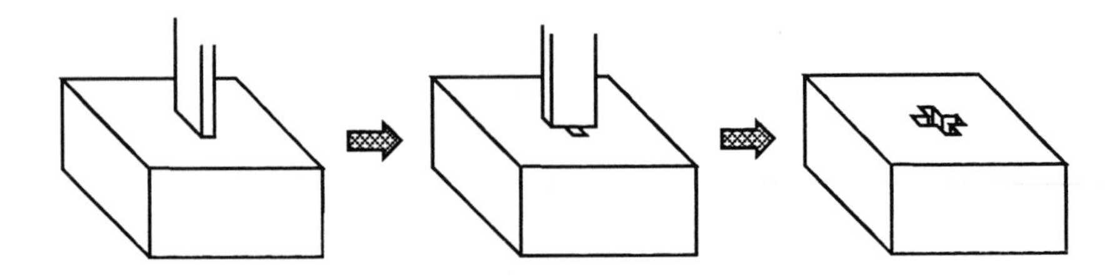

Fig. 9. Sequential operation for holes with a section including concave parts.
(By rotation of electrode.)

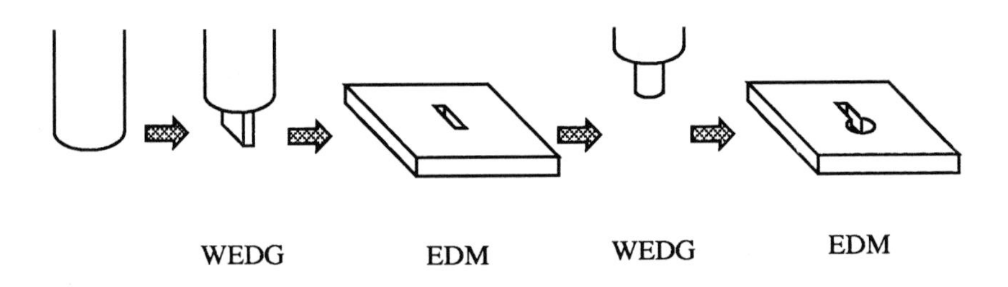

Fig. 10. Sequential operation for holes with a section including concave parts.
(By reformation of electrode.)

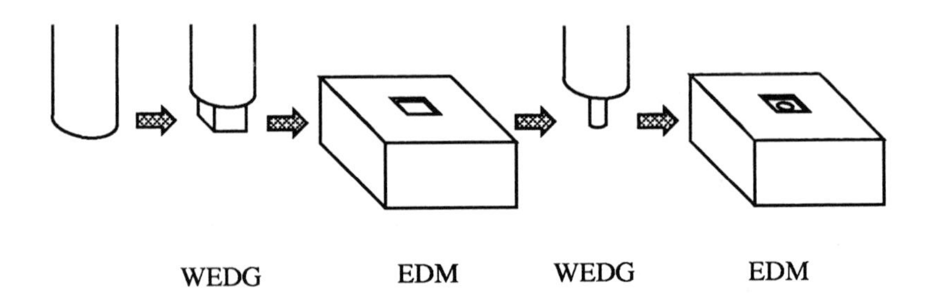

Fig. 11. Sequential cavity formation.

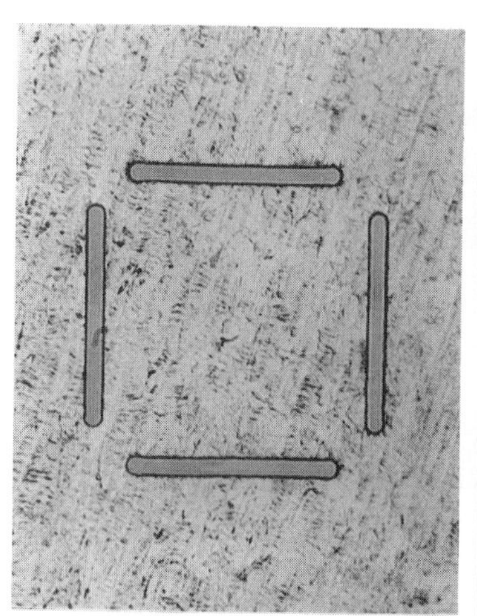

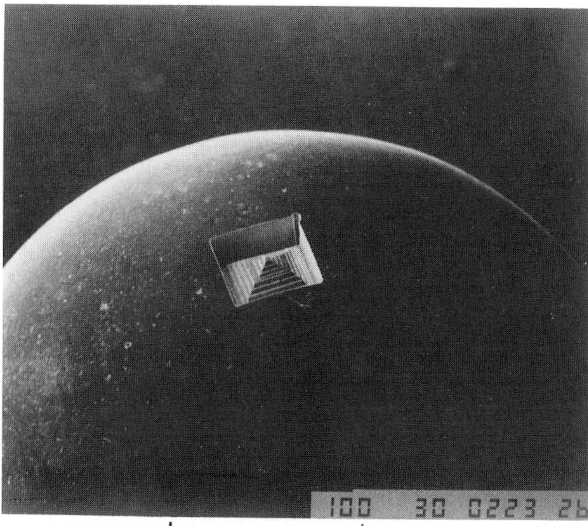

a. Arranged slits complex.

b. Micropyramid.

Fig. 12. Machining of complicated shapes by controlling the electrode path. (Matsushita Research Institute Tokyo Inc.)

However, a carefully prepared tool enables a cutting or shearing with an acceptable small force even for micromachining applications. This issue is important because EDM is usually applicable only for electroconductive materials, while mechanical machining does not care the electrical conductivity of the material. A typical example is the application in hole machining for print circuit boards, where the workpiece is of a composite material basically including an insulator substrate.

6.1. Microdrilling, micromilling[4]

Drilling is a simple and useful process in many applications. Microdrills are available commercially, rather easily down to ϕ40μm. However, there are some problems in using a commercial microdrill. One point is the quality of the cutting edge, and another is the alignment when it is chucked on a drilling machine.

Combination of drilling and tool (drill) making by WEDG is one of the solutions for these problem. A flat drill can easily be fabricated by WEDG as a slight modification of the example, Fig. 7g. The machine for this WEDG-microdrilling system can be with the same construction as Fig. 6. Since the prepared microdrill is used without rechucking, the errors such as tilt and eccentricity are perfectly avoided.

A WC alloy microdrill prepared by WEDG and the holes machined on the same machine are shown in Fig. 13. The workpiece is a stainless steel sheet.

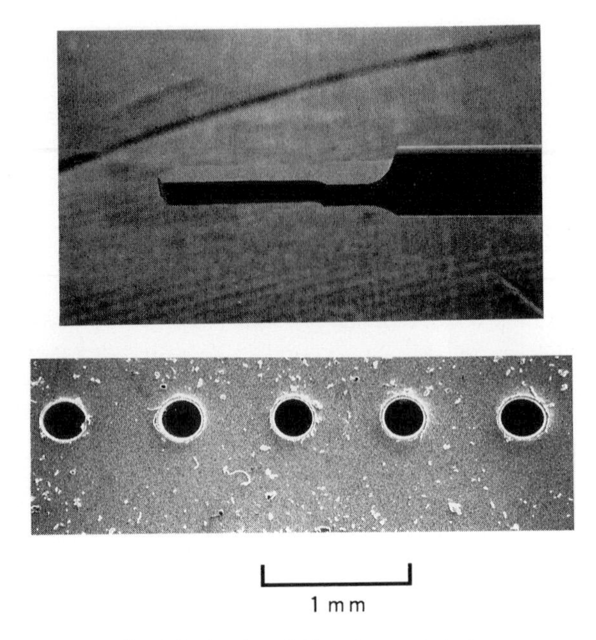

Fig. 13. A microdrill and sample holes.

Fig. 14. A sample groove machined with a microendmill.

By tilting the wire guide in WEDG, a twist drill can also be faabricated.

3.6.2 Micro endmilling

The choice of tools in such system shown in Fig. 6 includes an endmill. Thus, micro endmilling is available as easily as microdrilling. An example of machined groove is shown in Fig. 14.

6.3. Micro ultrasonic machining (MUSM)

A similar system to the above systems can be realized for ultrasonic machining (USM)[5]. By attaching rotational movement on the main shaft, or the actuator-horn setup, and a wire feed system, an USM machine can be equipped with WEDG function. In such system, the preparation and setting of microtools is much easier, compared with conventional USM machines, on which the chucking and alignment is a difficult task to maintain the precision for micromachining. Since USM is a suitable technique to machine brittle materials such as silicon, glass and ceramics, different application from MEDM and MMM will become practical by MUSM.

Fig. 15 shows some examples machined in a silicon plate by USM with on-the-machine tool making by WEDG. Very fine abrasive (grain size $\approx 0.5\mu m$) and a rather high frequency (45kHz) was applied for these examples.

6.4. Micropunching

Punching is very popular mechanical machining method for mass production. However, when a micro product is the target, the preparation of die set is very difficult. A delicate punch and a very small die hole make the adjustment of die set very difficult. Making such die set requires the high skill of the assembly technician and usually it takes a long time for adjustment, even if it is possible. Very often it is not possible at all to obtain a die set for punching holes with diameters of $100\mu m$ or less.

A unique system which does not include the adjustment process makes the micropunching practical[6,7]. Fig. 16 shows the scheme of this special punching process. This is a continuous process from tool making to punching. Essentially, it consists of four stages; (a) electrode making by WEDG; (b) die making by MEDM; (c) punch making by WEDG; and (d) punching. In practical applications, some extra stages such as stripper making will be inserted.

The most important concept in this system is that the positions of the punch and the die are known in the same coordinates throughout the process. Practically, the whole process takes place on the same equipment without any horizontal movement of punch and die. Therefore the punch and the die are at the correct position ready for punching, right after they are fabricated.

In many applications, it takes only one hour from the beginning of tool making to the first punching. In this system, regrinding of the punch which is more often necessary in micropunching is very easy. Just several minutes of interrupt by stage c refreshes the punch, and punching can be continued immediately.

Since no adjustment is necessary in this system, holes with irregular section are also punched without difficulty.

Some microhole examples produced by this process are shown in Fig. 17.

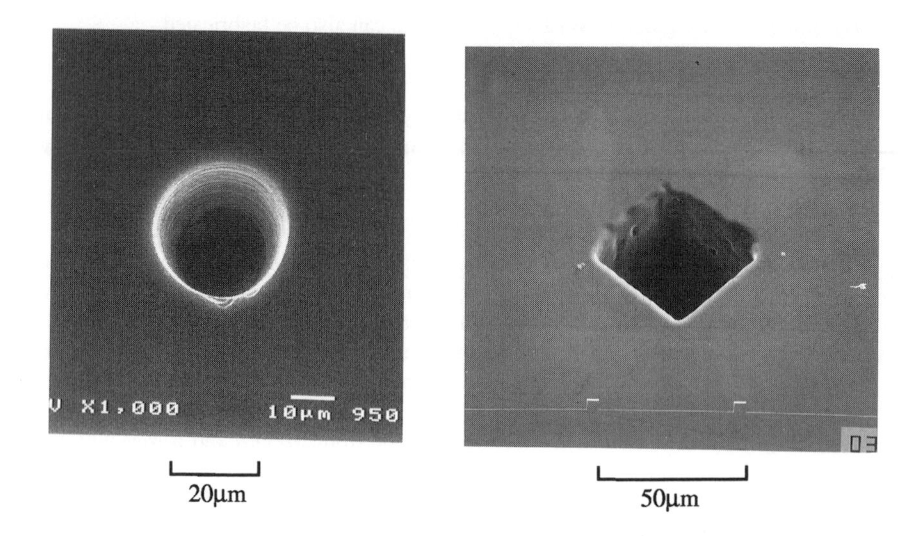

Fig. 15. Samples machined by micro ultrasonic machining.
(workpiece: silicon)

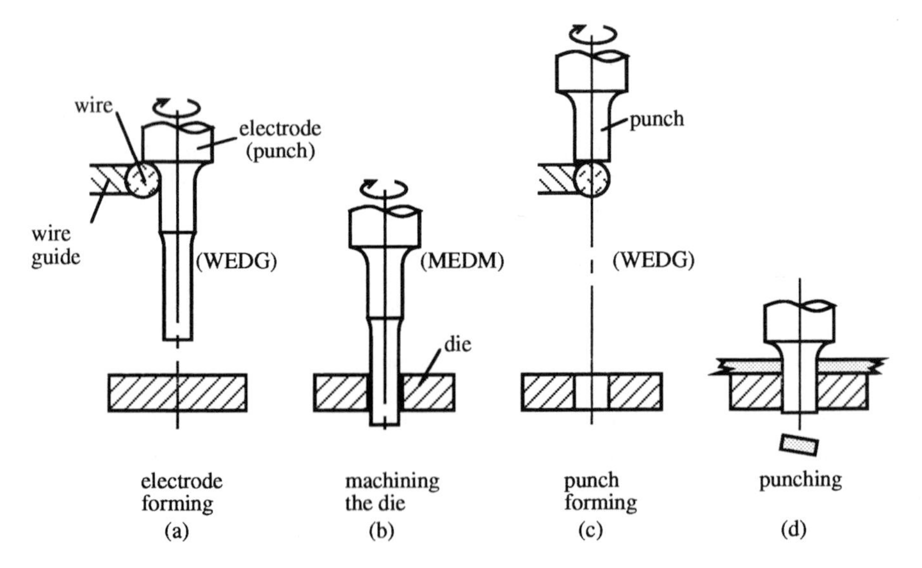

Fig. 16. Micropunching process including tool making.

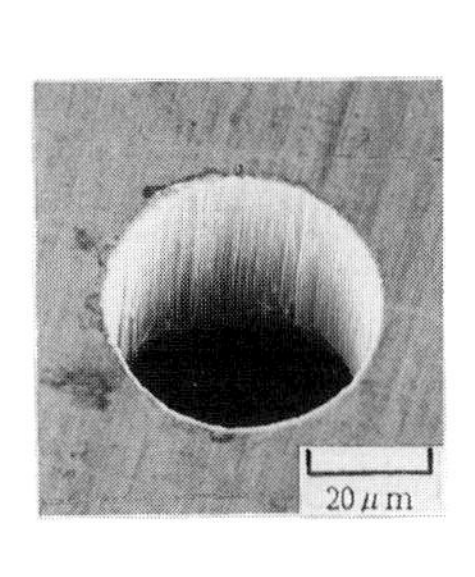

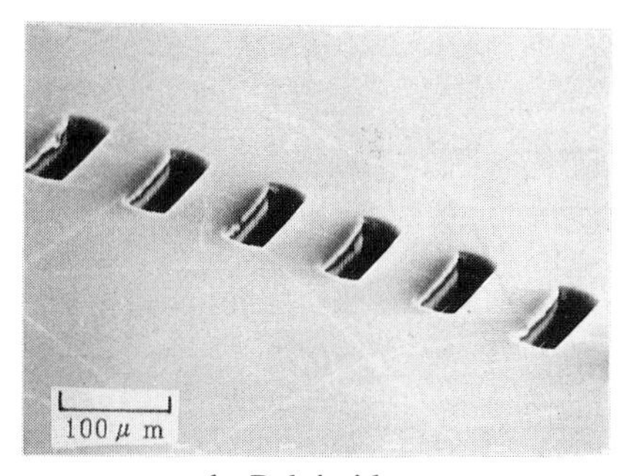

a. Phosphor bronze. b. Polyimide

Fig. 17. Sample holes by micropunching.

7. COMBINED METHOD FOR MICROMACHINING

7.1. Introduction

In some applications, combination of different techniques are effective to overcome the difficulties introduced by the miniaturization. An example applied for micronozzle fabrication[8-10] is introduced in this section.

For micronozzles, long and narrow holes are often requested. Moreover, non straight holes are desirable in many cases, for instance, a hole with a narrow opening and a wider leading channel. such a hole is often difficult to obtain because machining deep microholes and profiling inside of the holes are both difficult tasks by any micromachining processes.

7.2. Outline of a combined process

Fig. 18 shows a process which enables this difficult fabrication. the process is a combination of WEDG, wire electrochemical grinding (WECG) and electroforming.

A core is made first by WEDG (a). This core is designed to have the shape and dimension of the inside of the nozzle to be fabricated. To improve the smoothness of the surface, the core is finished by an electrochemical process (b). Since a wire electrode is used similarly to WEDG, this process is named wire electrochemical grinding (WECG). The gap between the wire and the core is set as large as to avoid any discharge. The electrolyte must have rather high resistivity, around 500kΩcm, because the gross removal is very small. After a core with smooth surface is prepared, the nozzle material is deposited on the core by electrochemical deposition or electrical plating (c), Outer shape of the deposited nozzle is formed, again by WEDG (d). Finally the core is mechanically removed (e).

When stainless steel is used as the core material, the core is easily removed up to the aspect ratio of 40 for $\phi 50\mu m$.

The first advantage of this process is that a large aspect ratio of the hole is easily obtained. Because, a long core, instead of a hole, is much more easily obtained when the diameter is very small.

Secondly, since the core is neither removed nor rechucked from a holder throughout the process, the concentricity between the inner and the outer shapes of the nozzle is ideally

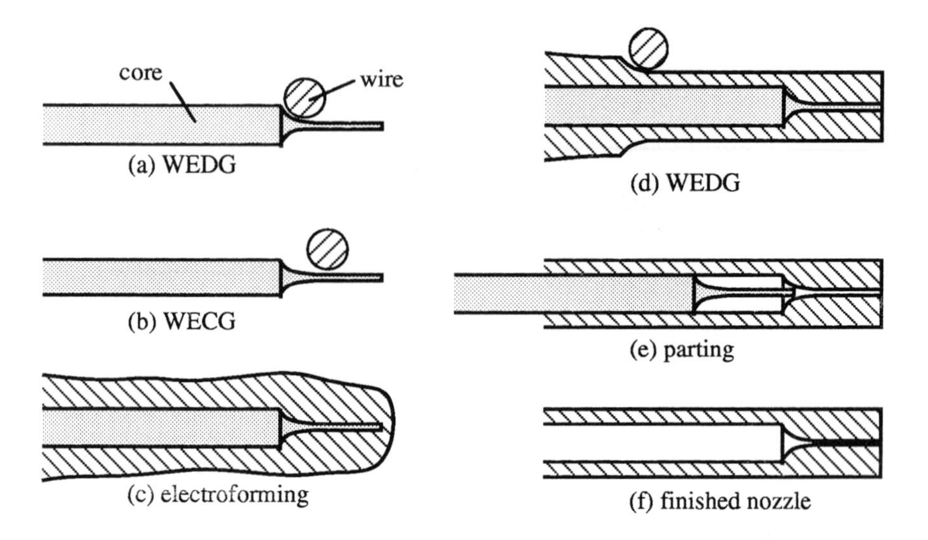

Fig. 18. A combined process for micronozzle fabrication.

maintained.

The third one from the viewpoint of practical applications is the wide choice of inner and outer shapes. Since the core and the nozzle outside are machined by WEDG, various inside and outside shapes are available.

In summary, the combination of three techniques, WEDG, WECG and electroforming, enables the fabrication of nozzles with,

(a) high aspect ratio,
(b) free design of inside and outside, and
(c) smooth inside wall.

7.3. Example

A typical sample product by this combined process is shown in Fig. 19. The nozzle material is Ni. Stainless steel is used for the core.

8. SUMMARY

The basic concept of the micromachining based on machine tool technology was discussed, and several important processes were introduced.

This type of process is suitable for micromachining three dimensional shapes and high aspect ratio holes and pins.

WEDG or wire electrodischarge grinding is a powerful method for precision machining of microtools and micropins.

Using microtools, micro EDM, microdrilling, micromilling, micropunching and micro ultrasonic machining are practically applicable for microproducts ranging down to tens of micrometers. By EDM, several micrometers is already in the practical range.

The combination of several machining techniques realizes more complicated fabrication. A process combining WEDG, wire electrochemical grinding and electroforming was introduced as an example which enables the fabrication of micronozzles with very high aspect ratio and variety of inside/outside shapes.

In the micromachining by machine tool technology, the choice of workpiece material includes metals, alloys, amorphous materials, glass, ceramics, monocrystal materials, silicon, diamond, plastics, composite materials, paper, wood, shells and biological materials.

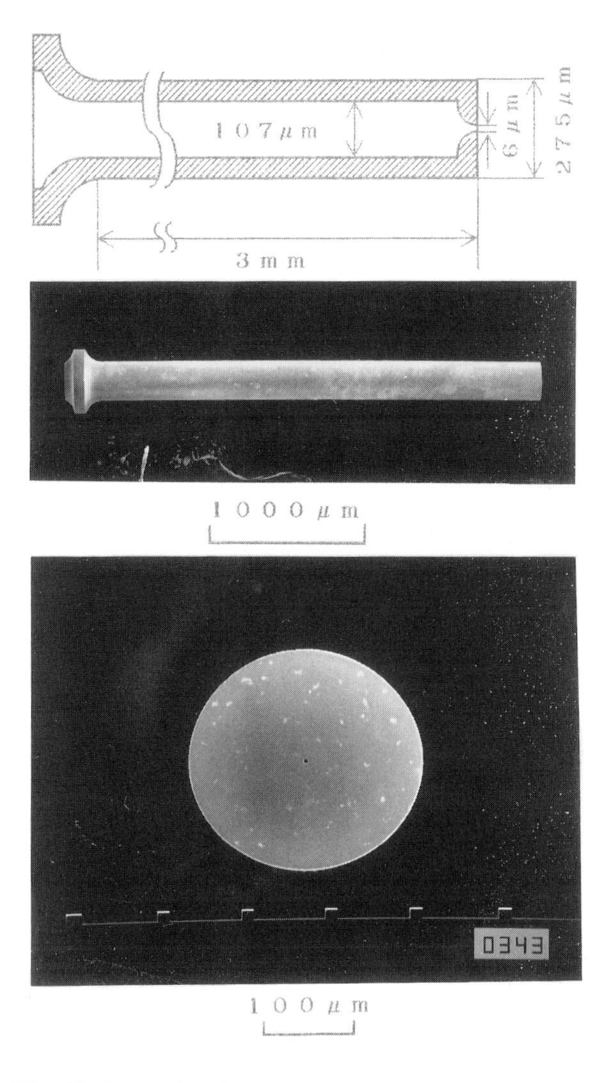

Fig. 19. A sample micronozzle produced by the combined system.

REFERENCES

1. T. Masuzawa and M. Fujino, K. Kobayashi and T. Suzuki, Annals of the CIRP, 34, 1 (1985) 431-434
2. T. Masaki, K. Kawata, T. Sato, T. Mizutani,K. Yonemoto, A. Shibuya and T. Masuzawa, Proc. of Int'l Symposium for Electromachining (ISEM-9) (1989) 26-29
3. T. Masaki, K. Kawata and T. Masuzawa, Proc. of 3rd IEEE Workshop on MEMS (1990) 21-26
4. M. Fujino, N. Okamoto and T. Masuzawa, Proc. of International Symposium for Electro Machining (ISEM XI) (1995) 613-620
5. X.-Q. Sun, T. Masuzawa, M. Fujino and K. Egashira, Proc of Asian Electrical-Machining Symposium '95 (1995) 31-36
6. T. Masuzawa, M. Yamamoto and M. Fujino, Proc. of Int'l Symposium for Electromachining (ISEM-9) (1989) 86-89
7. K. Wakabayashi, A. Onishi and T. Masuzawa, Bull., JSPE, ２４，　４ (1990) 277-278
8. C.-L.Kuo and T. Masuzawa, Proc. IEEE MEMS'91(1991) 80-85
9. C.-L. Kuo, T. Masuzawa and M. Fujino, Proc. of IEEE MEMS '92(1992) 116-121
10. T. Masuzawa, C.-L. Kuo and M. Fujino, Annals of the CIRP, 43, 1 (1994) 189-192

4. Tribological Aspects of Microsystems

K.-H. Zum Gahr

Research Center Karlsruhe
Institute of Materials Research I
and
University of Karlsruhe
Institute of Materials Science II
P.O. Box 3640, 76021 Karlsruhe
Germany

4.1 Introduction

In recent years there has been an increasing interest for using movable microscale structures in microsystems. Rotary or linear micromotors, microturbines, gears, sliders, miniature switches or microrobotic manipulators are fabricated by using technologies such as LIGA process, UV lithography, surface micromachining, laser ablation and others. In general, rotational structures with high aspect ratios offer greater driving torques which are necessary for overcoming friction between rotor and axle. Static friction determines the critical start-up conditions of a rotational structure and kinetic friction influences the speed and power output it will yield. Quite similar as in rigid data storage devices with a flying slider carrying a read/write head, even very small amounts of wear resulting in loose wear debris can cause the loss of the function of such devices or micromechanical components. Hence, friction and wear have been recognized as very important properties of micromechanical systems [1-5].

4.2 Tribology - Friction, Lubrication and Wear

Tribology is defined as the science and technology of interacting surfaces in relative motion and embraces the scientific investigation and practical application of the knowledge about all types of friction, lubrication and wear. Microtribology is specifically engaged in the study of phenomena and interaction of surfaces at atomic and molecular scales [6]. It is a highly multi-disciplinary discipline which uses knowledge of physics, chemistry, biology, mechanics, mathematics, materials science, engineering and others.

Friction is the resistance against the start (static friction) and continuing (kinetic friction) of relative motion between the contacting bodies in the direction of the common interface. It arises at the real area of contact, i.e. due to surface forces, physical, chemical and/or mechanical interactions on the scale of asperities or microasperities. Friction can be desirable (brakes) or undesirable (bearings) but wear is always undesirable if we do not consider some types of running-in wear.

Wear is the gradual change of geometry and progressive loss of material from the surface of a solid body owing to mechanical action. In general, wear of a solid body can be caused by a solid, liquid or gaseous counterbody or a mixture of these. Wear is the result of overstressing of the material in the immediate vicinity of the surface and/or of destroying a beneficial surface film (lubricant, chemical reaction layer). It is rarely catastrophic, but it does reduce operating efficiency of a mechanical system. Generation of wear debris, particularly in tribosystems with small clearances such as used in magnetic storage devices or micro mechanical systems, may be more serious than the actual dimensional changes of components. Friction and wear are related to each other in the sense that frictionless processes will not result in wear. On the other hand, increasing friction forces do not automatically imply increasing wear loss.

Separation of solid surfaces by lubricating films can be very effective for reducing friction and wear. Gases, liquids, pastes or solids are used as lubricants depending on the tribological system. Primary function of lubricants is to avoid solid/solid contact and absorb the dissipation of frictional energy. During relative motion all shearing processes should be constricted to the lubricant. An acceptable lubricant must exhibit high load carrying capacity and low resistance against shearing for obtaining low friction. In addition atoms or molecules of the lubricant should adhere strongly to the surface of the load carrying solid bodies owing to physical and/or chemical reactions. A lubricating film should be formed on the surface that can withstand the shear stresses encountered in the contact area.

Friction and wear are not intrinsic material properties but are characteristics of the tribological system [7]. The solution to a special friction or wear problem, maybe by using lubricants, other materials or a different design, depends upon the exact identification of the parameters of the tribosystem and the nature of the problem. Application of systems thinking or systems analysis can be very useful for describing tribological processes. The purpose of tribosystems is the transformation and/or transmission of "inputs" into "outputs" which are used technologically. Useful inputs and outputs may be classified in motion, work (mechanical, hydraulic, pneumatic, chemical, electric or thermal), materials or mass, and information. Undesirable inputs may be vibration, heat, material (dirt) and humidity or in general a chemically reacting atmosphere. Loss-outputs are friction and wear which can lead to undesirable outputs such as wear debris, heat, vibration or noise. Figure 1 shows a schematic functional and structural description of tribosystems in general. In systems thinking, the structural description is considered as internal and the functional description as external, meaning that it analyses the connection between the tribosystem and the rest of the micromechanical system. The structural description analyses the internal structure of a tribosystem, its elements such as solid body, counterbody, interfacial element and environment, their properties and their interactions. The structure of a tribosystem is, in general, changed with time through the action of friction and wear. This may result in a change of the functional behaviour of the system, either wanted (running-in) or unwanted (damage). The interfacial elements can be lubricants, adsorbed layers, tribochemical reaction products, dirt or wear debris. As a special case, the interfacial element may be absent. Related to the interfacial element, friction and wear processes are called unlubricated or lubricated.

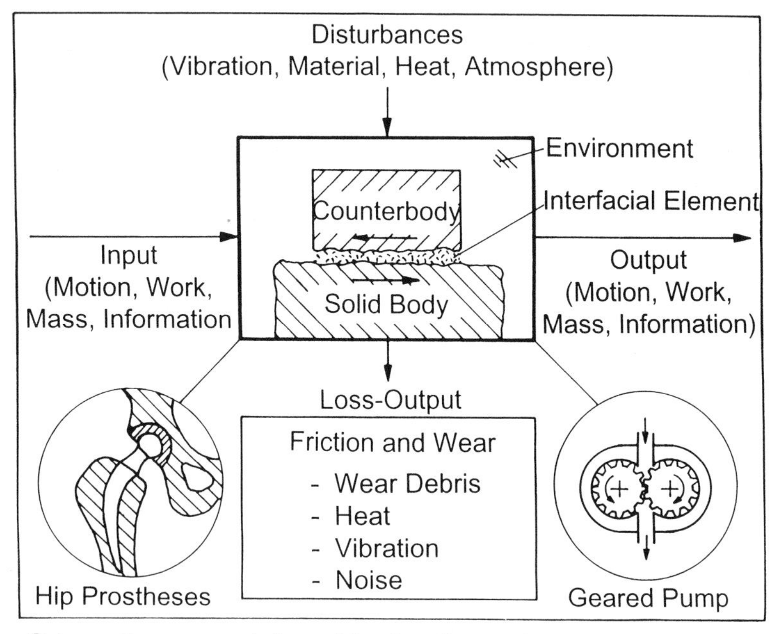

Fig. 1: Schematic representation of the function and structure of tribosystems.

Wear modes, e.g. sliding wear, rolling sliding wear, impact wear, or abrasive wear, are distinguished due to the kinematics, action and/or physical state of the counterbody. Wear mechanisms (Fig. 2) describe the energetic and material interactions between the elements of a tribosystem [7]. Abrasion causes removal of material due to scratching by a harder counterbody, e.g. mineral particles. Adhesion results in formation and breaking of interfacial adhesive bonds, i.e. cold-welded junctions. Surface fatigue leads to formation of cracks in surface regions due to stress cycles that result in the separation of material. Tribochemical reaction is characterized by formation of chemical reaction products as a result of chemical interactions between the elements of the system initiated by the tribological action.

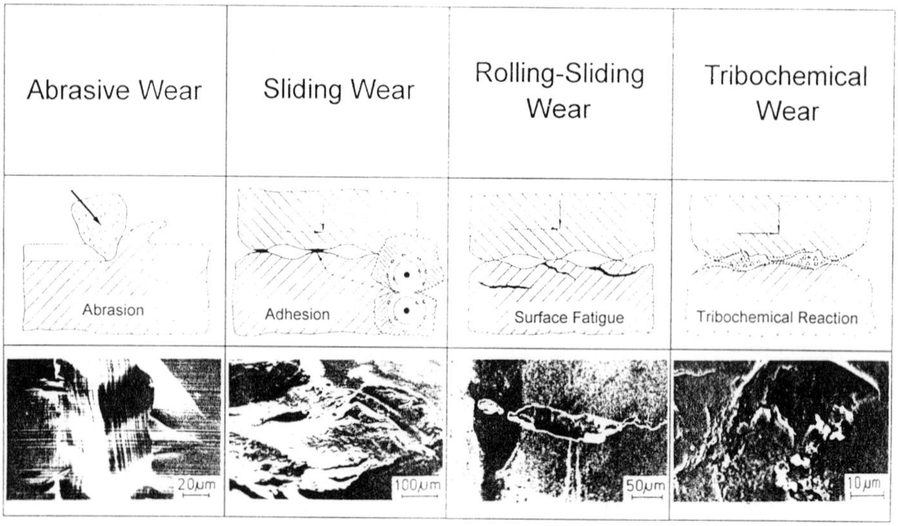

Abrasive Wear	Sliding Wear	Rolling-Sliding Wear	Tribochemical Wear
Abrasion	Adhesion	Surface Fatigue	Tribochemical Reaction
20μm	100μm	50μm	10μm

Fig. 2: Representation of wear mechanisms and resulting worn surfaces as a function of wear mode.

4.3 Surface Structure and Tribological Contact

Tribological interactions of a solid body and a counterbody are constricted to relatively thin surface layers or even to the scale of surface asperities in the case of very mild loading. It follows that tribological behaviour depends strongly on surface properties. With increasing miniaturization of micromechanical components, the ratio of surface to volume increases and hence the influence of the surfaces in contact on functional behaviour of the microsystem.

The term "surface" can be understood as the transition of a solid body into its environment. This includes instances where the structure is altered by changing the environmental conditions. Surfaces can be described by their topography and structure and also by their physical, chemical, mechanical and other properties. Figure 3 shows different features of surfaces depending on the scale. Surface topography typical for the fabrication technique used can be displayed by three-dimensional profile maps (Fig. 3a) recorded by a stylus profilometer. Surfaces contain irregularities or hills and valleys of different shape and size. In interpreting these profile maps on a scale of about 0.1 to 100 μm, it should be recognized that a substantially greater magnification has been used in the vertical than in the horizontal direction, resulting in a distorting effect on the profile. Technical surfaces are covered by environmentally initiated reaction layers such as oxides and/or adsorbates. Atoms or molecules adsorbed by physical and chemical mechanisms, usually up to a thickness of about 0.4 nm, change the physical, chemical and electrochemical properties of the surface. Beneath these layers there can exist surface zones highly deformed and of high defect density and containing internal stresses which results in mechanical properties far different from those of the bulk material (Fig. 3b). Scanning probe microscopy (STM, AFM, FFM) can be very effective for studying submicrometer or nanoscale structural details of solid surfaces or surface films (Fig. 3c). On the atomic scale (Fig. 3d) the surfaces of solids are formed by terraces, ledges, kinks, adsorbed atoms, impurity atoms and terrace vacancies according to the terrace-ledge-kink (TLK) model.

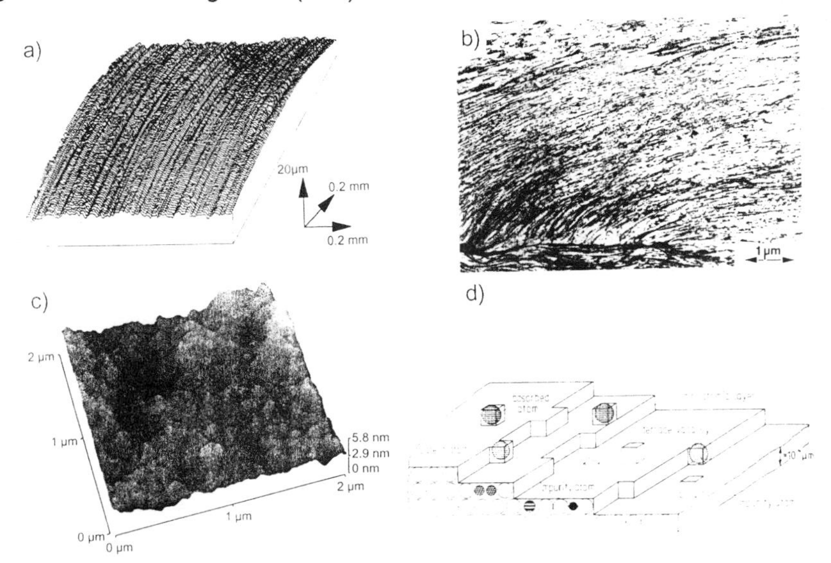

Fig. 3: Different aspects of surface structures (a) three-dimensional stylus profilometry of a ground surface of alumina ceramic, (b) highly deformed near surface structure of a metallic Cu-40Zn alloy, (c) atomic force micrograph (AFM) of the surface of a polished single-crystal Si (100) wafer and (d) terrace-ledge-kink (TLK) model of surface topography on an atomic scale.

The configuration of atoms at a free surface should be that with the minimum free energy. Ideally, the structure is the result of a compromise between the low energy arrangement of surface atoms in a perfect lattice structure and the energy involved in producing the facetted surface required to minimize surface energy. Real surfaces differ from these ideal arrangements owing to effects such as relaxation, reconstruction and the presence of intrinsic and thermally activated defects [8]. Relaxation and reconstruction refer respectively to changes in the interplanar spacing and in the crystalline arrangement of the atoms in the outermost surface layer as a result of differences between the interatomic forces at the surface and in the bulk material. Common defects on single-crystal surfaces are growth steps developed during solidification and growth from the liquid or vapour phase, cleavage steps (river patterns) on brittle inorganic materials, ledges and steps from dislocations emerging the crystal surfaces, segregated atoms, or etch pits due to chemical reaction of high-energy sites, e.g. where the dislocations emerge from the surface, with the environment or constituents of lubricants. Grain boundaries and phase boundaries are additional defects on surfaces of polycrystalline and multiphase materials, respectively. Figure 4 shows some examples of surface features on ceramic materials. Surface preparation by machining can produce plastically deformed surface layers (grooves and indentations) on ductile or cracks and pull-out of grains on more brittle materials. Porosity and preferred crystal orientations (textures) can be other features on surfaces.

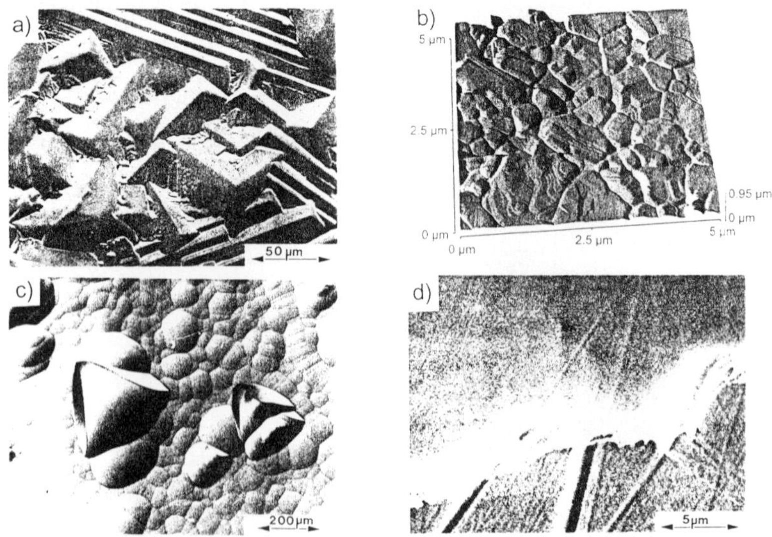

Fig. 4: Surface features on ceramic materials (a) growth steps of crystallites after laser remelting of Al_2O_3, (b) grain boundaries on Al_2O_3 after thermal etching, (c) etch pits on polished sapphire owing to chemical reaction with phosphoric acid and (d) chemical reaction film on anorganic (cordierite) glass.

Surfaces of metals exposed to air are covered with oxide and contaminant layers of a thickness of about 1 to 10 nm. Contaminant layers are built up by adsorbed gases, water molecules and hydrocarbons. Such layers reduce friction between solid surfaces in sliding contact but only at the beginning of relative motion because they are easily removed due to the weak physical rather than chemical bonds.

Properties of solid surfaces [7] can strongly be influenced by environmental effects (Fig. 5). Films of liquids such as organic acids can increase the mobility of dislocations, i.e. decrease the surface hardness (Fig. 5a). By contrast, oxide films can substantially harden the surfaces of metals, which has been confirmed on nickel covered by an oxide layer about 5 nm in thickness [9]. Adsorbed atoms of an aqueous film can change the ductility and fracture behaviour of surface crystals. The cohesive binding strength between atoms near the surface can be altered by the presence of active anions (Fig. 5b). The presence/absence of adsorbed anions favours/depresses brittle cleavage fracture, with an opposite effect on ductile rup-

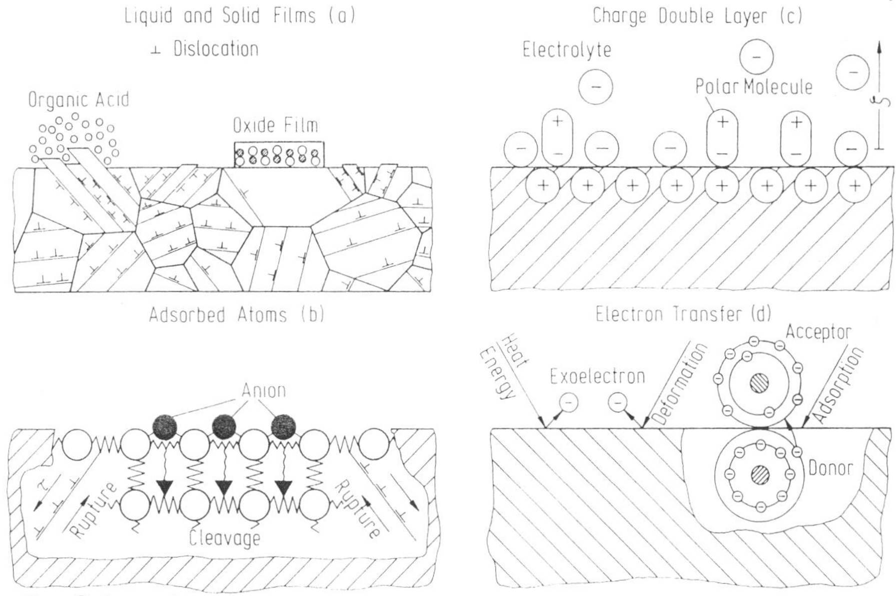

Fig. 5: Schematic representation of environmental effects influencing mechanical properties of solids.

ture. In general, there is an electrostatic potential difference associated with a double charge layer (Fig. 5c) at a metal/electrolyte interface. The Helmholtz double layer consists of an excess or deficiency of electrons at the solid surface and a layer of ions of charge opposite in sign at the interface to an electrolyte. The distribution of charge at a surface can be changed by chemical adsorption of charged molecules, for example polar water or impurity ions. The adsorbed molecules and the change in charge alter the mutual interaction between crystal defects, which is reflected in changes in near surface dislocation mobility and, macroscopically, in strength or hardness. The mechanical properties of solid surfaces can also be modified by electron transfer processes (Fig. 5d). Deformation and/or adsorption lead to changes in the binding energy and electronic state of the surface electrons. Processes such as exoelectron emission, triboluminescence and the formation of donor-acceptor complexes have been reported [10].

Technical surfaces are far from ideally smooth, and exhibit more or less roughness (Fig. 3a). The characteristics of surfaces are described by the arrangement, shape and size of individual elements such as asperities consisting of hills and valleys on a microscopic scale. Hence, the static contact between two solids is generally discrete and occurs on single "point" contacts. These contacts are elastically and/or plastically deformed and add up to the true area of contact which can be some orders of magnitude smaller than the apparent area of contact. During relative sliding of ductile materials such as metals the area of contact is increased owing to plastic deformation. The intimate contact between asperities on two mated clean solid surfaces leads to sticking together which is called adhesion. Surface forces involved in adhesion can be effective only over very short distances, in the range up to about 3 nm. Thus growth of the real area of contact during sliding becomes necessary for strong adhesion. A contaminant layer even one monolayer thick can substantially reduce the adhesion forces.

Figure 6 shows the change in tribologically relevant properties when the scale of view is reduced from the macroscopic to the atomic level. On the scale of a few mm, the local stresses can be calculated from the geometrical shape and size of the contact area and compared with bulk properties of the materials measured in conventional hardness or strength tests. The individual interfaces between the solids may be treated as micro-Hertzian contacts to a first approximation. Tribological behaviour is determined by the properties of a very small volume of material in the outermost surface layer alone in conjunction with environmental and loading parameters. As the volume stressed decreases, classical continuum mechanics becomes of doubtful applicability and surface properties can differ substantially from bulk properties of the solids. At the nano- or atomic scale, surface defects and surface films affect tribological contact and hence friction and wear. As a result of formation of secondary or primary bonds between micro-asperities on the contacting surfaces, individual atoms or groups of atoms may be detached from one surface and transferred to the other [11].

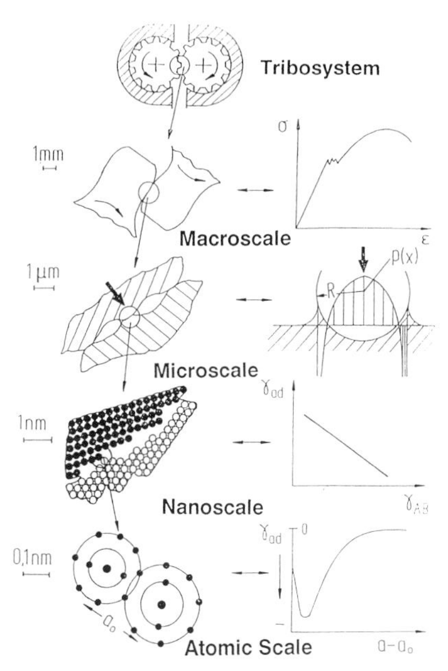

Fig. 6: Description of tribocontacts and relevant properties at various dimensional scales: p = contact pressure, γ_{ad} = work of adhesion, γ_{AB} = interface energy, ε = strain and σ = stress [6].

A simplified estimate of the tendency for adhesion can be obtained from a modified Dupré equation

$$\gamma_{ad} = \gamma_A + \gamma_B - \gamma_{AB} \tag{1}$$

where γ_{ad} is the work of adhesion, i.e. the energy required to separate the adherent surface areas and γ_A, γ_B and γ_{AB} are the surface free energies of the contacting materials (A,B) and of the formed interface (AB) respectively. The elastic and plastic deformations which accompany formation of adhesive junction are not taken into account in this model. It can be concluded that an increased tendency for adhesion (high values of γ_{ad}) will result from high values of the surface energies γ_A and γ_B and a low interface energy γ_{AB} between the mated surfaces. To a very rough approximation, metals and solids of covalent bonding have surface energies ranging from 1000 to 3000 mJ/m^2, solids of ionic bonding 100 to 500 mJ/m^2 and polymeric materials less than 100 mJ/m^2. In general, an amorphous structure such as a glass exhibits a lower surface energy than a crystalline structure, since the atoms can more easily arrange themselves to positions of low energy.

Fig. 7 shows some experimental results which support the equation (1). Adhesion between contaminated or oxidized metallic surfaces (Fig. 7a) is lower than between clean metallic surfaces [12-15]. Adhesion decreases with increasing free energy

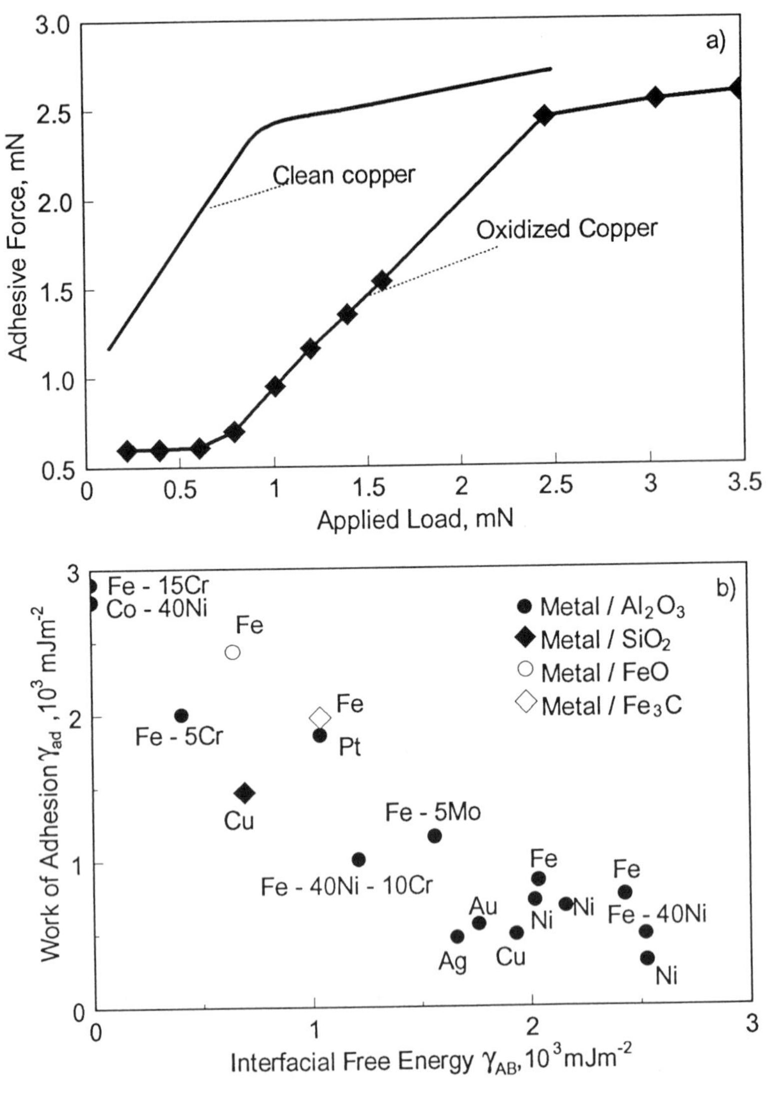

Fig. 7: Adhesion between metallic and oxidized surfaces (a) force to fracture adhesive junctions between iron (001) surface and copper in vacuum at 20 °C versus previously applied normal load, after data of D.H. Buckley [12] and (b) work of adhesion versus interfacial free energy between metals or alloys and nonmetals, after data of E.D. Hondros [7].

γ_{AB} of the interface. An influence of solute atoms such as Cr or Mo on adhesion is also exhibited. Increasing the Cr content in iron up to 15 % resulted in increasing adhesion in contact with Al_2O_3 (Fig. 7b).

Low coverage with nonmetal atoms (O, N and also C or H) may increase adhesion up to a maximum value, at about 0.5 monolayer for iron [16] (Fig. 8), but a further increase in adsorbed atoms leads to decreasing adhesion [17]. Pull-off forces (Fig. 9) for separating adhesive junctions between solid surfaces can substantially be reduced by contaminant layers such as adsorbed hydrocarbons [14] or water molecules at high humidity of ambient air [18]. The contaminant layers on the as-received surface (Fig. 9a) reduced the adhesion for all the Mn-Zn ferrite-metal couples. The contact radius was in the range of 1 to 2 μm.

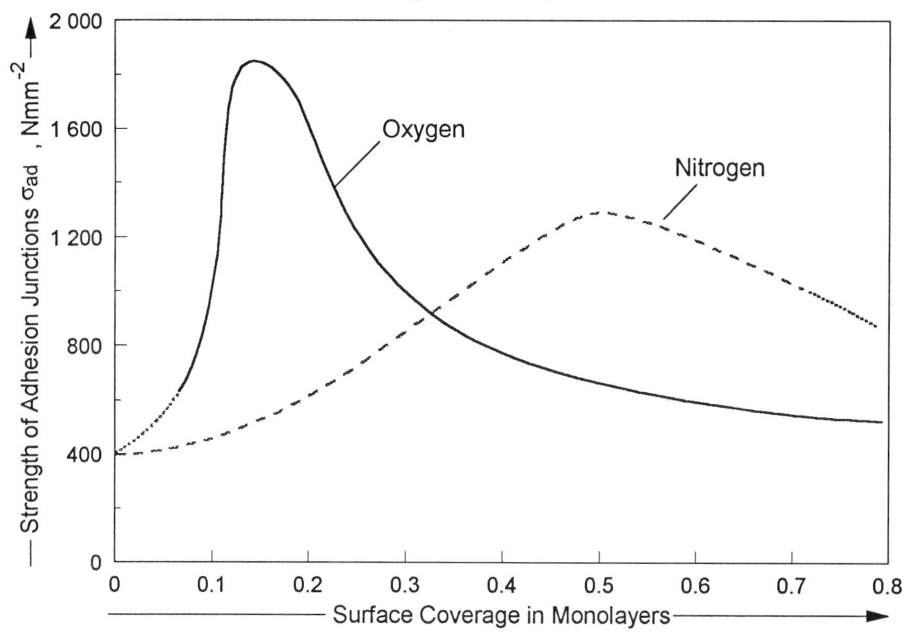

Fig. 8: Strength of adhesion junctions for oxygen or nitrogen coverage at the interface between two iron surfaces. Surface coverage in monolayers, i.e. number of nonmetal atoms per iron atom on the surface, after data of Hartweck and Grabke [16].

An atomic force microscope (AFM) was used for studying the attractive forces in Si-Si nanoscale contacts as a function of relative humidity in air (Fig. 9b). The pull-off force for separating the Si-tip from the Si (100)-wafer surface increased with humidity [18]. This can be explained by increasing forces with increasing humidity due to water molecules forming a capillary meniscus between tip and surface [19].

4.4 Friction and Wear

Tribological tests with real micromechanical systems are generally to be preferred for controlling functional behaviour and estimating service life but showing in practice a large scatter of results due to insufficiently known loading, surface and ma-

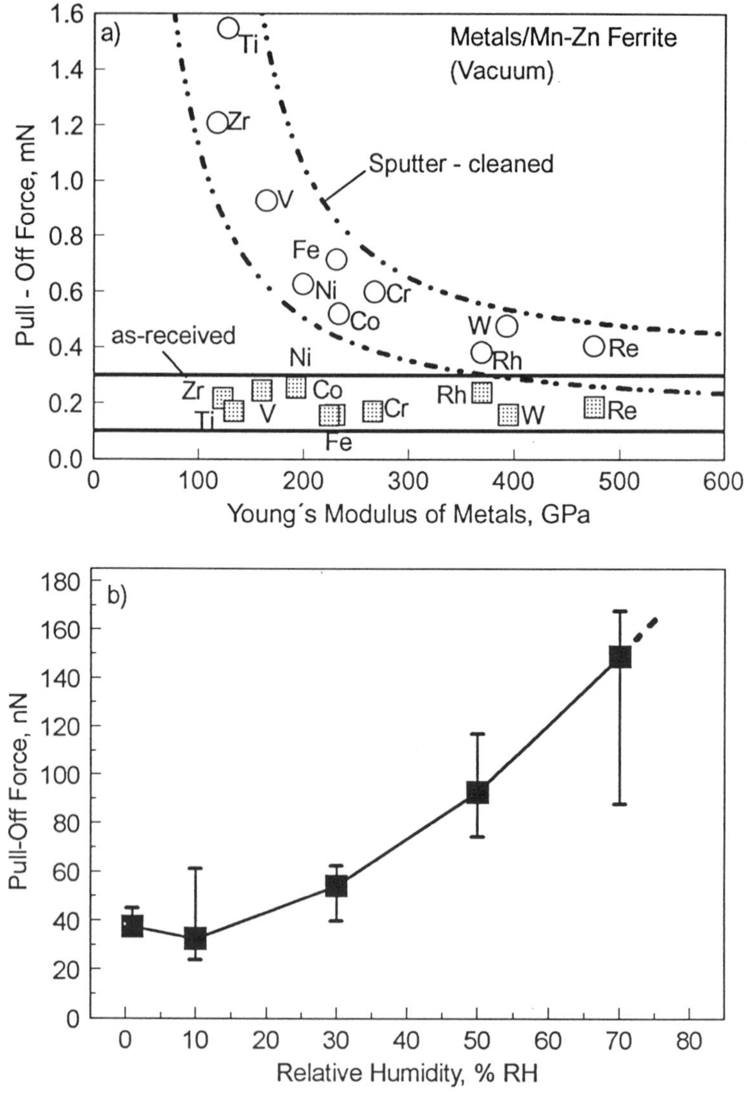

Fig. 9: Pull-off forces for breaking adhesive junctions (a) versus Young's modulus of metals in contact with polycrystalline Mn-Zn ferrite in vacuum for sputter-cleaned and as recieved surfaces after K. Miyoshi [14] and (b) versus relative humidity in air for silicon-silicon nanoscale contacts, after S. Franzka, K.-H. Zum Gahr [18].

terials parameters. Model tests using simplified geometries and strictly controlled, well-known loading, environmental, surface and materials properties, which are adapted to that of the real micromechanical system in service, are very important for understanding friction and wear mechanisms on a microscale and also for the design of a microsystem.

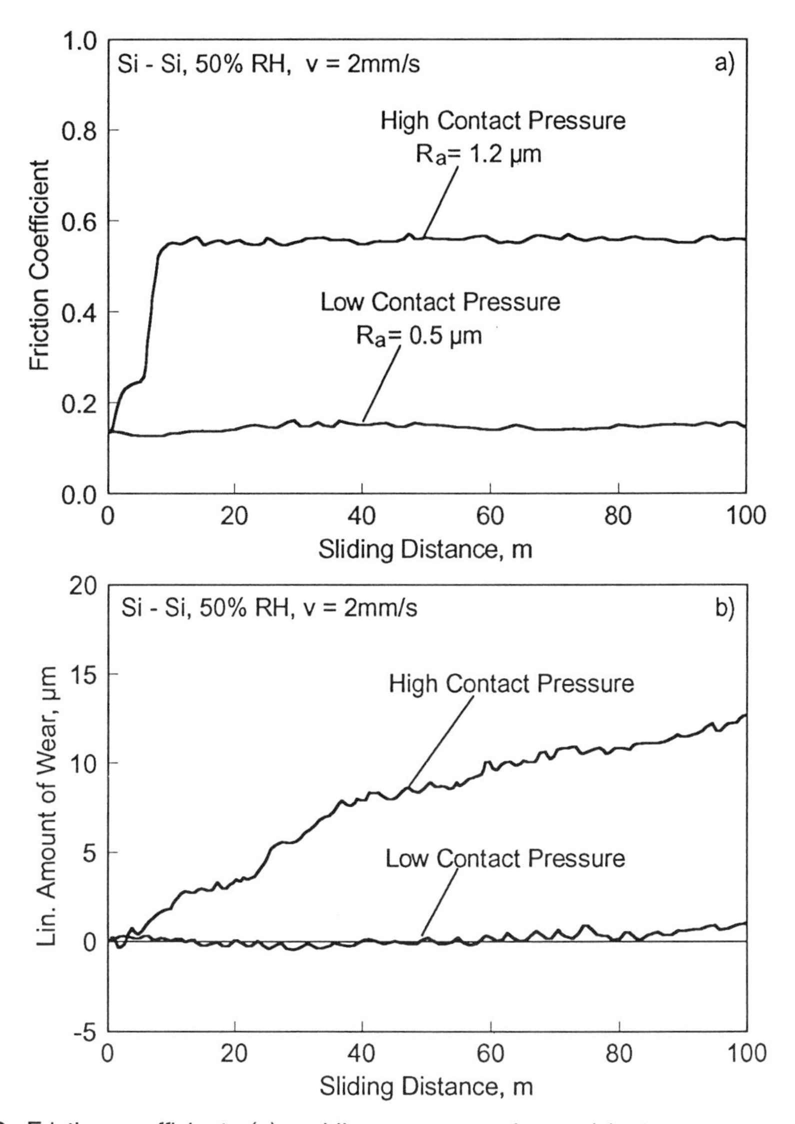

Fig. 10: Friction coefficients (a) and linear amount of wear (b) of self-mated single-crystal silicon versus distance during unlubricated, unidirectional sliding in air.

At some examples, tribological behaviour of materials relevant for micromechanical devices will be discussed in the following.

Silicon micromechanics is a widely used technique for producing movable components such as micromotors. Hence, characterization of tribological properties of silicon is essential in order to fabricate reliable microparts. Figure 10 shows results of friction and wear studies [20] of self-mated single-crystal silicon using pin-on-disk geometry with miniaturized specimens. The pin specimens had hemispherical tips of radii between 0.2 and 0.9 mm, were ground to R_a (c.l.a.) values between 0.5 and 1.2 μm and were loaded by a normal force of 100 mN. Unidirectional, unlubricated sliding tests were run in air at a speed of 2 mm/s, at room temperature and a relative humidity of 50 %. Small radii of curvature of the pins resulted in high applied contact pressure and in combination with high surface roughness (R_a = 1.2 μm) in high kinetic friction coefficients of about 0.55 to 0.60 (Fig. 10a). Under these experimental conditions severe wear damage occurred. In contrast, under mild loading conditions friction coefficients between 0.10 and 0.20 and extremely small wear (Fig. 10b) were measured after a sliding length of 100 m. From other studies [2], kinetic friction coefficients of self-mated Si couples between 0.25 to 0.35 have been reported.

Figure 11 shows worn surfaces of the Si-wafer disks [20]. Severe damage by grooving and formation of relatively large wear particles (Fig. 11a) were observed at high loading conditions according to Figure 10. At higher magnification,

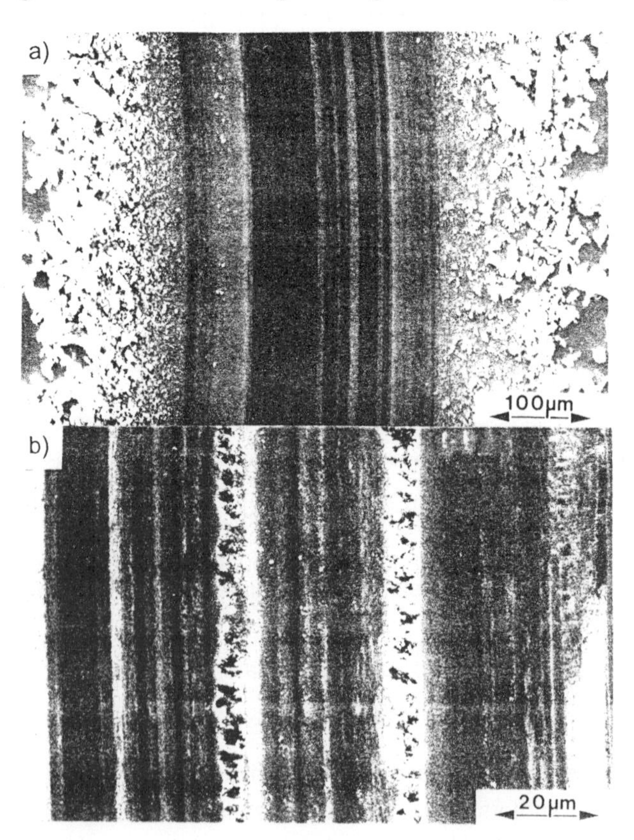

Fig. 11: Scanning electron micrographs of worn surfaces of single-crystal silicon after unlubricated, unidirectional sliding against Si pins under high contact pressure in air.

microcracking becomes visible on the wear path (Fig. 11b). In agreement with other studies [21,22], it can be concluded from the presented data, that under mild loading in air, i.e. low applied contact pressure and smooth surfaces in sliding contact, tribochemical reaction dominates the tribological interaction between the unlubricated, self-mated Si surfaces (Fig. 12). Surface layers of a thickness of some nanometers owing to oxidation (at lower humidity) and/or hydrooxidation (at greater humidity) and depending on the surface cleaning also contaminants such as hydrocarbons determine friction and wear. With increasing severity of loading (Fig. 12), a transition from tribochemical wear to abrasive wear occurs. The small friction coeffi-

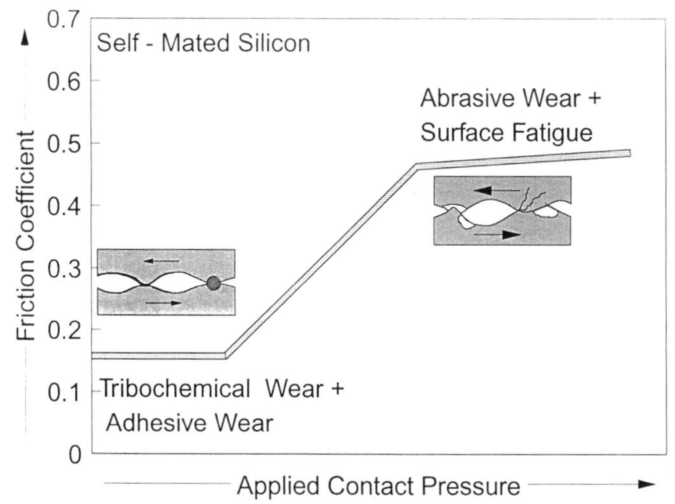

Fig. 12: Schematic drawing of friction coefficient and wear mechanisms of self-mated silicon during unlubricated sliding in air as a function of the applied contact pressure.

cients at the beginning of the tests (see Fig. 10a) display the influence of surface layers. Abrasive wear mechanisms include micromechanical interactions between surface asperities and destroying of tribochemical reaction layers and lead to a combination of microploughing, microcutting and microfatigue (Fig. 11b). With increasing loading an increasing tendency for microcracking occurs owing to the low fracture toughness of silicon wafers (about 0.7 MPa $\sqrt{m}$). Values of static friction coefficient can be above 1.0 (values of 4.9 ± 1.0 have been reported [23]) whereby these values are a strong function of surface conditions, environment, area of true contact and loading, i.e. of all factors which affect adhesion, capillary forces in high humidity or micromechanical interactions of surface asperities. In general, an influence of the size and shape of the contact area should be expected on both static and kinetic friction coefficient. An increasing true area of contact can increase adhesion forces and formation of junction due to microwelding of asperities of the mated surfaces, but as the result of running-in effects and enhanced smoothness of the surfaces the micromechanical interaction of surface asperities can be reduced and as the consequence the kinetic friction coefficient.

Compared with silicon micromechanics, the LIGA process [24] allows the use of a large range of materials for micromechanical components, e.g. metals, polymers and ceramics. At present, nickel and copper are the most important metallic materials used for LIGA micromechanical systems such as micromotors or microturbines. Figure 13 shows a microturbine produced from nickel by the LIGA process. The rotor, about 260 μm in diameter, is driven by a gas stream and rotates at speeds up to 30000 rev min^{-1}. Thin wear sheets are produced owing to frictional contact be-

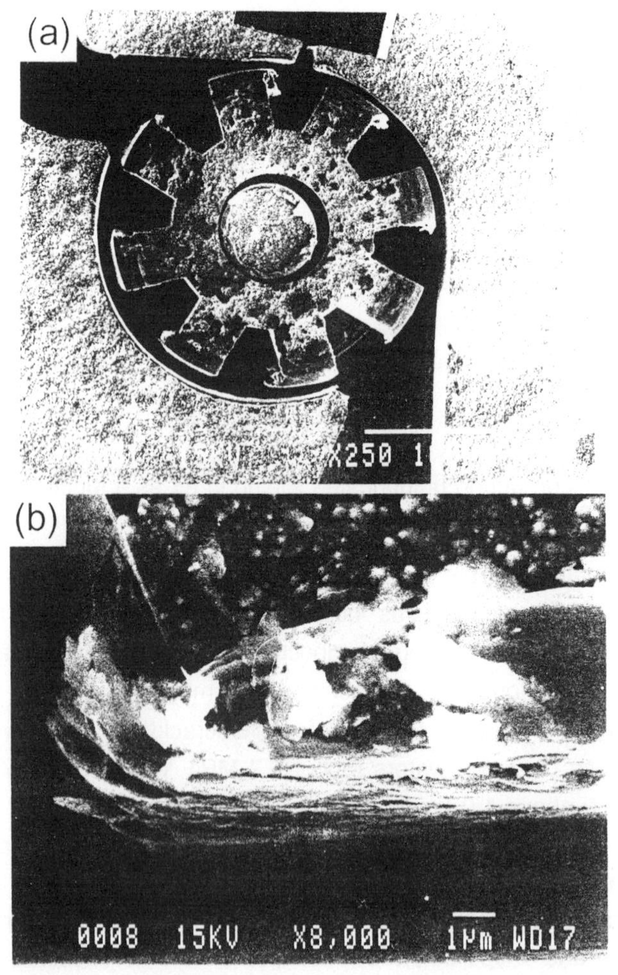

Fig. 13: Microturbine produced from nickel using LIGA [24] process (a) worn turbine and (b) wear sheets formed after running in air stream at 30000 rev min^{-1}.

tween the rotor and the housing. In this system, nickel/nickel contact occurs between rotor and axle or rotor and housing. Nickel/alumina contact prevails between the bottom side of the nickel rotor and the alumina substrate.

Studies [25] on real LIGA-processed rotors in air of relative humidity of 40 % and sliding speeds of 0.01 mm/s were carried out using different surface cleaning methods. Static friction coefficients of Ni against Al_2O_3 ranged from about 0.3 to 1.0 and of Cu against Al_2O_3 from about 0.5 to 1.4. The average surface roughness of the Al_2O_3 substrate was R_a = 75 nm, of the Ni rotors 30 and 260 nm and of the Cu rotors 50 nm, respectively. Values of the kinetic friction coefficients of these pairs ranged from 0.40 to 0.59 for Ni/Al_2O_3 and 0.46 to 0.64 for Cu/Al_2O_3 respectively. During these tests, copper showed a greater friction coefficient than nickel against alumina.

More fundamental studies [26] were carried out by using pin-on-disk geometry inside a microtribometer with miniaturized specimen sizes (about ∅ 200 μm of the pin) and fabricated by standard LIGA process. In the working range of the micromotor, no pronounced effect of sliding speed was observed on kinetic friction and amount of wear independently of the mating Ni/Ni or Ni/Al_2O_3. Figure 14 shows friction coefficient and linear amount of wear (summary of wear of Ni pin and Ni or Al_2O_3 disk) versus sliding distance up to 20 m. Friction coefficient of the Ni/Ni pairs increased continuously up to the quasi-stationary value of about 0.62, while friction coefficient of the pair Ni/Al_2O_3 increased to a maximum and decreased then slowly to a quasi-stationary value of about 0.56 at longer sliding distance. The Ni/Al_2O_3 pair exhibited a particularly high wear intensity (slope of the curve in Fig. 14b) at the beginning of the tests. This was caused by severe wear of the Ni pin owing to mechanical action of asperities on the harder Al_2O_3 surface (R_a = 50 nm) and Ni transfer to the alumina disk. After the initial high amount of pin wear, the Al_2O_3 surface was covered by nickel and both the frictional and the wear behaviour became similar to that of the self-mated nickel pairs. Wear particles and nickel transfer layers showed evidence of oxidation products. The ratio of oxide to metal hardness of NiO/Ni and CuO/Cu is about 1.6, i.e. the oxidation products act only moderately abrasive against the metals.

Relative humidity of the surrounding air and applied contact pressure were important parameters for tribological behaviour of the Ni/Ni and Ni/Al_2O_3 sliding pairs. Figure 15 shows the effect of relative humidity on the quasi-stationary values (after running-in, see Fig. 14a) of friction coefficient and amount of linear wear. Independently of the mating, friction coefficients and amount of wear decreased with in-

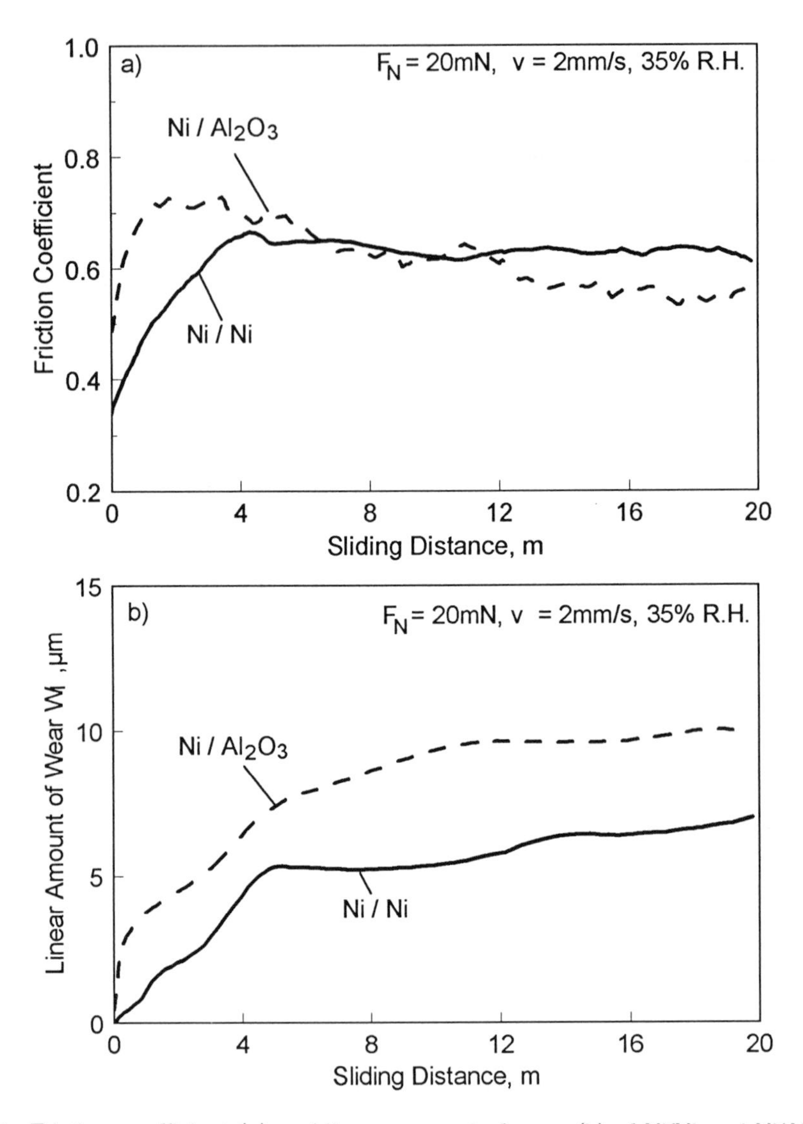

Fig. 14: Friction coefficient (a) and linear amount of wear (b) of Ni/Ni and Ni/Al_2O_3 pairs versus sliding distance during unlubricated, unidirectional sliding in air. After data of T. Bieger and U. Wallrabe [26].

creasing humidity. Friction coefficients of 2.0 [26] were measured for Ni/Ni pairs at vacuum (0 % RH). In agreement with this high friction coefficient, the greatest wear was measured for Ni/Ni in vacuum. Increasing humidity can lead to lubricating effects on a molecular level, reduced adhesion between the mated surfaces and formation of lubricating oxide or hydrooxide layers such as $Al(OH)_3$ on alumina. At high contact pressure and sliding speeds, spinels such as $NiAl_2O_4$ may also be formed. The effect of applied contact pressure is shown in Figure 16. Average fric-

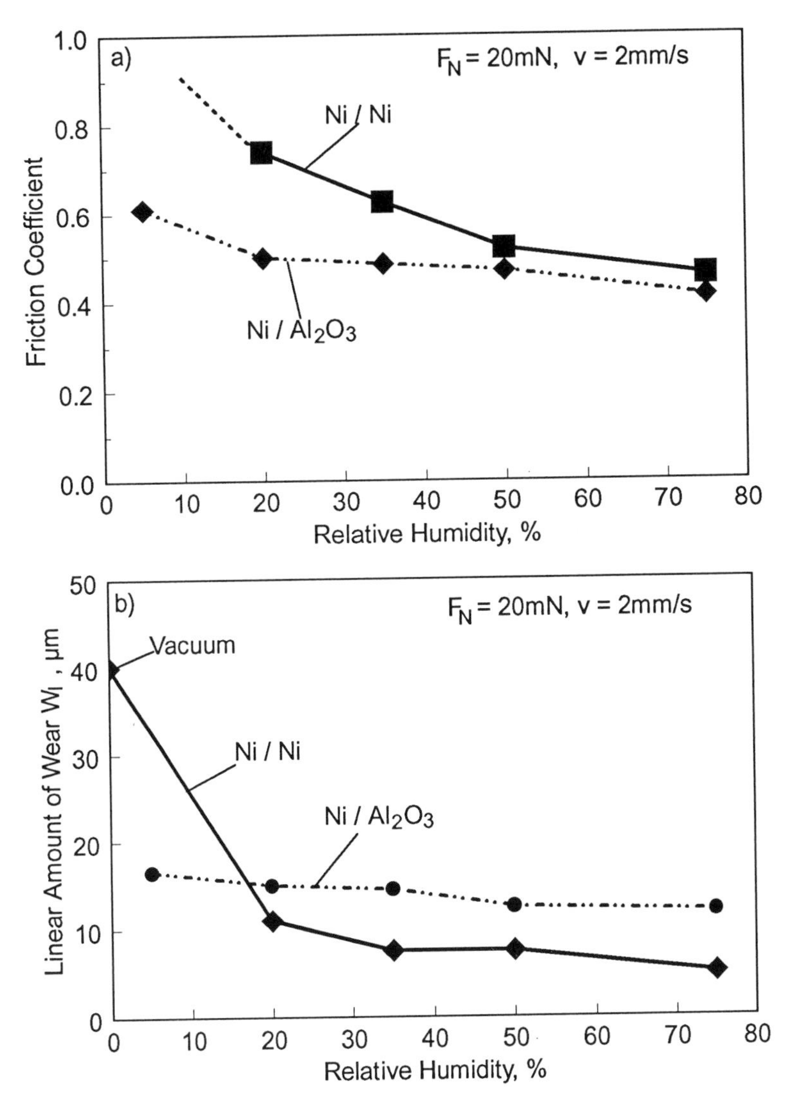

Fig. 15: Friction coefficient (a) and linear amount of wear (b) of Ni/Ni and Ni/Al_2O_3 pairs versus relative humidity during unlubricated, unidirectional sliding in air. After data of T. Bieger and U. Wallrabe [26].

tion coefficient of Ni/Al_2O_3 pairs decreased with increasing contact pressure from about 0.9 to less than 0.4. At low applied pressure, loose wear debris and islands of Ni transfer occurred, while at high contact pressure relatively thin and highly deformed transfer films covered entirely the wear path on the Al_2O_3 disk. Sheet like parts of the mutual transfer layers were finally detached from the alumina disk. On both types of sliding pairs, the amount of wear increased strongly after surpassing a critical value of contact pressure (Fig. 16b).

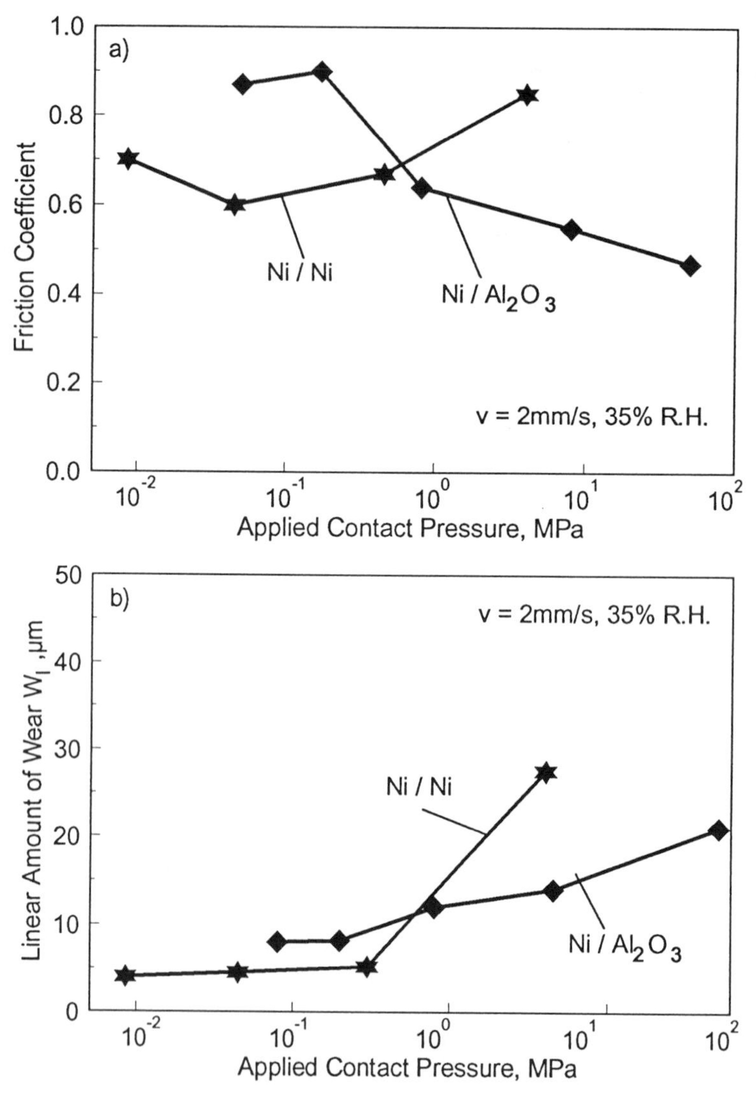

Fig. 16: Friction coefficient (a) and linear amount of wear (b) of Ni/Ni and Ni/Al_2O_3 pairs versus applied contact pressure during unlubricated, unidirectional sliding in air. After data of T. Bieger and U. Wallrabe [26].

Figure 17 shows a thin copper transfer film on an alumina disk and a polymer film (high density polyethylene HDPE) on a steel disk. In general, soft surfaces result in large areas of contact owing to large plastic deformations and hence to high friction. However, low coefficients of friction can be obtained by thin films of low hardness but lying on hard substrates. The hard substrate reduces the area of contact and avoid excessive surface deformations while the soft film reduces the shear resistance during relative sliding. Polymers exhibiting a smooth molecular profile such as high density polyethylene (HDPE) or polytetrafluorethylene (PTFE) show excellent sliding properties owing to the ease with which the long chain molecules shear across each other [27].

Fig. 17: Scanning electron micrographs of transfer layers (a) copper film on alumina surface and (b) polymer (HDPE) film on steel surface after unlubricated, unidirectional sliding contact of Cu/Al_2O_3 and HDPE/steel pairs.

4.5 Materials for Reliable Micromechanical Systems

It is generally recognized that friction and wear are not intrinsic materials properties but are characteristics of the system in question (tribosystem) and can substantially be influenced by design, manufacturing and/or mounting quality of the components.

Rolling contact may be more favorable than sliding contact. Friction and wear result from the intimate contact of solid surfaces on a microscale (Fig. 6). Hence, surface roughness plays an important role during unlubricated sliding. Mechanical interactions of surface asperities depend on surface roughness. For solid surfaces of large differences in hardness, abrasion resulting in grooving of a softer ductile or microcracking of a brittle surface can be reduced by lowering surface roughness. Enhanced smoothness of the mated surfaces may increase the true area of contact and promote adhesive forces. Wear particles or dirt can cause abrasive wear or the loss of the function of the whole system. Misalignment of the components leads to locally overstressing or uncalculable forces.

Separation of the two mated solid surfaces by liquid or solid films can reduce adhesion and abrasion and hence friction and wear, if we consider that in a simplified form the friction coefficient is given by

$$\mu = \mu_{ad} + \mu_p \qquad (2)$$

where μ_{ad}, μ_p represent the adhesive and ploughing contribution to friction, respectively. In this case, friction force F_R is more or less determined by the shear resistance τ_f of the film and the true area of contact A

$$F_R = \tau_f \cdot A$$

Under boundary lubrication, asperity interactions contribute to the friction force additionally. While the shear resistance of a liquid film depends on viscosity and may be influenced by film thickness, temperature and contact pressure, the true area of contact depends on surface roughness, contact load and hardness of the mated solid surfaces. Friction values on the microscale are frequently much smaller than those on the macroscale. Microcomponents sliding under very low loads should show very low friction and wear.

Use of special lubricants is not possible in many micromechanical systems but in fluidic systems the fluid may be used for avoiding solid/solid contact of the relative moving parts. Liquid lubricants can cause problems owing to the surface tension of the liquids. Films of adsorbed gases or hydrocarbons can substantially reduce friction but this effect is mainly constricted to the initial period of running.

Modelling of ultra-thin films with respect to tribology has been received increasing attention during the last years. According to this [28], solid-like films exhibit stick-slip motion, amorphous-like films exhibit high friction due to the molecular interdigitations and entanglements occurring between the two surfaces, and liquid-like films exhibit low, viscous-like friction with smooth sliding. High sliding speeds make the film more solid-like and low speeds make it behave more like a liquid. For boundary lubrication of microsystems, organized, dense molecular layers of long-chain molecules are considered on the solid surfaces in contact. Such monolayers and thin films can be produced by Langmuir-Blodgett deposition and by chemical grafting of molecules into self-assembled monolayers [28,29]. Langmuir-Blodgett films are bonded to the solid surface by weak van der Waals forces and self-assembled monolayers are chemically bonded via covalent bonds. The chain length of the latter molecules and the terminal linking group can be modified that makes these molecules interesting for boundary lubrication of microsystems. Branched-chain molecules of complex fluids and polymers may be better lubricants than straight-chain molecules because the irregularly shaped, branched molecules remain in the liquid state even under high loads [28]. Octodecyl (C_{18}) compounds based on aminosilanes on an oxidized silicon led to lower friction coefficient (μ = 0.018) and greater durability than Langmuir-Blodgett films of zinc arachidate adsorbed on a gold surface coated with octadecylthio ($\mu \approx 0.03$) [28]. Diamond-like carbon films received also a lot of attention owing to their lubricating properties and high wear resistance [30-32].

As mentioned above, many micromechanical systems have to run insufficiently lubricated or even unlubricated. Hence, improved tribological performance needs a proper materials selection and control of materials properties by considering microstructural elements such as textures, grain size and shape, segregated atoms, second phases, porosity etc. These can be a function of fabrication methods of the microcomponents, e.g. galvanic deposition during LIGA processing or PVD (physical vapour deposition) techniques for thin film actuators.

Adhesion, abrasion, tribochemical reaction and surface fatigue are the main wear mechanisms [7]. Figure 18 shows different mechanisms of interactions during unlubricated sliding contact of metals, ceramics and polymers without considering abrasion owing to rough surfaces of very different hardness.

According to the equation 1, the tendency for adhesion γ_{ad} can be reduced by great interfacial energy γ_{AB}, i.e. by mating materials of different main groups, e.g. metals or ceramics with polymers. Hardness and crystal structure of self-mated metals (Fig. 19) should influence the strength of adhesive bonds (adhesion coefficient is equal to the ratio of the force necessary to break the adhesion junction to the normal loading force with which the surfaces were initially compressed). Increasing hardness results in a decreasing coefficient of adhesion, in general [33].

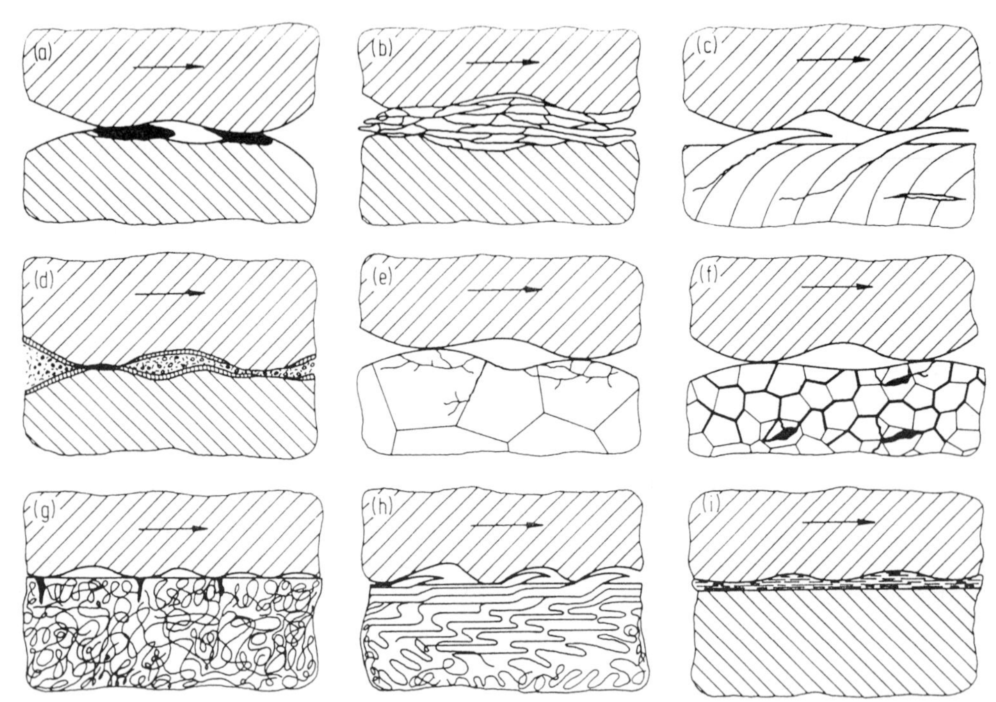

Fig. 18: Mechanisms of surface interactions during unlubricated sliding contact of metals, ceramics and polymers: (a) formation of adhesive junctions, (b) formation of third body layers due to mutual material transfer, (c) flakes and cracks cause by plastic deformation and/or fatigue of metals, (d) formation of tribochemical reaction layers on metals and ceramics, (e) microcracking during mechanical interactions of microasperities on ceramics, (f) cracking at grain or phase boundaries in brittle ceramics or metals, (g) cracking in amorphous polymers, (h) stretching, reorientation and pull-out of molecular chains in partially crystalline polymers and (i) interfacial layers consisting of strongly oriented molecules or molten polymers.

The strength of adhesive bonds increases from close-packed hexagonal over body-centered cubic (bcc) to face-centered cubic (fcc) metal lattices (Fig. 19). Both the number of available slip systems and hardness of the metals influence the area of contact. With increasing area of contact, the adhesion forces are increased for a given γ_{ad}.

Studies on polymer/polymer sliding pairs [34,7] show that friction coefficient increased exponentially with the work of adhesion (Fig. 20). From literature [7] it is also known that wear intensity of mated polymers can linearly increase with γ_{ad}. For metallic materials, hardness is one of the most important factors which affect friction and wear. Observations on many materials combinations demonstrate, however, that the effect of hardness is very complex [35,7]. Hardness of metallic materials can be increased or decreased by work-hardening or work-softening owing to

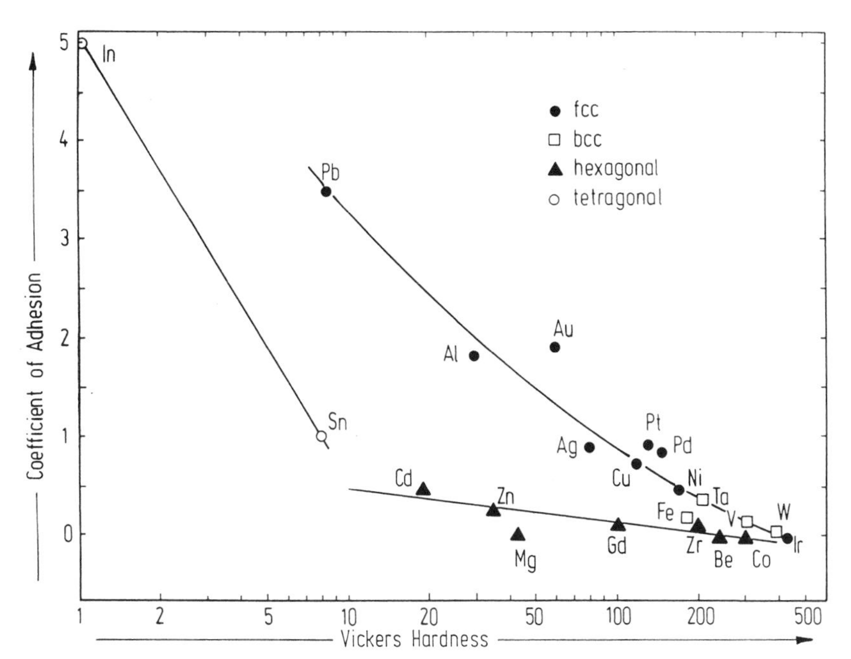

Fig. 19: Coefficient of adhesion plotted against the hardness of metals with different crystal structure. After data of M.E. Sikorski [33].

plastic deformation during sliding contact. Hardness or yield strength of metals can substantially be increased by alloying or by precipitation hardening. Second phases with different chemical properties compared with the matrix can also reduce adhesion to the mated solid surface [6]. Cu-Sn, Cu-Be, Cu-Cr, Cu-Ni or Cu-Si can show greater wear resistance in dry sliding contact against steel than pure copper. Considering the LIGA process, favourable materials for microcomponents should be producible by electrochemical (or chemical) deposition. From this aspect Cu, Ni, Co and some alloys play a role. Co offers a lower adhesion coefficient according to Fig. 19 due to its hexagonal structure. Chemically deposited Ni-8P exhibits hardness values of about 500 to 600 HV and up to about 300 °C a roentgen-amorphous or nanocrystalline structure. Aging at 400 °C results in a hardness increase up to about 1000 HV owing to precipitation of Ni_3P. For reducing friction coefficient in self-mated sliding pairs, soft dispersoids such as PTFE particles can be added. PTFE (polytetrafluorethylene) dispersoids show lubricating effects [36] and friction coefficient can drop to the half.

Embedding of soft or hard dispersoids for improving friction coefficient (soft phases such as PTFE or graphite) or wear resistance (hard phases such as SiC or diamonds) may be of limited usefulness for micromechanical systems due to the miniaturized size of the components and hence the necessary fineness of microstructural elements of the materials applied. Ion-implantation and surface

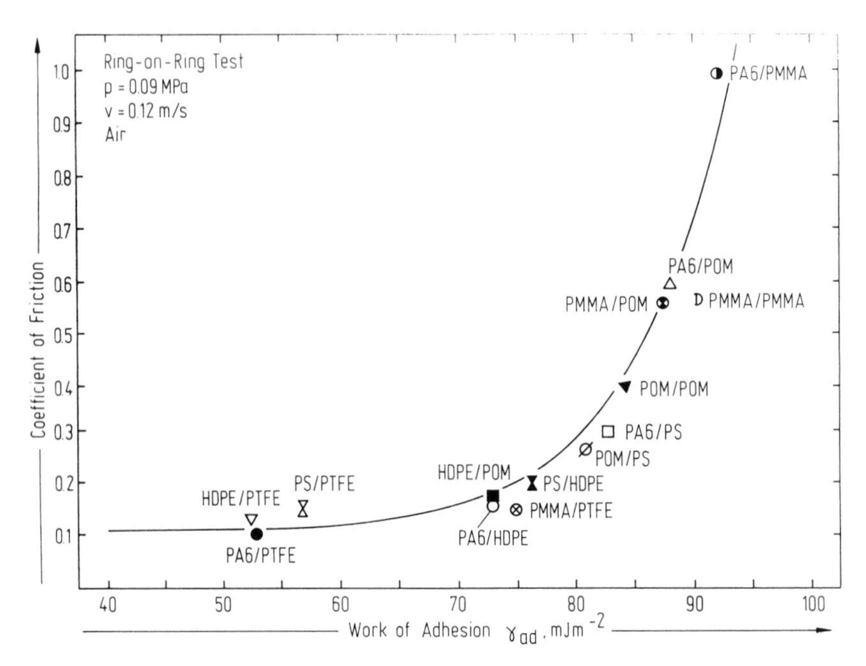

Fig. 20: Coefficient of friction measured by using a ring-on-ring test at ambient temperature, versus the work of adhesion of different polymer/polymer sliding pairs. After data of G. Erhard [34].

coatings produced by PVD or CVD deposition techniques can be very effective for improving tribological behaviour, but it may also be constricted to special applications due to processing difficulties.

Many concepts well recognized for reducing friction and wear on a macroscale are not directly transferable to the microscale of micromechanical components of sizes in the range of about 500 µm or even less and complex geometries involved.

4.6 Summary

Tribology of micromechanical systems has particularly to consider surface effects owing to the miniaturized size of the components. Static friction is very important for the starting of the system while kinetic friction determines the necessary driving power during operation. Friction and wear or lifetime of a system are related to each other in the sense that frictionless systems are also without wear. But increasing friction coefficient does not inevitable mean increasing wear.

The running-in period of a system can be affected by adsorbed films on the mated surfaces and also by surface or assembling defects which can be even out by initially high wear or it results in the loss of the function. Materials and their properties influence friction and wear but also other factors such as design or surface finishing are very important for tribological performance. Friction and wear are systems properties but not intrinsic properties of the mated materials. Microstructure of the materials used for the components can substantially differ from that of macroscale components due to the miniaturized size and special fabrication methods used. It is now recognized that many materials properties show scale effects. Hence, results of tribological studies on macroscale specimens or components cannot easily be transferred to submicron or even nanoscale components.

Literature

[1] Yu-Chong Tai and R.S. Müller: Frictional study of IC-processed micromotors. Sensors and Actuators, A 21 - A 23 (1990) 180 - 183.

[2] K.J. Gabriel, F. Behi, R. Mahadevan and M. Mehregany: In situ friction and wear measurements in integrated polysilicon mechanisms. Sensors and Actuators, A 21 - A 23 (1990) 184 - 188.

[3] K. Noguchi, H. Fujita, M. Suzuki and N. Yoshimura: The measurements of friction on micromechatoronics elements, in 4th Int. Workshop IEEE MEMS, Nara, Japan 1991, pp. 148 -153.

[4] U. Wallrabe, P. Bley, B. Krevet, W. Menz and J. Mohr: Design rules and test of electrostatic micromotors made by the LIGA process. J. Micromech. Microeng., 4 (1994) 40 - 45.

[5] U. Beerschwinger, D. Mathieson, R.L. Reuben, S.J. Yang, R.S. Dhariwal and H. Ziad: Tribological Measurements for MEMS applications. Proc. Micro System Technologies, Berlin 1994, pp. 115 - 124.

[6] K.-H. Zum Gahr: Microtribology. Interdisciplinary Science Reviews, 18 [3] (1993) 259 - 266.

[7] K.-H. Zum Gahr: Microstructure and Wear of Materials. Trib. Series 10, Elsevier, Amsterdam 1987.

[8] G.E. Rhead: Atomic mobility at solid surfaces. Int. Mater. Rev., 34 (1989) 261 - 276.

[9] M.D. Pashley, J.B. Pethica and D. Tabor: Adhesion and micromechanical properties of metal surfaces. Wear, 100 (1984) 7 - 31.

[10] Y. Enomoto and K. Saito: Triboemission of continuum X-ray from ceramics, in Wear of Materials 1991, K.C. Ludema and R.G. Bayer (eds.), ASME, New York 1991, pp. 223 - 228

[11] U. Landman, W.D. Luedtke, N.A. Burnham and R.J. Colton: Atomistic mechanisms and dynamics of adhesion, nanoindentation and fracture. Science 248 (1990) 454 - 461.

[12] D.H. Buckley: Adhesion of metals to a clean iron surface studied with Leed and Auger emission spectroscopy. Wear, 20 (1972) 89 - 103.

[13] D.H. Buckley: Surface effects in adhesion, friction, wear and lubrication. Trib. Series 5, Elsevier, Amsterdam 1981, p. 293.

[14] K. Miyoshi: Properties of solid surfaces, in Handbook of Micro/Nano-Tribology, B. Bhushan (ed.), CRC Press, Boca Raton 1995, pp. 81 - 107.

[15] D. Maugis: Adherence of solids, in Microscopic aspects of adhesion and lubrication, J.M. Georges (ed.), Trib. Series 7, Elsevier Amsterdam 1982, pp. 221 - 247.

[16] W. Hartweck and H.J. Grabke: Effect of adsorbed atoms on the adhesion of iron surfaces. Surface Science, 89 (1979) 174 - 181.

[17] H. Mishina: Chemisorption of diatomic gas molecules and atmospheric characteristics in adhesive wear and friction of metals. Wear, 180 (1995) 1 - 7.

[18] S. Franzka and K.-H. Zum Gahr: Microtribological studies of unlubricated sliding Si-Si contact in air using AFM/FFM. Tribology Letters (1996), in press.

[19] M. Binggeli and C.M. Mate: Influence of capillary condensation of water on nanotribology studied by force microscopy. Appl. Phys. Lett., 65 (4) (1994) 415 - 417.

[20] G. Kolbe: Research Center Karlsruhe, Institute of Materials Research I, unpublished data.

[21] B. Bushan, B.K. Gupta, M.H. Azarian: Nanoindentation, microscratch, friction and wear studies of coatings for contact recording applications. Wear, 181 - 183 (1995) 743 - 758.

[22] E. Zanoria, S. Danyluk and M. McNallan: Effects of length, diameter and population density of tribological rolls on friction between self-mated silicon. Wear, 181 - 183 (1995) 784 - 789.

[23] M.G. Lim, J.C. Chang, D.P. Schultz, R.T. Howe and R.M. White: Polysilicon microstructures to characterize static friction. Proc. IEEE MEMS 1990, IEEE New York 1990, pp. 82 - 88.

[24] J. Mohr, C. Burbaum, P. Bley, W. Menz and U. Wallrabe: Movable microstructures manufactured by the LIGA process as basic elements for microsystems, in Micro system technologies 1990, H. Reichl (ed.), Springer Verlag, Heidelberg, 1990, pp. 529 - 537.

[25] U. Beerschwinger, S.J. Yang, R.L. Reuben, M.R. Taghizadeh and U. Wallrabe: Friction measurements on LIGA-processed microstructures. J. Micromech. Microeng., 4 (1994) 14 - 22.

[26] T. Bieger and U. Wallrabe: Tribological investigations of LIGA-microstructures. Microsystem Technologies, (1995) in press.

[27] K. Marcus and C. Allen: The sliding wear of ultrahigh molecular weight polyethylene in an aqueous environment. Wear, 178 (1994) 17 - 28.

[28] B. Bushan, J.N. Israelachvili and U. Landman: Nanotribology: friction, wear and lubrication at the atomic scale. Nature, 374 (1995) 607 - 616.

[29] E. Meyer, R.M. Overney and J. Frommer: Lubrication studied by friction force microscopy, in Handbook of Micro/Nano Tribology, B. Bhushan (ed.), CRC Press, Boca Raton 1995, pp. 223 - 241.

[30] S. Miyake, T. Miyamoto and R. Kaneko: Microtribological improvement of carbon film by silicon inclusion and fluorination. Wear, 168 (1993) 155 - 159.

[31] A. Grill: Review of the tribology of diamond-like carbon. Wear, 168 (1993) 143 - 153.

[32] S. Suzuki, T. Matsuura, M. Uchizawa, S. Yura and H. Shibata: Friction and wear studies on lubricants and materials applicable to MEMS. Proc. IEEE 4th Workshop on MEMS, Nara, Japan 1991, pp. 143 - 147.

[33] M.E. Sikorski: The adhesion of metals and factors that influence it. Wear, 7 (1964) 114 - 162.

[34] G. Erhard: Sliding friction behaviour of polymer-polymer material combinations. Wear, 84 (1983) 167 - 181.

[35] D.A. Rigney: The roles of hardness in the sliding behaviour of materials. Wear, 175 (1994) 63 - 69.

[36] T. Sakamoto, O. Tahano and M. Nishira: Friction of electroless nickel-PTFE composite coating. Jap. J. of Tribology, 38 (1993) 521 - 530.

5. Silicon Microsensors

P.J. French

Laboratory for Electronic Instrumentation, DIMES, Faculty of Electrical Engineering, Delft University of Technology, Postbus 5031, 2600GA Delft, The Netherlands.

5.1 INTRODUCTION

As the title suggests a microsensor is simply a miniaturised sensor and often refers to silicon as the base material. This chapter will concentrate on micromechanical sensors which includes both micromachined devices and sensors designed to measure a mechanical parameter.

Silicon has long been known to have excellent mechanical properties although the full potential of the material was not realised until micromachining techniques enabled the fabrication of true mechanical structures. The first types of structures were fabricated using bulk micromachining techniques, where the silicon wafer is etched to leave free standing mechanical structures. More recently surface micromachining techniques have received considerable attention. In this case the mechanical structures are fabricated in thin films deposited on the silicon wafer surface.

If the microsensor is to be integrated with electronics, it is necessary to consider the compatibility of the two fabrication processes. In the case of bulk micromachining, compatibility problems generally arise from the incompatibility of the processing with the clean room environment and this can often be solved by performing the micromachining at the end of the processing. In the case of surface micromachining there are several considerations. The first is the thermal budget, since the micromechanical layers may require annealing. The second is non-planarity of the wafer surface caused by the deposition of multiple layers. The last major consideration is the sacrificial etching. In surface micromachining it is usually necessary to selectively remove one of the layers to leave a free standing structure and this layer is known as the sacrificial layer. At this stage it is important to ensure that the etching required to remove this sacrificial layer does not attack the mechanical layers or the electronics. In many cases this etchant will also attack the electronic components. In recent years alternative micromachining techniques have been developed. One which has attracted considerable attention is the LIGA process and this is dealt with in chapter 2.

For all these structures the mechanical properties of the materials are an important consideration. These properties, for commonly used materials are listed in table 5.1.

Bulk micromachined structures have been used for a wide variety of structures for many years. The include, thermal of flow devices where thin membranes are formed to enhance the temperature difference [1-2], pressure sensors [3], accelerometers where a micromachined mesa is supported by thin beams of epitaxial thickness [4-5], micro-pumps [6-7] and optical devices [8]. In some cases the micromachined structures have been combined with electronic feedback systems [9]. All the above mentioned structures require several additional processing steps after the completion of the electronics processing.

These include the etching described in this section and also, in many cases some form of silicon or glass bonding which will be dealt with in section 5.2.6.

In microelectronics terms these bulk micromachined structures are usually relatively large. With the push towards yet smaller devices came the development of surface micromachining. Surface micromachining is based on the deposition of thin films on the surface of the wafer and removing one or more layers to release the mechanical structures.

Table 5.1. Properties of the most important materials in micromachining.

	Melting point (°C)	Thermal expansion ($x10^{-6}/°C$)	Density g/cm^3	Young's modulus ($10^{11}Pa$)	Yield strength (10^9Pa)
Si	1415	2.5	2.4	1.3-1.69	6.9
SiN	1900	2.8	1.48	2.43	14.0
SiO_2	1610	0.5	2.27	0.73	8.4
Al	660	25	2.70	0.70	0.17

Probably the first references to surface micromachining were in the mid 60's where a resonating metal gate was suspended over an MOS device [10-11]. In the 1970's there were several attempts to fabricate surface micromachined devices where the part of the epitaxial layer was removed to leave free standing structures. An example of this was the deflectable aluminium coated oxide mirror [12].

In the 1980's came the first of the micromachining processes using entirely chemical vapour deposited layers (CVD) [13-14]. In this case polysilicon and oxide were used as the mechanical and the sacrificial layers, respectively. This early work showed the potential of this new process and the second paper [14], demonstrated examples of working gears and cranks.

Surface and bulk micromachining both have advantages and disadvantages. A recent development which has tried to take the advantages from both sides while minimising the disadvantages, is epi-micromachining. These devices are fabricated in the substrate but only in the top few microns.

5.2 FABRICATION

Micromachining is a technique for fabricating microstructures usually by silicon etching techniques. The chemical etching possibilities of silicon was discussed in a series of papers over several years by Schwartz et. al. [15-18]. Bulk micromachining possibilities using orientation dependent etching was cover in studied by Bean [19] and the mechanical potential by Petersen [20].

If the micromachined structures are to be integrated with the electronics it is necessary to establish the process limitations and whether additions to the standard process are to be permitted. The devices can be integrated with CMOS [21-22] or bipolar [23] electronics. It is important to establish a strategy in terms of how the additional steps required for the micromachining can be inserted into the standard process. Such a discussion is given in [24].

5.2.1 Bipolar technology

A typical bipolar process contains npn and pnp transistors which can be orientated vertically or laterally. The most commonly used devices are the vertical npn and the lateral pnp transistor. Bipolar circuitry has been widely used for both digital and analogue electronics. Although bipolar technology has lost its dominance in the field of digital electronics it remains an important technology for analogue circuits. In this section the process sequence of an example bipolar process is described. The basic process sequence for vertical npn and lateral pnp transistors is given in figure 5.1. The start material is a p-type (100) wafer. The n+ buried layer is formed by ion implantation of antimony followed by an anneal and drive-in. An n-type epitaxial layer is then grown on the wafer resulting in the structure shown in figure 5.1a. Bipolar technology requires that each transistor be isolated and therefore a deep p+ diffusion is made which must make a p-type contact with the substrate. This is followed by the deep n+ diffusion which makes a contact to the buried layer. The resulting structure is shown in figure 5.1b. This is the last of the long high temperature process steps and therefore extra thermal step after this point should be added with caution. The most critical steps in the bipolar process are the formation of the base and emitter, which are formed by ion implantation. The base is usually formed by two implantations with a higher dose being used to ensure a low resistance contact. The emitter can be formed directly by ion implantation or via a polysilicon emitter [25], yielding the structure shown in figure 5.1c. Finally contact window opening and aluminium contact formation are used to complete the device fabrication as shown in figure 5.1d.

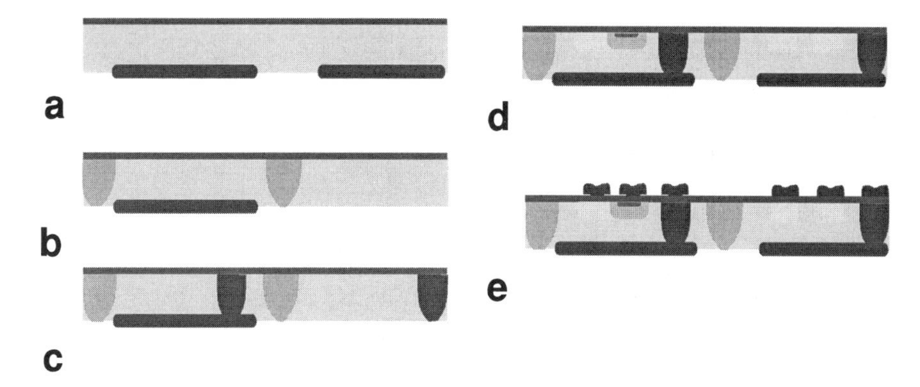

Figure 5.1 Simplified bipolar process

5.2.2. CMOS processing

An NMOS process uses enhancement and depletion type n-channel transistor to form the driver and load, respectively, in a simple inverter. The CMOS process brought together the enhancement devices of the NMOS and PMOS devices, and these two transistors can be connected in series to form a simple inverter. Since both the n-channel and p-channel transistor must be fabricated in the same substrate, a diffused well is formed in the substrate.

The operation of this inverter relies on the positive value of threshold voltage of the n-channel device and the negative value for the p-channel device. Typical values of the threshold voltages are, $V_{Tn} \simeq 1V$ and $V_{Tp} \simeq -1V$.

The start material is a p-type substrate on which a thin oxide layer and a nitride layer are deposited and patterned to define the borders for device isolation [26]. The nitride film prevents oxidation of the substrate and only the exposed areas are oxidized as shown in figure 5.2a. This LOCOS (LOCal Oxidation of Silicon) oxide is usually preceded by a boron channel stopping implant. The nitride is removed after LOCOS and the n-well is to be formed resulting in the structure shown in figure 5.2b. Subsequently, the gate oxide, typically 500-1000 Å thickness, is thermally grown. Then a polysilicon layer is deposited and patterned to form the gate. Doping takes place either in-situ or after deposition by $POCl_3$ or ion implantation after deposition. This results in the structure shown in figure 5.2c.

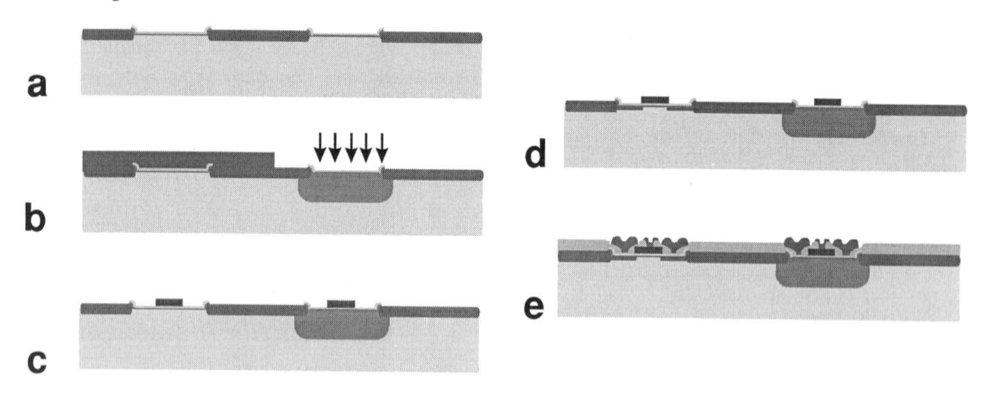

Figure 5.2 Simplified CMOS process

The threshold voltage of the n-channel transistors is adjusted using an unmasked boron implantation, through the gate and the gate oxide. Masking is not needed as the dose is relatively low compared to the source/drain and n-well implants. The self-aligned n^+ source and drain in the n-channel transistor is formed, while masking the p-channel transistors, and similarly for the p^+ source and drain in the p-channel transistors are formed, resulting in the structure shown in figure 5.2d. A dielectric layer is then deposited to enable isolation between the gate polysilicon (first polysilicon) and a second polysilicon or metal. Integrated capacitors are usually formed using poly-poly capacitors with an oxide dielectric material. Finally contact holes are etched and the second polysilicon or aluminium interconnect is deposited and patterned resulting in a structure as shown in figure 5.2e.

Figure 5.2 represents a simplified version of the CMOS process, although it shows the main steps. As with bipolar processing, it is important to establish where in the process flow additional step may be added. Any changes in substrate or additional thermal steps required for the sensor, may significantly effect the characteristics of the CMOS transistors and should therefore be added with caution.

5.2.3. Backside bulk micromachining

Bulk micromachining involves etching wells in the silicon substrate leaving, for example, membranes, mesa's or beams. Silicon has excellent mechanical properties and lends itself well to micromachining. The mechanical properties of the material are discussed in detail by Petersen [20] and crystal orientation dependent etching, by Bean [19].

Anisotropic wet etchants use the fact that some crystal orientations etch much slower than others. For example using a (100) wafer will yield the etch cavity shown in figure 5.3. In this case the etching has stopped on the <111> plane. This effect is due to the energy required to remove an atom from the surface and the atom spacing revealed to the surface. The large differences between etch-rates of silicon planes can present problems when etching corners and this will be discussed below.

Figure 5.3 Basic bulk micromachined structure.

When considering the most suitable etchant for compatible bulk micromachining thermal budget is not a consideration. There are several suitable etchants and each, must be fully characterised in terms of etch-rate, masking materials and compatibility with circuit processing/clean rooms. Etchants commonly used are Hydrazine-water, EDP (ethylenediamine-pyrocatechol-water), KOH (potassium hydroxide) and TMAH (tetramethylammonium hydroxide) [20,27,28]. All characteristics are dependent upon the concentration, and temperature of the etchant and only commonly used examples are quoted here. The first two etchants, hydrazine and EDP have the disadvantages of being highly toxic and therefore difficult to handle. The most commonly used etchant is KOH. This is relatively easy to handle but the potassium deposits of the silicon surface is not fully compatible with a cleanroom environment. The etching should therefore preferably be performed as a post-cleanroom-processing step. Recently, there has been increasing interest in TMAH as an alternative. The wafers can be cleaned more easily after etching and can therefore be returned to the cleanroom for further processing if necessary. It has been found to have similar etching properties to KOH although to etch stop on the (111) plane is slightly inferior.

The etchant KOH, or KOH with IPA [29], is probably the most commonly used etchant. One disadvantage is the relatively high etch-rate of silicon dioxide. Silicon nitride is therefore a commonly used masking layer. If plasma enhanced CVD (PECVD) nitride is used the whole bulk micromachining processing can be performed after the electronics processing has been completed.

Bulk micromachining involves etching through the whole thickness of the substrate and stoping the etching when the required membrane thickness is reached. There are three techniques commonly used and these are time-stop, p+ stop and electrochemical etch-stop. The first of these is the most simple and is used for mechanical sensors on the market. It does however require precise control of the etch-rate and has limited accuracy. The second etch-stop available with some etchants is a high boron etch-stop (>$2.5\times10^{19}cm^{-3}$). Etchants such as KOH and EDP have been found to have a considerably reduced etch-rate for highly p-doped material. EDP has an extremely sharp cut-off, and more suitable than KOH as shown in figure 5.4 [30]. However, a standard bipolar process does not have a suitable high p-doped buried layer, and therefore would have to be added. Furthermore, care must be taken to ensure that such a highly doped layer does not effect other devices. The third technique is the electrochemical etch-stop [31]. This has the advantage of high accuracy and it is compatible with bipolar electronics. The disadvantage is that the etching set-up is more complicated. The basic electrochemical etching set-up is shown in figure 5.5. In the case of KOH etching the masking material is usually nitride. The etching of silicon does not require a current flow from the platinum electrode. This is however required for the etch-stop. The epi-layer biased as the anode and the platinum as the cathode. When the sample is immersed in the etchant the etching begins. In the case of an electrochemical etch-stop, the electrical potential at the etch plane when the epitaxial layer is reached ensures that the etching does not proceed. Thus the electrochemical etch-stop provides a highly reliable method to control the thickness of a beam or membrane. The mask on the backside is of silicon nitride and required special alignment techniques. Two possible techniques for easing this problem are a) forming a special alignment marks at the beginning of the process [32] and b) to form a potential pattern in the substrate, via electrodes on the silicon surface [33].

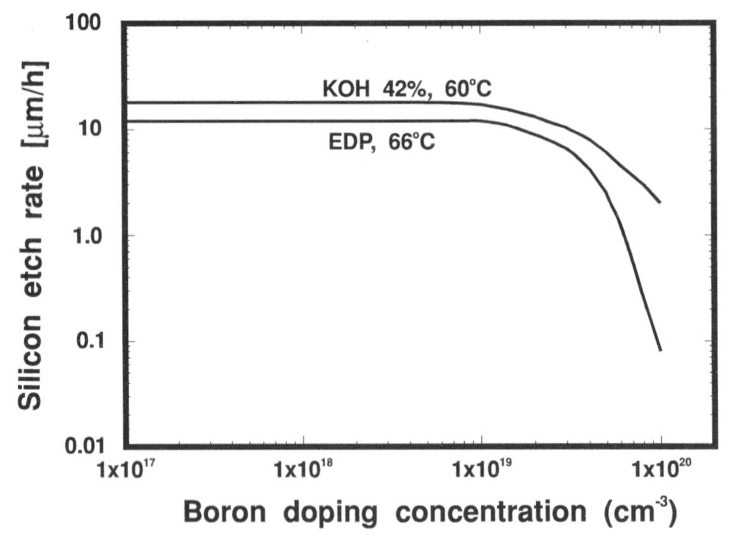

Figure 5.4 Etch-rate as a function of boron doping concentration for EDP and KOH.

In many cases it is necessary to etch through the epitaxial layer in selected areas which can be achieved by either by ensuring certain epitaxial layers are not passivated by isolating them from the main epitaxial areas, or by plasma etching from the front-side.

All the etchants operate at low temperature and can be masked using low temperature deposited layers. They are all therefore compatible with the electronics in terms of thermal budget. Such structures have been incorporated in both bipolar [5] and

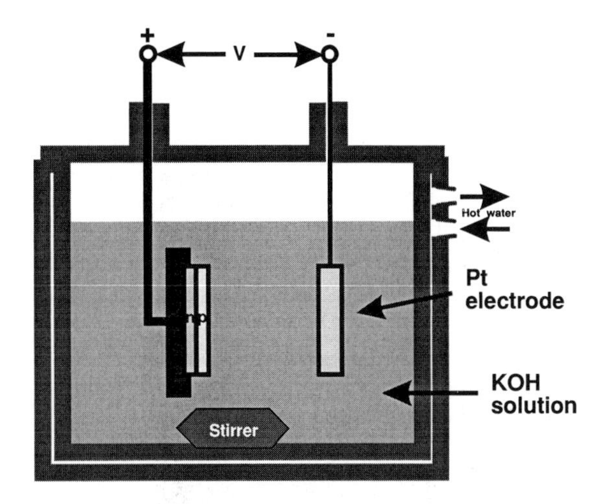

Figure 5.5 Basic electrochemical etching set-up.

CMOS [21] processes. A further example of the flexibility of these processes is the combination of bulk micromachining, electronics and piezoelectric zinc oxide (ZnO) in the fabrication of surface acoustic wave devices [34].

Figure 5.6 Three designs for corner compensation.

An important design consideration is corner compensation. When the etchant meets a convex corner the plane revealed may have a high etch-rate, resulting in the removal of the entire corner [35-36] and for smaller structure the possible removal of the whole mesa. Various designs have been developed to compensate for this effect and an example of these is shown in figure 5.6 [36]. The principle of the method is to add structures to the corners which are rapidly under-etched. If the extension is large enough the desired corner will be reached just as the etching reaches the epitaxial layer. If the etching is continued beyond this point the corner will be removed as before.

5.2.4. Surface micromachining

One of the advantages of surface micromachining on silicon is that, due to its inherent small size, it is possible to either include more electronics on the same chip or decrease the die area. To achieve intergration, a micromachining process has to be combined with a CMOS or bipolar process, without influencing the characteristics of the electronic devices. Many compatibility problems must therefore be solved, before the process is ready to be used by the designers. A major consideration in the development of a process is the thermal budget as any additional high temperature step can have a considerable influence on the bipolar characteristics [37]. A bipolar process is able to withstand limited additional thermal processing, such as 800-900°C. If higher temperatures are used after the formation of the base and emitter, the increased diffusion can seriously effect the transistor characteristics.

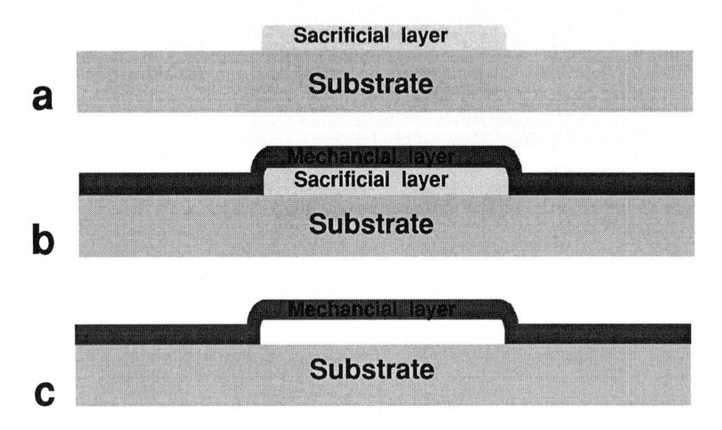

Figure 5.7 Basic surface micromachining process; a) deposition and patterning of sacrificial layer, b) deposition and patterning of the mechanical layer and c) removal of the sacrificial layer.

A simple single layer surface micromachining process is shown in figure 5.7. The sacrificial layer is first deposited and patterned, figure 5.7a. The desired mechanical layer is deposited, annealed (if necessary) and patterned figure 5.7b. The final step is the sacrificial etching, where the layer is etched through access holes leaving a free standing structure, figure 5.7c. The additional steps, such as contact windows and metallisation will be discussed in a later section. An example of a micromachined membrane is shown in figure 5.8.

The layers used in surface micromachining are usually deposited using chemical vapour deposition (CVD). The two main techniques are low pressure CVD (LPCVD) and plasma enhanced CVD (PECVD). The basic layout of a typical LPCVD system is shown in figure 9. The main parameters are the deposition gas flows, the system pressure and the deposition temperature and these parameters can have a significant effect on the properties of the deposited layers.

If these micromachined structures are the only structures to be fabricated on the wafer then the process can be optimised for the micromachining. If, however, the micromachined devices are to be combined with electronics a number of problems may arise. The first is the thermal budget, since the micromechanical layers may require annealing. The second is non-planarity of the wafer surface caused by the deposition of

Figure 5.8 SEM photograph of a surface micromachined membrane.

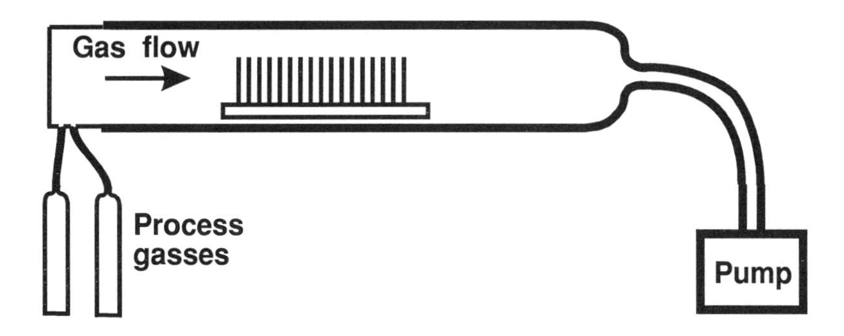

Figure 5.9 Schematic of a typical LPCVD system.

multiple layers. The last major consideration is to ensure that the etching required to remove the sacrificial layer does not attack the mechanical layers or the electronic components. Initially a decision has to be made whether the surface micromachining steps should precede or follow the thermal processing of the electronics. If the surface micromachining is done before the electronics, this has the advantage that the characteristics of the electronic devices are not degenerated by any further processing. However, the mechanical characteristics of the surface micromachined structures may suffer from the many high temperature steps necessary in bipolar processing. Furthermore, the poor planarity of the wafer after processing of the surface micromachined structures, may cause problems with defining small structure due to inhomogeneity of the resist layer.

The requirements for good compatibility with standard electronics implies that the surface micromachining process must either follow the electronic processing, or be inserted at a late stage, where the last annealing step is suitable for both the electronics and the micromechanical structures.

It is essential to establish a process where the characteristics of both the mechanical and electrical components can be maintained. Furthermore the designer must know the limitations in terms of additional steps which can be added, and both these considerations are discussed below. Taking the example of bipolar there are two options. The first is to insert the micromachining processing before the emitter anneal and to use this anneal for both the emitter and the micromachined layers. The second option is to add the micromachining module after the completion of the bipolar, but before contact opening and aluminium. In this case the total thermal budget available for the micromachining is low as the emitter junction depth is extremely sensitive to any additional steps. In both cases it must be established whether the sacrificial etching is performed before or after aluminium formation.

The requirements for good compatibility with standard electronics implies that the surface micromachining process must either follow the electronic processing, or be inserted at a late stage, where the last annealing step is suitable for both structures.

5.2.4.1 Sacrificial layers

For surface micromachining, the sacrificial layer must be remove to leave free standing structures. There is a wide range of both sacrificial and mechanical layers available to the designer, where the sacrificial layer depends upon the required mechanical layer. Some of these layers such as polysilicon and nitride, can be incorporated into a standard silicon process. Other layers, such as Tungsten, molybdenum, polyimide, TiNi or ZnO must be added to the end of the process. Several materials available in standard processing have been quoted in the literature, such as:

Aluminium [38], polysilicon [39] and oxide (using SOI [40], SIMOX [41], LOCOS [42] and deposited oxide [43].

each requiring a specific etchant. This section will concentrate on deposited oxide as a sacrificial layers which are removed by wet etching in HF. There are various types of oxide, all of which have advantages and disadvantages. These include thermal oxide, and deposited oxides. Oxides are usually deposited using low pressure chemical vapour deposition (LPCVD) or plasma enhanced chemical vapour deposition (PECVD). In this section and LPCVD phospho-silicate glass (PSG) is described. This has a deposition temperature compatible with the electronics, provided the deposition is performed before aluminium deposition [44]. PSG is deposited using O_2, SiH_4 and PH_3 which results in a phosphorus doped glass. It is the phosphorus doping which enhances the etch rate which, along with its low deposition temperature makes it highly suitable as a sacrificial material.

The high etch-rate of the PSG is mainly due to the presence of phosphorus, but one potential problem is structural modification during subsequent thermal processing [43]. The main effects are: shrinkage and etch-rate reduction.

The considerable increase of both etch-rate and underetch-rate with increasing phosphorus doping is shown in figure 5.10. Both vertical and under etch-rate, show an exponential dependence on the phosphorus content. After annealing (850°C/30 minutes in dry nitrogen) the etch-rate reduces, although values over 50 Å/sec. can be achieved for PH_3/SiH_4 flow ratios larger than 21%. The lateral underetch-rate in 20% HF are presented together (for comparison, the vertical etch-rate of thermal oxide in 1% HF is 0.3 Å/sec. and its underetch-rate is 0.3 μm/min.).

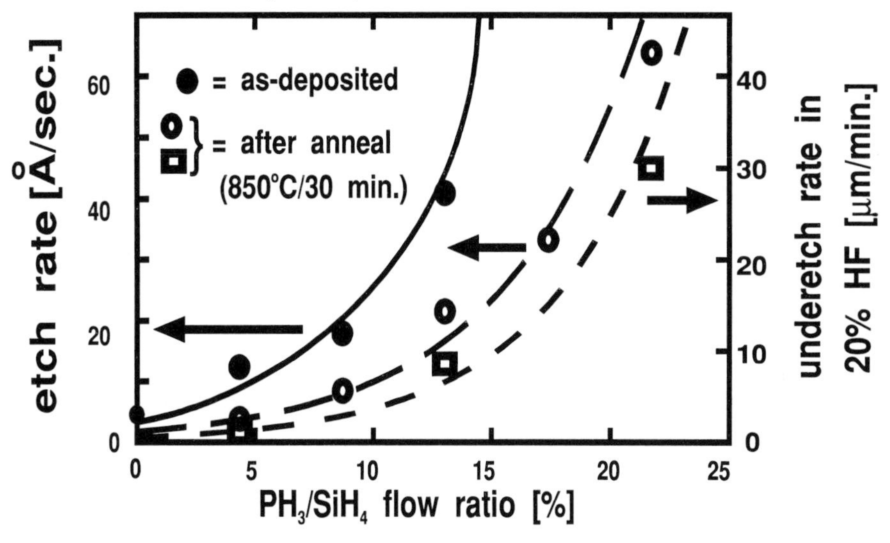

Figure 5.10 Under-etch-rate of PSG in HF as a function of PSG phosphorus content.

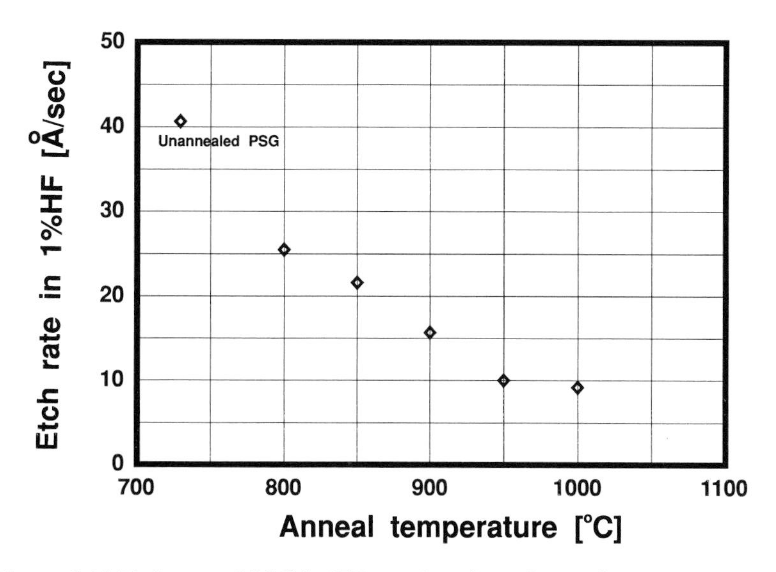

Figure 5.11 Etch-rate of PSG in HF as a function of anneal temperature.

The modification induced by the anneal is extremely rapid, the shrinkage and the vertical etch-rate reach stabilized values after the first few minutes of annealing, even at an anneal temperature of 850°C. The reduction in etch-rate is dependent on anneal temperature as shown in figure 5.11. Once annealed, however, further annealing at the same, or lower temperature have no additional effects of the PSG. Therefore provided the PSG is annealed before further processing the film remains stable.

5.2.4.2 Mechanical layers

Polysilicon is a material frequently used for both electronics and sensors. In the field of bipolar electronics it has been applied to polysilicon emitter, and in MOS as a polysilicon gate. Since the piezoresistive effect in polysilicon was first studied in the early 1970's [45], there have been many studies of its electrical [46-47] and piezoresistive properties [48-49]. With the development of surface micromachining techniques has come an increasing interest in the mechanical properties of the material [50-55].

There are two processing steps after deposition of the mechanical layers which may require high temperatures. These are to activate doping (for ion implanted polysilicon) and a stress anneal. It is therefore essential that the mechanical properties of all the micromechanical layers are characterised in order to develop a low temperature fabrication process. The mechanical strain can be measured by the buckling technique [56] but this has the disadvantage of requiring an array of structures and accuracy of the measurement is dependent upon the number of structures. A second technique, used for measuring tensile strain measures directly the displacement of the intersection of a two beams [57]. This is however only useful for relatively high strain levels. A simple technique to measure both compressive and tensile strain using a pointer structure as shown in figure 5.12 [58]. Here a wide range of strains can be measured with a single structure.

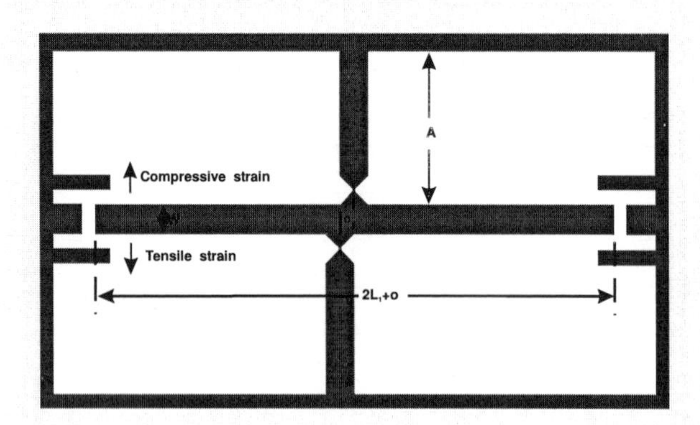

Figure 5.12 Strain measurement structure

A method frequently used for the relaxation of residual stress in polysilicon films is a high temperature (>1000°C) anneal for an hour or more [53]. This yields the desired mechanical characteristics but is obviously not compatible with bipolar electronics. It is furthermore advantageous to be able to use film thicknesses commonly used in standard processing.

Processing parameters have a considerable effect on the electrical and mechanical properties of polysilicon [49]. The important processing parameters include; deposition temperature, *in-situ* [51] annealing, doping and post implant annealing.

Deposition temperature is known to have a strong effect on both the electrical and mechanical properties of polysilicon. Above a certain temperature, which depends on both pressure and gas flows, the films are deposited as polysilicon with fully formed grains and grain boundaries. Below this temperature a semi-amorphous film is deposited. In this case, grains are formed but separated from each other by an amorphous layer. A sharp transition exists for strain as shown in figure 5.13. These films were deposited using a pressure of 150mtorr and a silane flow of 45sccm in a 15cm diameter tube.

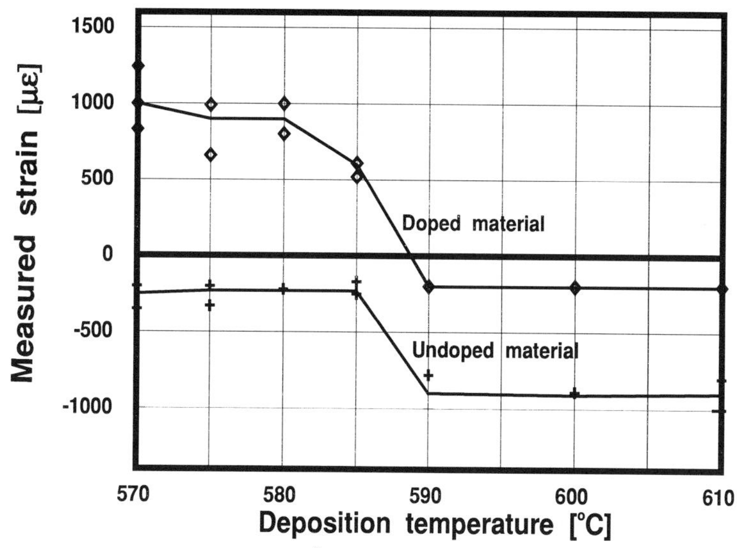

Figure 5.13 Strain levels in 4000Å thick polysilicon films as a function of deposition temperature. Doped films were implanted with phosphorus at $1x10^{15}cm^{-2}$ and annealed at 850°C for 30 mins.

As shown above, undoped and unannealed films are in compressive strain for all deposition temperatures. A low temperature, 600°C, *in-situ* anneal, first proposed by Guckel et. al. [54], as short as 30 minutes is sufficient to have a significant effect on the stress levels in the films resulting in a change from a compressive strain of ~-300με to a tensile strain of ~+1000με for undoped films. Doped films are less influenced by the use of the *in-situ* anneal. The measured strain for these films is shown in figure 5.14 [55]. If the grain boundary is not formed during deposition, low temperatures can be used to remove any compressive strain. If a polysilicon structure is formed during deposition, such a low temperature anneal cannot be used as temperatures above 1000°C are required to significantly change the structure. It is therefore the formation of the grain boundary and not the grain size which is the main contributing factor in determining the stress levels.

A further consideration is the stress profile through the thickness of the film. If there is a significant difference in mechanical stress through the film this may result is bending up, or down, of single ended beams. The film thickness can also have an influence on the strain levels for these polycyrystalline materials. Figure 5.15 shows the

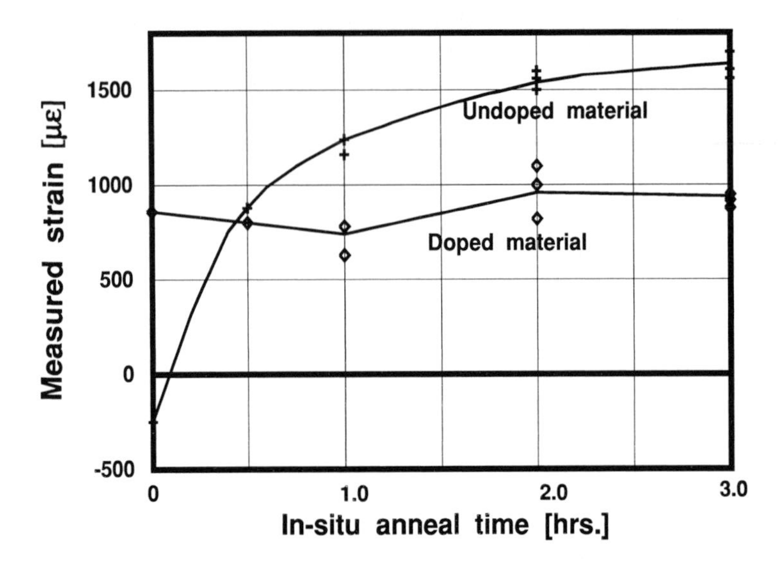

Figure 5.14 Film strain levels as a function of anneal time for an anneal temperature of 600°C. Implantation for doped samples as figure 5.13.

strain for film thicknesses ranging from 2000Å to 7500Å. The implanted energy for each film was chosen to yield a range approximately in the middle of the film. This graph further shows a greater effect for the undoped polysilicon than for doped, since the implantation and recrystallisation of the film has reduced the differences through the film.

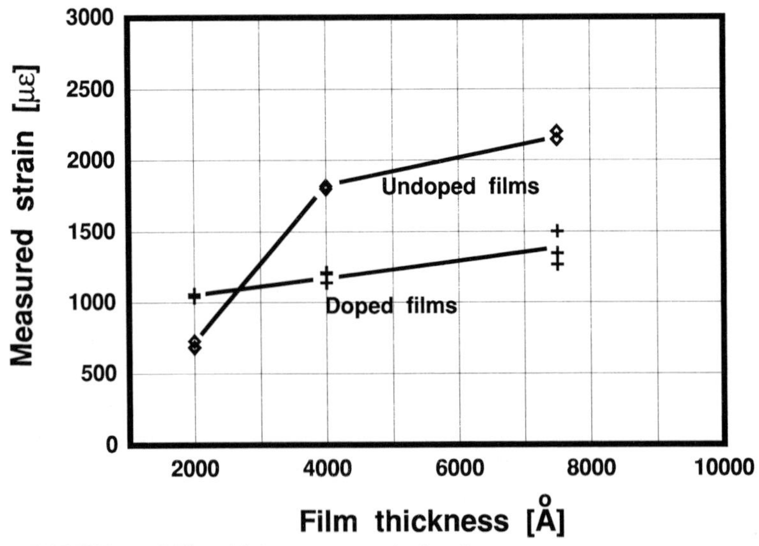

Figure 5.15 Effect of film thickness on strain levels.

The above experiments show how through optimisation of two processing parameters, deposition temperature and *in-situ* anneal, a process can be developed with the desired mechanical properties while maintaining a low thermal budget. Although, as shown above, the *in-situ* anneal is not essential for the doped films, its incorporation ensures suitable mechanical properties for both doped and undoped films with a single standard process.

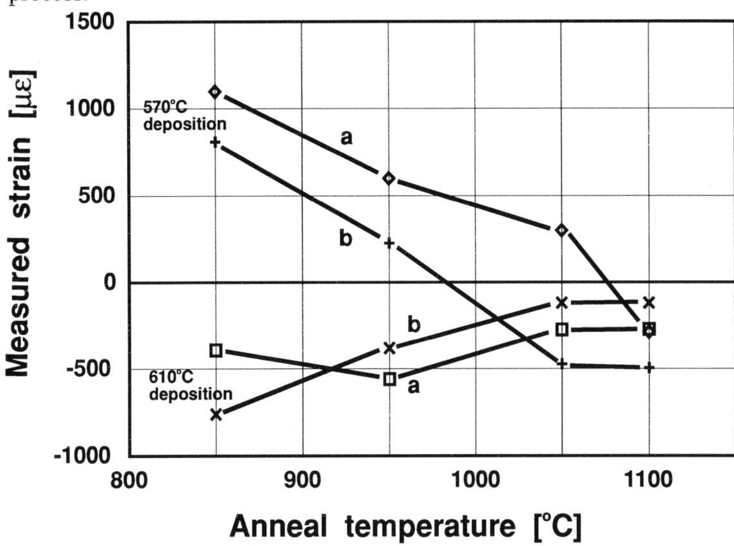

Figure 5.16 Effect of post implant anneal temperature on strain levels.

If the micromachined device is to be combined with other devices, a further degree of freedom is the stage in the process flow where the polysilicon deposition can be placed while maintaining satisfactory mechanical properties. For films deposited as semi-amorphous, strain levels reduce with increasing anneal temperature, crossing over to a compressive strain at anneal temperatures above 1000°C. For depositions above the transition temperature, the films were in compressive stress for all anneal temperatures, although the strain approached zero as the anneal temperature was increased. Figure 5.16 shows the measured strains for films deposited at 570°C (below the transition) and 610°C (above the transition) as a function of anneal temperature after doping by phosphorus implantation at a dose of $1x10^{15}cm^{-2}$, using both thermal oxide and PSG as a sacrificial material. This figure shows that if the deposition parameters are not optimised, high-temperature annealing is indeed required to remove the compressive stress.

The second main material used as a micromechanical layer is silicon nitride which is a mechanically strong material with low optical absorption. The stress levels in the material, as with polysilicon are highly dependent upon both deposition parameters and subsequent thermal processing. Deposition can be performed by either LPCVD or PECVD. Plasma enhanced systems operate at temperatures compatible with aluminium and therefore it is possible to use PECVD for fabrication of micromechanical layers after the aluminium has been deposited. If, however, this is not required, LPCVD is a suitable option. This description will concentrate on LPCVD deposited films.

The films are usually deposited using diclorosilane (DCS) and ammonia (NH_3) usually at a temperature range of 750°C to 850°C and pressures of 80mtorr to 300mtorr.

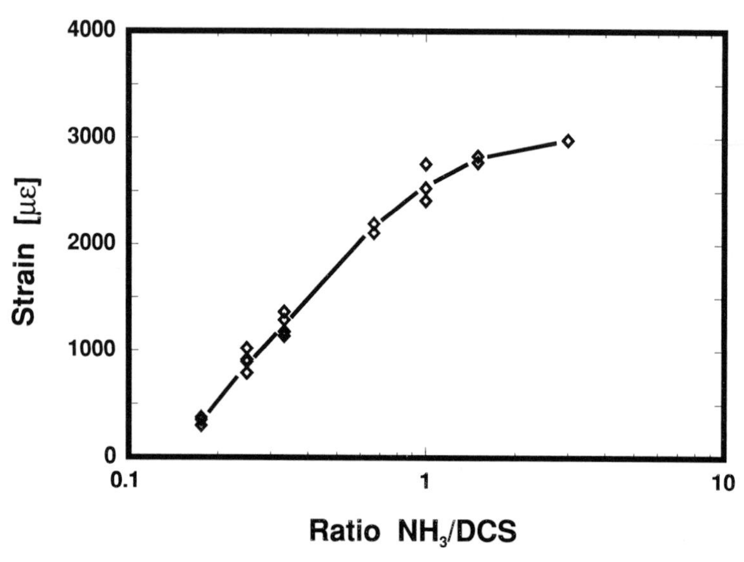

Figure 5.17 Effect of process gas flow ratio on strain levels in silicon nitride.

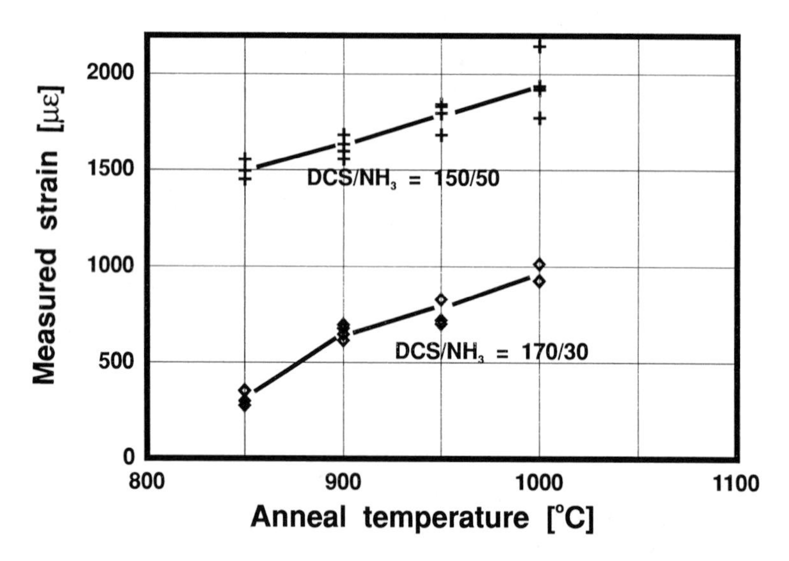

Figure 5.18 Effect of post deposition anneal temperature on strain levels in silicon nitride deposited with a gas ratio of 30/170 (see figure 5.17). Anneal time was 45 mins.

Within the deposition parameter range described above the mechanical strain can vary from 100με to 3600με, tensile. Higher deposition temperatures have been used to yield lower stress [59] but this would create an addition to the thermal budget and is therefore undesirable. Figure 5.17 shows the effect of gas flow ratio on the strain levels for a deposition temperature of 850°C and a pressure of 150mTorr.

All the above strain measurements are for nitride films which have had no further thermal treatment above the deposition temperature. Further annealing can have a considerable effect on the mechanical properties of the films as shown in figure 5.18 [60]. This considerable effect of anneal temperature is a possible limitation for this material.

5.2.4.3 Wafer planarity

Each time a layer is deposited and patterned on the surface of the wafer a step is created, which if several layers are deposited can create two problems; 1) step coverage of upper layers and 2) uniformity of resist.

Step coverage for polysilicon or nitride is not usually a problems but aluminium is known to have poor step coverage, and if necessary the step shape can be modified [60]. Alternatively, this problem can be overcome by using a staircase structure where the aluminium climbs one layer at a time. The second problem is that of resist spinning. Once large steps have been formed on the wafer, subsequent masking steps will have to use a thicker resist to ensure sufficient cover on the outer part of the wafer. This can ensure reasonable resist cover but limits the feature size that can be defined, and is therefore not suitable for defining small contact windows.

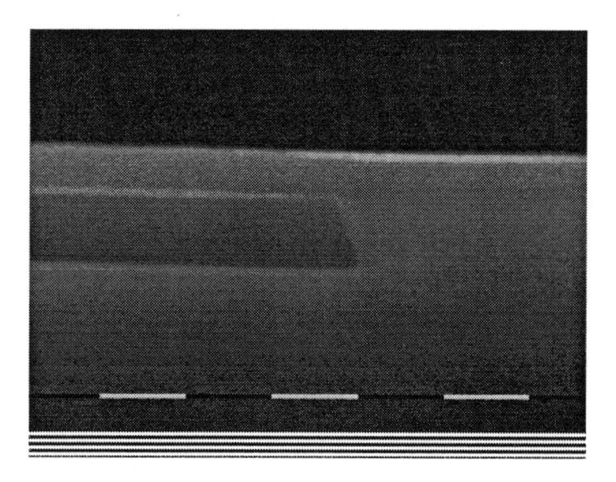

Figure 5.19 SEM photograph of surface micromachined beams using the SEG planarisation technique.

One possible solution to the problems described above is to planarize the wafer to fill cavity area with oxide. Both PSG and BPSG have been used for such purposes since during annealing they begin to flow thus reducing the step. The advantage of BPSG is its relatively low reflow temperature [62], however the etch-rate is considerably lower than that of PSG. An alternative technique is to use a plasma etch where the etch-rate of PSG is the same as the resist [61], since the resist is spun on and yields a planar surface. This does however require thicker PSG layers and accurate back-etching.

Rather than etching back, selective epitaxial growth (SEG) provides the opportunity of growing the epitaxial layer around the PSG [63] and an example of a resulting structure is given in figure 5.19. Provided the PSG areas are aligned in the [100] direction the epitaxial growth yields a planar surface on which the mechanical layers can be deposited. The same epitaxial layer can also be used for the electronics [64].

An alternative technique is the LOCOS process. In order to achieve a planar surface a recess should be etched after patterning of the masking oxide and before the oxidation. After the oxidation the nitride may be remove to leave a planar surface.

5.2.4.4 Sacrificial etching

Sacrificial etching is the step where the sacrificial layer is removed from under the micromechanical layers to produce free standing structures. It is important to ensure that the etch selectivity between sacrificial and mechanical layer etch rates is high and the electronics is protected. Using high HF concentrations yield a high under etch-rate and therefore large structures can easily be etched. The sacrificial etching mechanism is also an important consideration and there have been several recent studies on this topic [65-67]. The etch-rate can be enhanced adding HCl to the HF solution [68]. If, as mentioned above, the electronics is protected by resist, buffered HF (BHF) is the most suitable etchant. An alternative solution is to use a plasma deposited nitride layer to protect the electronics. This would have to be sufficiently thick to withstand the HF etching. The remaining layer would have to removed using plasma etching after sacrificial etching. This may, however, present significant problems with over-etching of mechanical layers during the removal of the nitride. An alternative etch is the pad-etch which is able to etch oxide without significantly attacking the aluminium [69]. Although the etch-rate is lower than HF, the electronics is able to withstand longer etch times.

An additional problem is that of sticking. When structures are released they remain free standing while remaining in the etchant. If the structures are cleaned in water and allowed to dry, capillary forces will pull the structures down to the substrate. This can create severe problems for fabricating surface micromachining devices and there have been several studies into the effect of sticking [70-71]. Therefore considerable care must be taken in the drying process. This process is further complicated when resist is present. There are two main techniques. The first is to immerse the sample in alcohol and dry the structure on a hot plate. The alcohol, with its low surface tension is able fill the cavity and the rapidly evaporates at the elevated temperature. An alternative technique is know as freeze dry. The technique makes use of sublimation to release the structures. This can be achieved by lowering the pressure of a sealed chamber containing the wafer [72], but this technique, although effective, has the disadvantage of being complicated. A simpler freeze dry technique involves the use of cyclohexane, which has a freeze temperature of +7°C [73]. After etching the samples are rinsed in water (with water or alcohol to maintain a wet surface), and then the resist is removed in acetone. The sample is emersed in cyclohexane and then reduced to a temperature of -5°C in a nitrogen flow. The cyclohexane sublimates and the sample is gradually brought back to room temperature.

5.2.5 Epi-micromachining

There are two limiting factors with back-side bulk micromachining; 1) the dimensions are usually large and 2) alignment is difficult and has limited accuracy (although recent improvements in back-side mask alignment techniques has greatly reduced this problem). If the bulk micromachining can be performed from the front-side of the wafer, this alignment problem is automatically removed and the dimensions can be reduced.

Membranes and beams can be produced on the surface by applying the wet etchant, such as KOH described above, to the front side of the wafer. First the front side must be protected with a nitride layer. In this case a PECVD layer must be used as it is used to

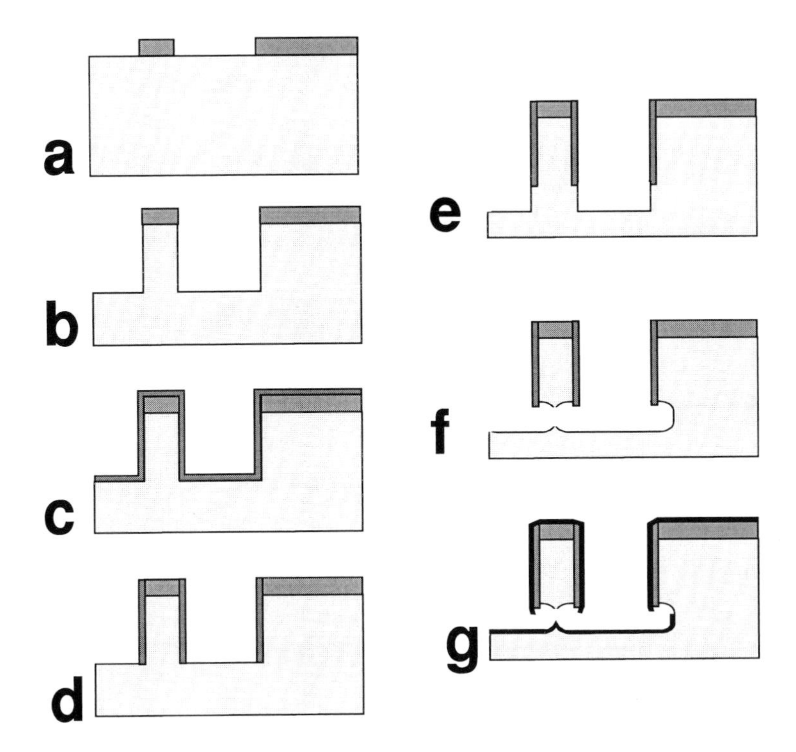

Figure 5.20 Process sequence of SCREAM micromachining technique [78]

protect the aluminium and therefore the temperature must be kept low. The bulk micromachining steps can be added to the end of the standard process. Use can be made of layers deposited or grown on the surface of the wafer, where the sacrificial layer is the underlying silicon [21]. Alternatively, both the mechanical and the sacrificial material can be silicon. Both these techniques have been applied to such structures as flow sensors and resonators. In recent years there has been increased interest in the combination of silicon microelectronics and optics. Front side bulk micromachining has been used to fabricate channels for fibres [74]. By careful consideration of the orientation dependence of the etching vertical mirrors and beam splitters can be fabricated [75]. Combinations of wet and dry etching on the front side of the wafer have also been applied to the fabrication of micro-tips [76-77].

If deep trenches are to be fabricated it is preferable to perform the etch at the end of the process (i.e. after aluminium deposition). This can be done using plasma etching of which two techniques have been proposed recently. The first, known as SCREAM uses a combination of anisotropic and isotropic plasma etching [77]. The basic process sequence is shown in figure 5.20. After patterning the oxide (figure 5.20a) trenches are etched which define the sidewalls of the structure (figure 5.20b). The sidewalls are protected by an oxide deposition (figure 5.20c) and etch-back using plasma etching as shown in figure 5.20d. A further plasma etch etches the trench beyond the sidewall protection (figure 5.20e) and finally the structure is under-etched using an isotropic etch, resulting in the structure shown in figure 5.20f.

The second technique, know as SIMPLE, can be achieved using a single etch [79]. This process makes use of a plasma which etches low doped material anisotropically and n-type material above a threshold of about $1x10^{20}cm^{-2}$. The basic process sequence is shown in figure 5.21. This technique has been used to fabricate micromachined devices and a cross-sectional SEM photograph of the final structure is shown in figure 5.22. A high aspect ratio can be achieved, where the height of the device is determined by the epitaxial thickness. The width of the structure shown here is 2μm. This technique has the advantage of simplicity, is performed after the aluminium deposition and is fully compatible with the electronics circuitry.

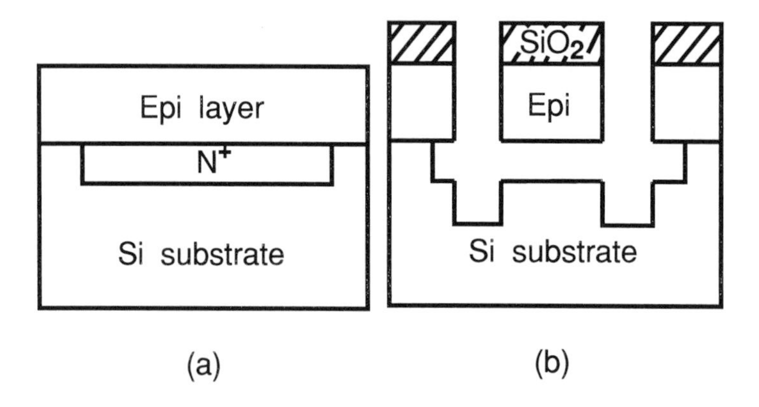

Figure 5.21 Process sequence for SIMPLE technique [79].

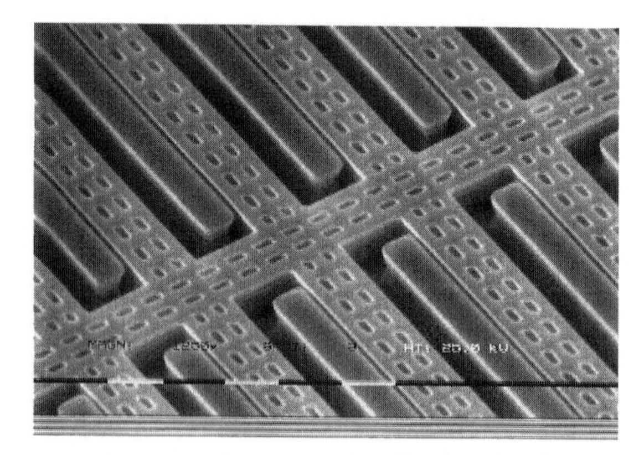

Figure 5.22 Example of micromachined structure using the SIMPLE process.

A further technique is to use a porous silicon layer as a sacrificial layer [80]. This etching set-up is very similar to the KOH etching, although in this case the etchant is HF. A positive voltage is applied to the platinum electrode and negative to the back of the wafer. The resulting current flow enhances the formation of holes on the surface which result in pores. The high surface area of this materials results in rapid etching, after porous formation, in KOH. This makes the material highly suitable as a sacrificial material. A simple process sequence is given in figure 5.23. In this case a buried layer is made porous and later removed to leave a micromechanical layer formed in an epitaxial layer.

An alternative technique is to start with a wafer with a buried sacrificial layer, such as a SIMOX wafer [41]. SIMOX wafers have a buried oxide layer above which is a single crystal silicon layer. These substrate have successfully been applied to micromachining. This has the advantage of process simplicity and high crystal quality of the mechanical layers. However, the higher costs of the start material may present problems for some applications.

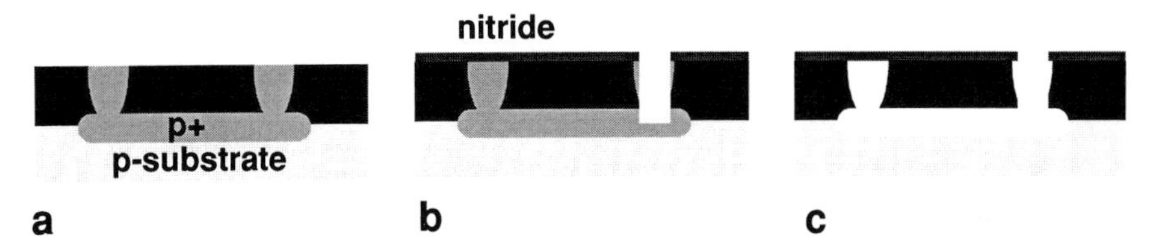

Figure 5.23 Process sequence for porous silicon sacrificial layer technique, a) standard bipolar processing, b) plasma etching through the epi to buried layer and c) formation of porous buried layer and removal in KOH to form the free standing structures.

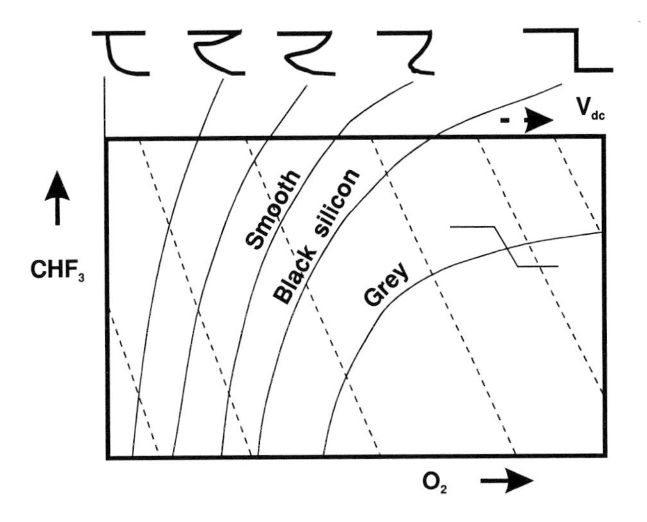

Figure 5.24 Control of etch profile by gas flow variations [81].

Plasma etching has found increasing application for micromachining. The profile of the etched cavity and the selectivity over underlying layers can be controlled by gas composition [81]. In deep plasma etching high aspect ratio and accurate control of the profile of the etched hole is required. In some applications it may be required to create a

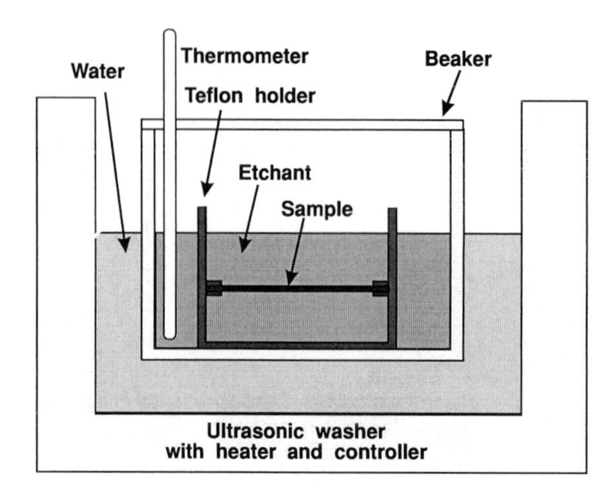

Figure 5.25 Etch set-up for ultrasonic agitation etching [82].

taper or undercutting. This can be achieved by adjusting the ratio of etching gasses. An example of profile for an $SF_6/O_2/CHF_3$ chemistry. is given in figure 5.24 [81]. A range of structures can be fabricated using this variation in etch chemistry. An alternative technique to achieve deep groove is to use ultrasonic agitation. The etching set-up is shown in figure 5.25 [82]. The process has been combined with silicon to glass bonding as shown in figure 5.26 to fabricate a lateral accelerometer [82].

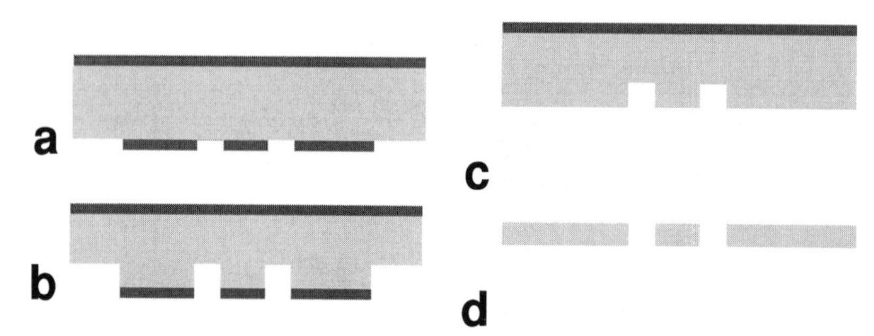

Figure 5.26 Fabrication process for a lateral accelerometer using the mechanical structures fabricated by ultrasonic agitation [82].

5.2.6 Wafer-to-wafer bonding

A further option used to create sealed cavities with bulk micromachined devices is to bond an additional chip, either silicon or glass, to the sensor chip. Bonding is also frequently used to obtain silicon on oxide devices. There is a range of bonding techniques applicable to micro-mechanical sensors and these are [83]:

Eutectic bonding
Polyimide bonding
Thermocompression metallic bonding
Ultrasonic welding
Laser welding
Low temperature glass bonding
Epoxy bonding
Non-uniform press bonding
Room temp. compression metallic bonding
Seam welding
Anodic bonding
Fusion bonding.

A common example of glass to silicon bonding is for bulk micromachined accelerometers [84-85]. In this case the glass plate serves several functions:

1) To provide a seal and the desired damping
2) In the case of capacitive read-out, a metal plate on the glass serves as one plate of the capacitor.
3) As overload protection.

The bonding is achieved by anodic bonding which requires an elevated temperature <450°C and high voltage 1000V. This technique is commonly used with fully processed wafers. The temperature is low enough to present no problems for the aluminium. Since the voltage is applied between the glass and the wafer, there is no problem of high current through the aluminium, and thus heating through high currents. The bonding set-up for anodic bonding is shown in figure 5.27.

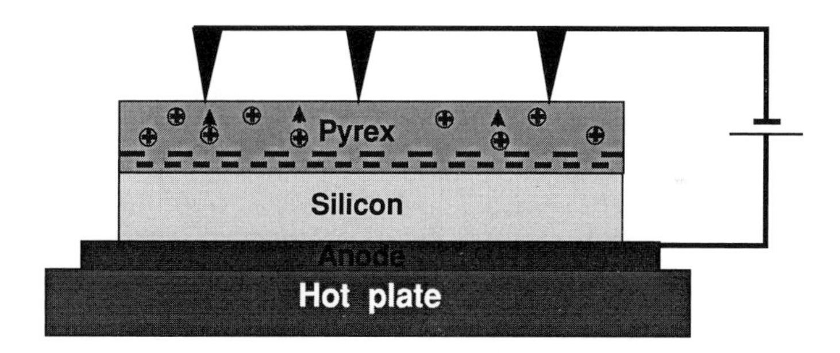

Figure 5.27 Bonding set-up for anodic bonding

Silicon direct wafer bonding is a technique for fusing wafers by heat and no additional layer is required to aid adhesion. Both wafers should be polished and hydrophilic (thus containing a high density of OH groups on the surface - this can be achieved by immersing in fuming nitric acid). The wafers are brought together at room temperature by either electrostatic of mechanical means. The wafers are then annealed to increase the strength of the bond. Early techniques used high temperatures and high pressures to bond the wafer [85]. Although temperatures of above 1000°C the short time required does not result in excessive out-diffusion of dopants. However, the mechanical properties may be changed and wafers with aluminium cannot be used. Lower temperatures can be used but this results in a lower bond strength.

Table 5.2 Comparison of bonding techniques [91].

Method	Materials	Intermediate Layers	Temperature [°C]	Surface preparation	Selective bonding
Anodic Bonding	Glass-Si Si-Si Si-Metal/glass	 Pyrex sputtered Al, W, Ti, Cr	>250 >300 300-500	Voltage 50-1000V	Lithography Etching, lift-off
Silicon fusion bonding	Si-Si SiO_2-SiO_2		700-1000	Standard cleaning	Lithography etching
Low temp. glass bonding	Si-Si SiO_2-SiO_2	Na_2O-SiO_2 and other sol-gel mat.	200-400	spin coating	
		boron glass	>450	CVD implantation	Lithography etching
Low temp. direct bonding	Si-Si SiO_2-SiO_2		200-400	Plasmas treatment wet surface activation (dip)	
Eutectic bonding	Si-Si	Au,Al	379,580	Sputtering electroplating	lift-off, etching
Welding	Si-Si	Au,Pb-Sn	300	Evaporation Sputtering	lift-off, etching
Adhesive	Si-Si Si glass SiO_2-SiO_2 SiN-SiN	Adhesives Photoresist	Rt-200	Spin-coating	Lithography

Low temperature wafer-to-wafer bonding has been achieved using a an intermediary layer such as such a polymer glue, low temperature glass [86], or gold [87]. Bonding with a glue is a simple technique, but has several drawbacks. Mainly reproducibility and contamination due to outgassing are the main disadvantages. However, this technique has been successfully applied to the fabrication of SOI devices [88]. Boron doped glass has been applied to low temperature bonding. Boron doped glass softens at relatively low temperatures [89]. Bonding takes place at 400°C and 20-200V and takes about 10 mins. The problem with boron doped glass is the sensitivity to phosphorus contamination. This increases the temperature required to reflow the glass. A recent alternative is a sodium silicate glass which bonds at 200°C for 2 hours [90] yielding high bond strength. Alternatively, materials such as gold have been used for a through wafer bonding process [87]. This process uses a 300Å titanium / 1200Å gold layer combination deposited on an oxidised wafer. The titanium is used to improve the adhesion. Bonding was performed at temperatures between 350°C and 400°C with a 100g distributed weight. This low-temperature process is thus, in terms of thermal budget, compatible with standard processing although the use of gold means that, to avoid contamination, the bonding must be performed at the end of the process. A comparison of 6 bonding techniques is given in table 5.2 [91].

There are a range of structures in the literature which use silicon-to-silicon bonding and some examples are given in [92-94].

5.3 PERFORMANCE OF MICROMECHANICAL SENSORS

Silicon sensors are able to yield a wide range of outputs. In the case of piezoresistive accelerometer this signal is a change in resistance. For the same measurand (i.e. acceleration) the output signal may be a change in capacitance. In many cases the signal will be converted into either an analogue voltage change or a digital signal. In some cases the signal from the sensing element itself is in a frequency or digital output. An example of a frequency output is to use a ring oscillator where the switching time of each element is modulated by the measurand [95]. The flip-flop sensor is a device which give a percentage of one's and zero's to represent the measurand. A counter is then used to give a direct digital output [96-97].

Alternatively a surface acoustic wave may be propagated in a thin film where the delay time is modulated by the measurand [98]. This basic principle can be applied to a wide range of sensors. It is important that the output signals are compensated for the unwanted effects such as:

Offset	Systematic error in output.
Offset drift	Change in offset with time.
Hysteresis	Difference in rising and falling output.
Sensitivity-drift	Change in sensitivity with time
Ageing	Change in characteristics with time
Cross-sensitivity	Sensitivity to more than one parameter.
Non-linearity	Output not directly proportional to the measurand.
Noise	Output contains a random signal.

The ability to compensate for these inaccuracies is an important factor determining the quality of the device. There are two types of cross-sensitivity for mechanical sensors. The first is the sensitivity to other directions which is of particular importance to accelerometers. The second is the influence of other parameters such as temperature. For both these problems there are two approaches. The first is to try to minimise the effect of the additional parameter and the second is to measure the parameter and then to perform the compensation with the electronics.

5.3.1 Read-out principles

The sensing element responds to a mechanical parameter by being deformed or displaced. The change must be converted into a signal which can eventually be read out. There is a wide range of read-out possibilities and these are described below.

5.3.1.1 Piezoresistive Read-out

Piezoresistors have been used extensively for accelerometers and pressure sensors. The piezoresistor is placed on the surface of the support beam or membrane. The location of the resistor is important to ensure the maximum stress and thus the maximum output. An example of a cross-section of the read-out for an accelerometer is shown in figure 5.28. This uses piezoresistors in both compressive and tensile stress which yields opposite changes in resistance. The strain, ε, at a vertical distance, h, from the central plane of the a beam of length, l, is given by:

$$\varepsilon = \frac{12(x-l/2)hD}{l^3} \tag{5.1}$$

where D is the deflection. The maximum values of strain are therefore theoretically achieved at the ends of the beam and this is where the resistors must be placed. This equation also shows that the strain increases as the beam thickness is increased. The strain sensitivity of the resistor is defined by:

$$G = 1 + \nu_x + \nu_y + \frac{\pi E}{\varepsilon} \tag{5.2}$$

where π is the piezoresistive coefficient, E the Young's modulus, ε, the strain experienced by the resistor and $1+\nu_x+\nu_x$ is the contribution due to volumetric changes. In the configuration given in figure 5.28 it is straightforward to place the resistors in compressive and tensile stress and if placed in the bridge configuration shown in figure 5.29, the output can be further increased. In the case of a pressure sensor based on a thin membrane the bridge is constructed by placing the resistors parallel and perpendicular to the direction of stress as shown in figure 5.29. The piezoresistive coefficient is highly anisotropic and quite different for n and p-type material. The longitudinal and transverse piezoresistive coefficients are given by:

$$\pi_l = \pi_{11} + 2(\pi_{12} + \pi_{44} - \pi_{11})(l_1^2 m_1^2 + l_1^2 n_1^2 + m_1^2 n_1^2) \tag{5.3}$$

$$\pi_t = \pi_{12} + (\pi_{11} - \pi_{12} - \pi_{44})(l_1^2 l_2^2 + m_1^2 m_2^2 + n_1^2 n_2^2) \tag{5.4}$$

where π_{11}, π_{12} and π_{44} are the piezoresistive coefficients and l_i, m_i, and n_i are the direction cosines related to the principal axes [99]. The values of the piezoresistive coefficient for n and p type material are given in table 5.3 [100].

Table 5.3. Piezoresistive coefficients for n and p-type material.

	π_{11}	π_{12}	π_{44}
n-type	-102.2	+53.4	-13.6
p-type	+6.6	-1.1	+138.1

In the case of a (100) wafer, p-type material yields a maximum value for both longitudinal and transverse piezoresistance in the [110] direction, which is parallel to the flat. N-type material, on the other hand, yields a minimum value for this direction. For

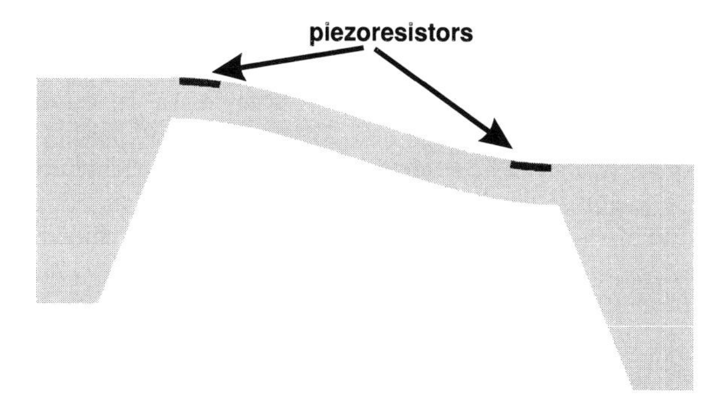

Figure 5.28 Cross section of the readout of a piezoresistive output accelerometer.

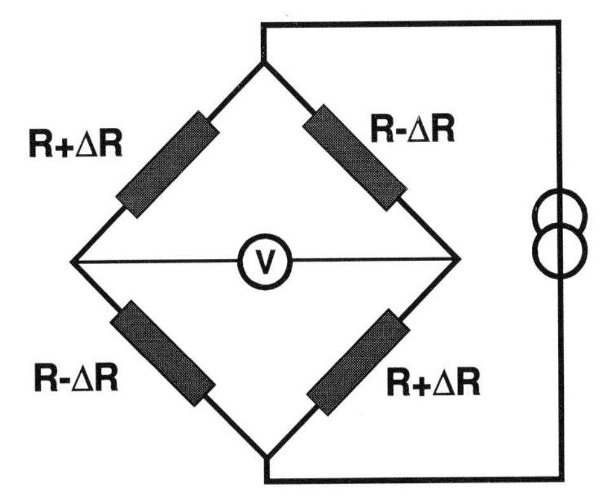

Figure 5.29 Wheatstone bridge configuration for a strain guage.

more information on the piezoresistance effect see [99]. Polysilicon, which is also commonly used as a piezoresistor, contains many orientations, and there are many additional factors to be considered. This is dealt with in detail in [49].

5.3.1.2 Piezojunction Read-out

The technique is basically the same as the piezoresistive except that the stress in the material is measured by changes in the characteristics of the p-n junction [101-102].

5.3.1.3 Electromagnetic Read-out

In this configuration coils are placed on the moving part of the device and the encapsulation. An alternating magnetic field is generated by one coil and this induces a voltage in the second coil. This generated voltage is proportional to the distance between the two coils.

5.3.1.4 Capacitive Read-out

In this case the capacitance is usually measured between the mass and the encapsulation [104]. The high temperature coefficient found with the piezoresistive effect has been avoided and the temperature effects arise from changes in Young's modulus and thermally induced packaging stress. In the case of surface micromachined devices it is therefore necessary to maximise the area of the capacitive plates and this is limited for practical reasons. This problem is usually solved by using a finger structure [105].

5.3.1.4 Thermal Read-out

An alternative read-out technique is thermal. The basic structure of a thermal accelerometer is shown in figure 5.30 [106]. The heat generated by the heat source is dissipated through the air or through the thermopile. If the mesa approaches the heat source the heat being dissipated through the air will be modulated thus changing the temperature difference across the thermopile. Thus the movement of the mesa can be read-out as a change in temperature.

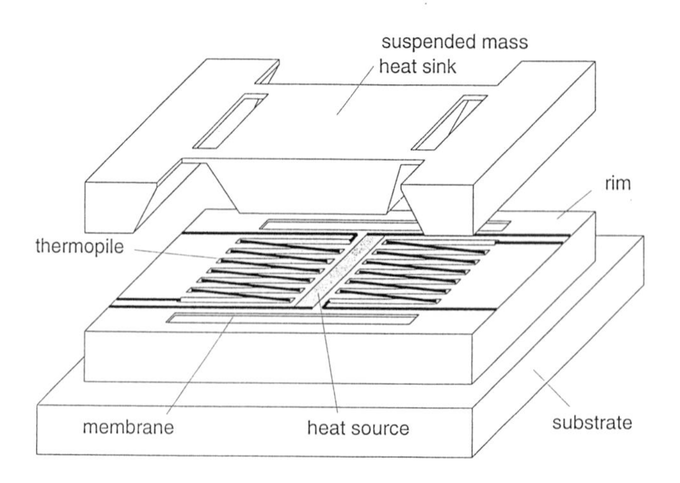

Figure 5.30 Thermal read-out based accelerometer.

Thermal read-out is also commonly used for flow sensors. In this case the heat generated by a resistor on the chip is carried by the flow. This results in a temperature difference between the up stream and down stream portions of the chip whcih can be measured simply with an integrated thermopile.

5.3.1.5 Resonant frequency Read-out

A resonator can be applied to a range of micromechanical sensors. The resonator must first be driven into resonance and this is achieved either electrothermally [107] or electrostatically [108]. The resonance is modulated by an induced strain and this is read out by means of piezoresistance, capacitance or optics. Examples of three resonator structures is shown in figure 5.31.

In the case of electrothermal activation a heating resistor is usually placed on a

Figure 5.31 Three examples of resonator structures.

beam or membrane. Taking the example of a double clamped beam, the driving moment is proportional to the power dissipation which is given by ($P=V^2/R$):

$$(V_1+V_0\sin(\varpi t))^2=(V_1^2+\frac{V_0}{2})+2V_0V_1\sin(\varpi t)-V_0^2\cos(2\varpi t) \tag{5.5}$$

and this will force the beam into resonance if the activation frequency is equal to the resonant frequency. If the beam is in the fundamental resonant mode then the frequency is given in terms of the length, L, and the thickness, d by [109]:

$$f_0=0.615\frac{d}{L^2}\sqrt{\frac{E}{(1-\nu^2)\rho_{Si}}} \tag{5.6}$$

where ρ_{Si} is the density of silicon. Under stress this frequency is modulated, yielding a frequency f_{load}:

$$f_{load}=f_0\sqrt{1+\frac{3L^2\sigma}{\pi^2Ek^2}} \tag{5.7}$$

The resonant frequency of a single clamped beam and a membrane are given in equations 5.8 and 5.9, respectively [109].

$$f_0=0.162\frac{d}{L^2}\sqrt{\frac{E}{\rho_{Si}}} \tag{5.8}$$

$$f_0=1.654\frac{d}{a^2}\sqrt{\frac{E}{(1-\nu^2)\rho_{Si}}} \tag{5.9}$$

where a is the side length of the membrane.

An alternative to a mechanical oscillator is an electrical oscillator. A simple ring oscillator has a oscillation frequency determined by the delay time of each inverter. Using MOS transistors as both electrical element and strain sensor is one technique to achieve such a device [110]. Alternatively, modulation of the power supply can be used to control the delay time. I^2L gates have a delay time which is strongly dependent upon the supply current and this effect has been applied to a frequency output strain sensor [95].

5.3.1.6 Optical Read-out

The optical read-out technique can be used with resonators as mentioned above. Alternatively the optical read-out can be achieved by interference patterns or the shutter type [111-112]. The mechanical movement casts a shadow on the sensing area yielding a photo-current proportional to the movement. In [112] this has been applied to an angle of rotation sensor.

5.3.1.7 Electron-tunnelling Read-out

This technique uses the emission from a sharp tip. The tunnelling occurs only at very small distances so in the case of the accelerometer shown in figure 5.32 a feedback loop is used to control the position of the mesa. It is therefore the voltage required voltage to maintain this position, and therefore the tunnelling, that is the measure of the acceleration [113].

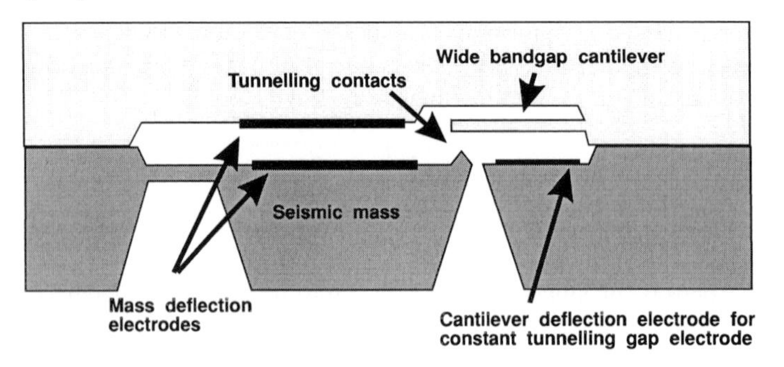

Figure 5.32 Electron-tunnelling read-out for a bulk micromachined accelerometer.

5.3.1.8 Piezoelectric Read-out

Silicon is not a piezoelectric material and therefore additional depositions are required if a piezoelectric sensor is to be fabricated. Materials such as ZnO can be deposited as a post-processing step. Piezoelectric materials generate a voltage when stressed and this can be used as a direct measure of the acceleration or pressure [114]. An alternative feature of piezoelectric materials is that this process may be reverse (i.e. an applied voltage may generate a stress in the material). Surface acoustic waves can then be generated in these materials where the detection is simply a reversal of the generation. The parameter measured here is the delay from generation to detection [115]. The layout of such a structure is shown in figure 5.33.This delay is highly sensitive to stress. This opens many opportunities for mechanical sensors and actuators.

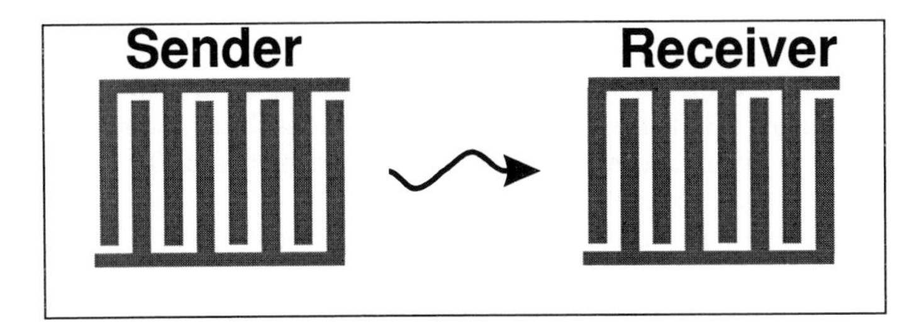

Figure 5.33 Surface acoustic wave based read-out scheme.

5.3.2 Readout electronics

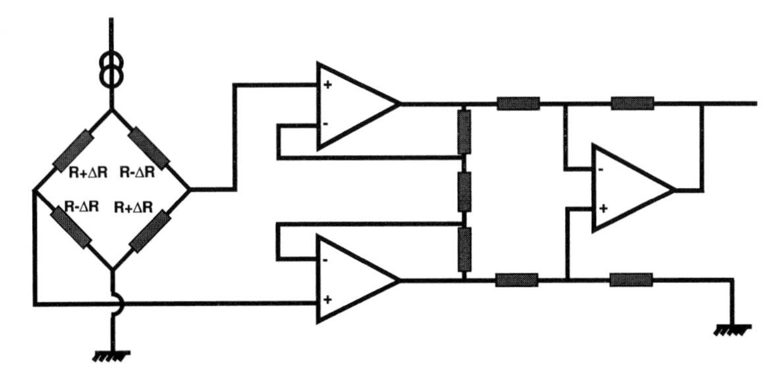

Figure 5.34 Simple amplifier for a Wheatstone bridge read-out scheme.

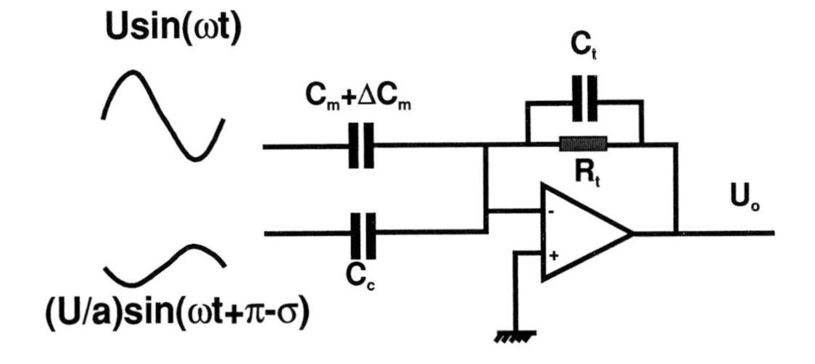

Figure 5.35 Simple amplifier for a capacitive read-out scheme.

As shown above there are various read-out techniques. For each of these the signal requires further handling. In most cases it is important to amplify the signal since many sensors yield low signal changes. In the case of a piezoresistive device a simple amplifier can be applied and this is shown in figure 5.34 [114].

For the case of capacitive read-out an alternative amplifier is required. The charge amplifier shown in figure 5.35 [115], uses a reference capacitor with a signal out of phase with the capacitance to be measured. If this phase difference is 180° the amplifier is extremely sensitive to changes in capacitance.

The two examples shown in figures 5.34 and 5.35 are the most simple examples of amplifier circuit and contain no compensation for cross-sensitvity or other parasitic effects.

5.4 EXAMPLES OF MICROMECHANICAL SENSORS

There are many micromechanical sensors fabricated in silicon and already on the market. In this section examples of these sensors, fabricated using the various techniques described above, are given.

In general the mechanical sensor measures a displacement or strain induced by the measurand. The simplest structure is probably the pressure sensor, which comprises a thin membrane which is distorted as a result of applied pressure. As described above there are three main micromachining technologies for fabricating micromechanical sensors and a wide range of read-out techniques.

5.4.1 Accelerometers

Accelerometers are forming an increasingly important sector of the sensor market and there is a wide range of structures available. In order to achieve sensitivity to acceleration a mass must be incorporated into the sensor. This can be achieved by simple backside etching mask. An additional etch is used in this case to form thin support beam in place of a membrane to further enhance the sensor properties. A linear accelerometer can be defined by the following specifications:

-Range a_{max}: $[ms^{-2}]$
-Bandwidth f_{max}: [Hz]
-Mass m : [kg]
-Sensitivity S_z : [V/g]

Negative effects such as cross-sensitivity can be defined by:

-Linear S_y, S_y: $[ms^{-2}]$
-Rotational S_{xy}, S_{xz}, S_{yz}: [m/rad]
-Non-linearity NL : [%/fso]

and two important considerations to define the operation of the device are:

-Overload capability
-Damping

The simplest model of an accelerometer is a mass and a spring, as shown in figure 5.36a. In the DC case the displacement is defined as a balancing of the force due to the displacement of the mass, m, and the spring constant, *k*. The displacement, *z*, is therefore given by:

$$z = \frac{ma}{k} \tag{5.10}$$

and the sensitivity is simply:

$$S_z = \frac{m}{k} \tag{5.11}$$

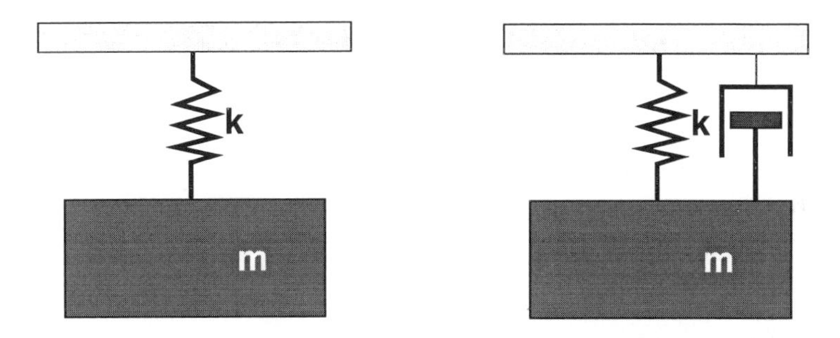

Figure 5.36 Simple model of an accelerometer; a) without damping and b) with damping.

This is a completely undamped system and will suffer severe oscillations. Therefore, in the AC case damping must be introduced. A simple model of an accelerometer comprising a mass, spring and damper is shown in figure 5.36b. There are therefore three forces acting on the mass, those of the acceleration itself (ma), the spring (F_s) and the damping (F_d). The mass reaches its stable position when:

$$ma = F_s + F_d \tag{5.12}$$

The force, F_d, can be given in terms of the damping force, D by:

$$F_d = -D\frac{dz}{dt} \tag{5.13}$$

The three forces acting on the mass must therefore equal the external force applied. Therefore the dynamic motion of the mass is given by:

$$-ma_{ext} = ma + F_s + F_d \tag{5.14}$$

$$ma = \frac{md^2z}{dt^2} + \frac{kdz}{dt} + Dz \tag{5.15}$$

The response of the mass to an acceleration can be given in the frequency domain by:

$$\frac{z}{a} = \frac{m}{ms^2 + Ds + k} \tag{5.16}$$

The damping factor, ξ, of the system can be given in terms of the damping, D, spring constant, k, and mass, m by:

$$\xi = \frac{D}{2\sqrt{km}} \tag{5.17}$$

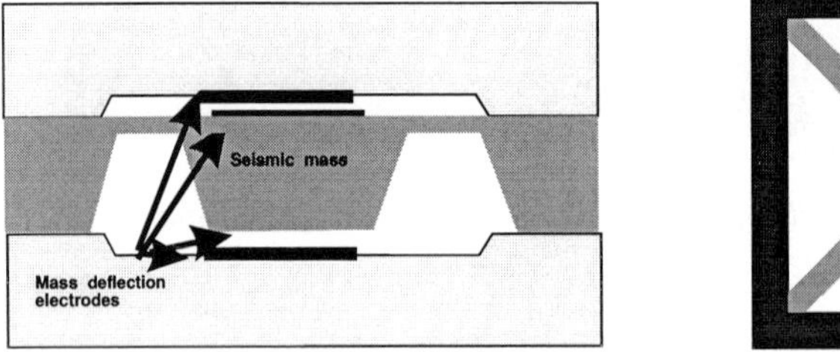

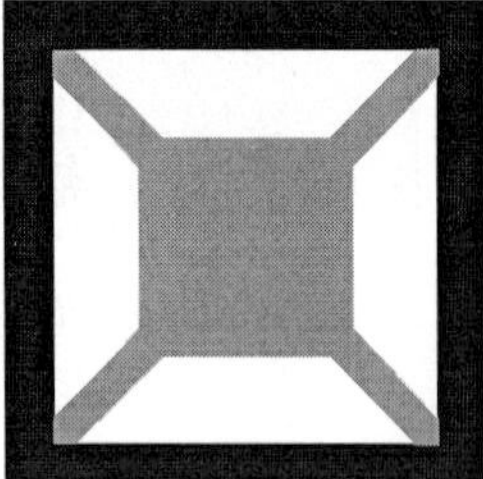

Figure 5.37 Vertical accelerometer structure.

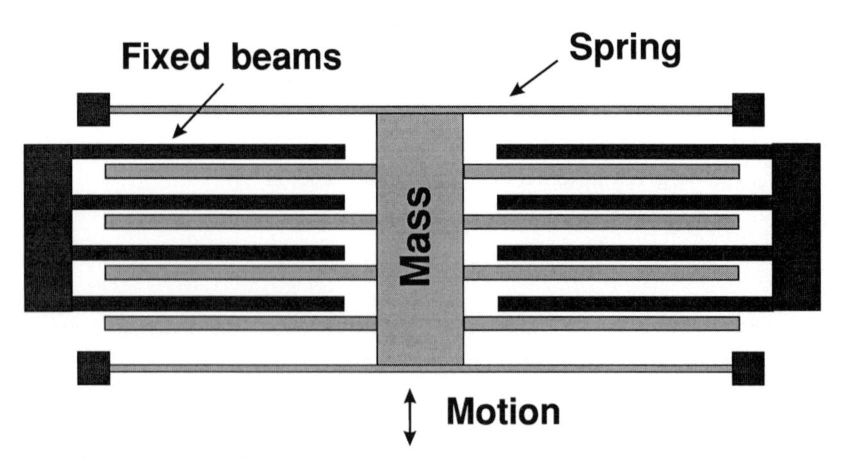

Figure 5.38 Lateral accelerometer.

As described above there are various read-out techniques and can be fabricated in three technologies. Backside bulk micromachined devices are usually used for vertical acceleration sensors and an example is shown in figure 5.37 surface and frontside micromachined devices, on the other hand, have been more frequently applied to lateral accelerometers such as shown in figure 5.38. This device has a capacitive read-out and thus the finger structure is used to increase the sensing area. A further improvement in the sensitivity of the device is to operate in resonator mode as applied by [104].

5.4.2. Gyroscope

There has been considerable interest recently in the possibilities of manufacturing micromachined gyroscopes. A silicon based micromechanical gyroscope was presented by Greiff et. al. in 1991 [118]. This device was based on two resonators , one resonating on the other, as shown in figure 5.39. In this case the resonators were excited electrostatically at a frequency of 100kHz. A second structure was proposed by Maenaka et. al. and was based on a tuning fork [119]. A third type of gyroscope is similar to a conventional lateral accelerometer, as shown in figure 5.38 [120]. This device is sensitive to rotation in the x direction shown in the figure.

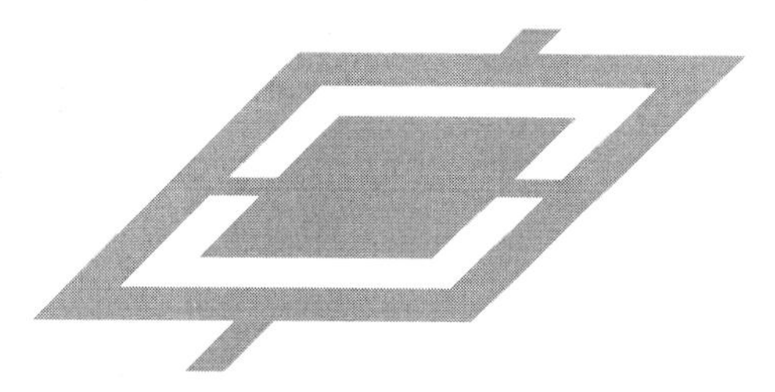

Figure 5.39 Double resonator based gyroscope.

All vibrating structures have two, preferably perpendicular oscillations. The Coriollis induced vibration is always much smaller than the reference vibration. The rotation of quasi-vibrating plates and vibrating beams is shown in figure 5.40 [121].

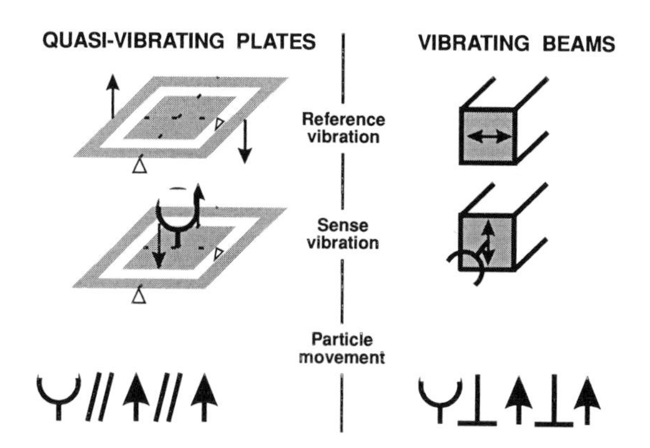

Figure 5.40 Vibration modes of quasi-vibrating and vibrating gyroscopes [121].

5.4.2. Pressure sensors

Figure 5.41 shows cross-sectional and planar views of a pressure sensor. In this case the epitaxial layer forms the membrane which, as shall be discussed later, has a well controlled thickness. The readout of these devices can be either capacitive, where the capacitance between the membrane and a fixed plate is measured, or piezoresistive, where the strain in the membrane is measured by a resistor. The example given in figure 5.41 is piezoresistive. Once again a surface micromachined device usually uses a capacitive readout. The sensing membrane has a sealed cavity, where the sacrificial layer has been removed and the inner-pressure is the sensor reference pressure. This basic structure is shown in figure 5.42. A further technique is to use piezoelectric materials. This can either be applied directly as a piezoelectric element, thus yielding a voltage as a function of strain, or as a surface acoustic wave. The second of these techniques is shown in figure 5.43.

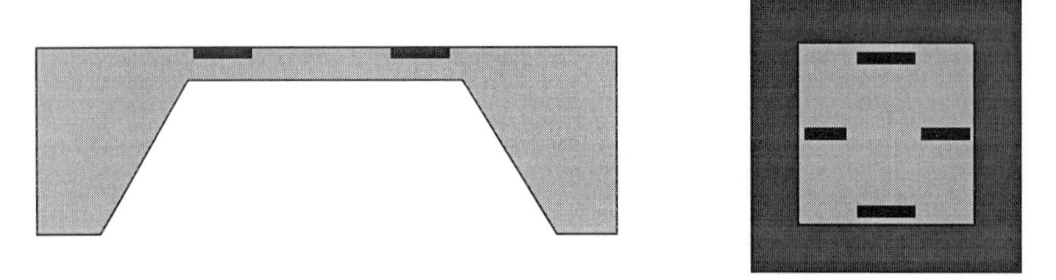

Figure 5.41 Cross sectional and planar views of a basic pressure sensor.

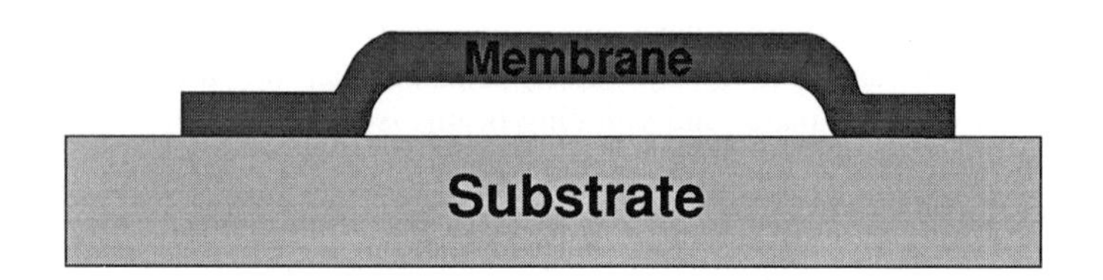

Figure 5.42 Surface micromachined pressure sensor.

In all these cases the pressure is measured as a function of the deformation of a membrane. Taking the example of a circular diaphragm of radius, R and thickness, t, with a pressure difference of Δp. The radial stress, σ_r, at a distance, r, from the centre of the diaphragm is given by [122]:

$$\sigma_r = \left(\frac{3\Delta p z R^2}{4t^3}\right)\left[(1+\nu)-(3+\nu)\left(\frac{r^2}{R^2}\right)\right] \qquad \textbf{(5.18)}$$

where z is the vertical co-ordinate with the centre of the membrane being zero. Similarly the tangential stress σ_t by:

$$\sigma_t = \left(\frac{3\Delta p z R^2}{4t^3}\right)\left[(1+\nu)-(1+3\nu)\left(\frac{r^2}{R^2}\right)\right] \tag{5.19}$$

where ν, is Poisson's ratio. It can be seen from equations 5.18 and 5.19 that at r=0 (the centre of the membrane) the two stresses are equal, as would be expected. r=R the difference is maximum and this is where the resistors should be placed, for piezoresistive based sensors. For capacitive readout the deflection of the membrane is the most important and this is given by:

$$D = \frac{3\Delta p(1-\nu^2)(R^2-r^2)^2}{16Et^3} \tag{5.20}$$

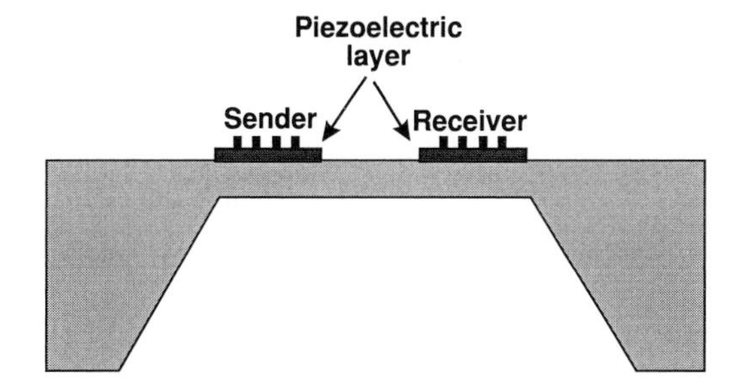

Figure 5.43 Surface acoustic wave based pressure sensor.

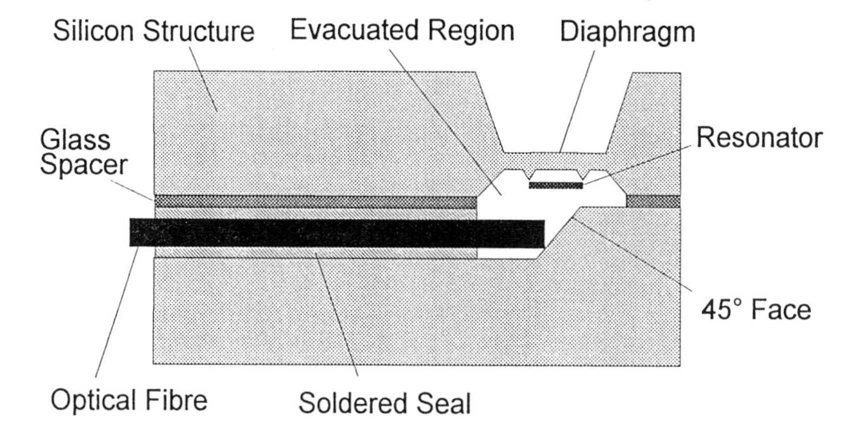

Figure 5.44 Resonator based pressure sensor (Courtesy of Lucas Industries).

An alternative to a piezoresistive or capacitive read-out is to use a resonator with an optical read-out. Such a structure is shown in figure 5.44, where the deflection of the membrane modulates the resonant frequency of an attached beam [74]. An SEM photograph of the beam and fibre interface is shown in figure 5.45. This device, which has been developed by Lucas Industries, is particularly suitable for harsh environments with high temperatures and high electric fields.

Figure 5.45 SEM photograph of the resonator with fibre (courtesy of Lucas Industries).

5.4.3. Flow sensors

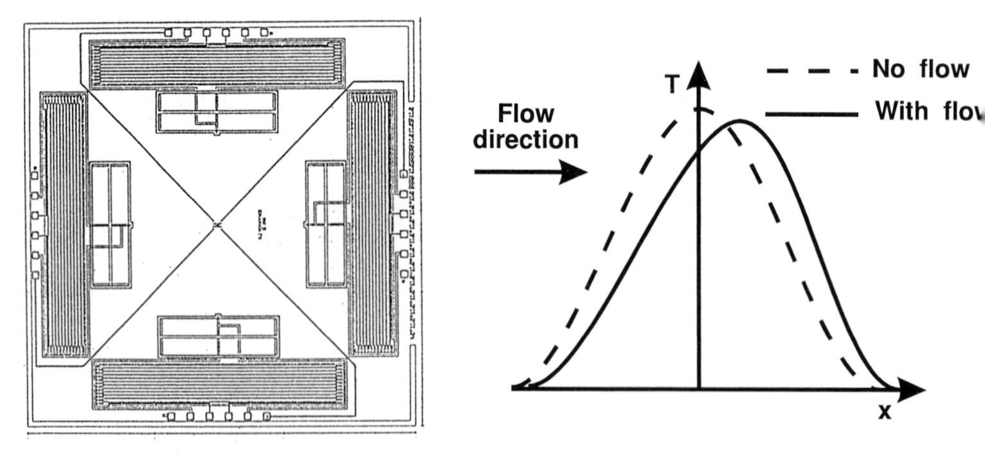

Figure 5.46 a) Layout of thermo-pile based sensor, b) temperature distribution across the chip..

A flow sensor, although measuring a mechanical parameter usually has no moving parts. The layout of a thermal based flow sensor is shown in figure 5.46a [123]. This is a two dimensional flow sensor containing 4 thermopiles. The chip is heated with a resistor and a temperature profile across the chip is measured. A fluid flow will yield a higher temperature downstream (therefore a lower temperature difference) than upstream. This temperature profile is shown in figure 5.46b [124].

5.4.4. Force sensors

Recent advances in etching techniques has open new possibilities in surface analysis. A micro-tip construction, as used for the accelerometer, figure 5.33, has found a range of applications in surface analysis. It has been applied to scanning tunnelling microscopes (STM) and Atomic force microscopes (AFM). The difference is that STM measures contours of constant electron density, by maintaining a constant tunnelling current and AFM measures the mechanical deflection of the probe. These two structures are shown in figure 5.47 [124].

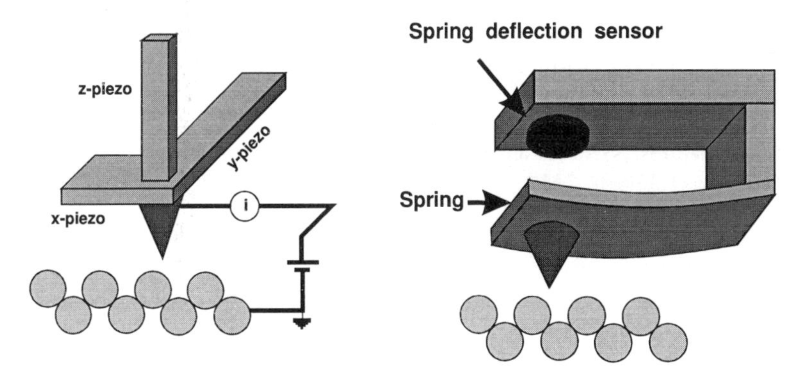

Figure 5.47 Two micro-tip based microscopes.

5.5 CONCLUSIONS

In the above sections both the electronic and the micromachining processing has been described. An important consideration when developing a sensor with in-built signal processing is to first establish from where the compatibility problems may arise. In the case of bulk micromachining the problem mainly arises potential contamination. This is usually solved by adding the additional steps to the end of the process after which the wafers are not returned to the clean room. One solution to this problem is to use etchants which do not contaminate, such as TMAH. In the case of surface micromachining care must be taken to ensure a) the sacrificial etchant is not allowed to attack the electronics and 2) the total additional thermal budget does not adversely effect the electronics.

It is essential to establish a process where the characteristics of both the mechanical and electrical components can be maintained. Furthermore the designer must know the limitations in terms of additional steps which can be added, and both these considerations are discussed below.

Silicon micromachining has presented the designer with many new opportunities for developing new microsensors. The first main breakthrough for silicon mechanical sensors was the pressure sensor which remains a major world market. With the explosion of interest in micro-accelerometers from the automobile industry has shown the silicon is able to meet this new challenge. The field of resonators has further opened the possibilities for pressure sensors and accelerometers and also gyroscopes for which a new large market awaits.

REFERENCES

1] A.W. van Herwaarden and P.M. Sarro, "Floating membrane thermal vacuum sensor", Sensors and Actuators, 14, (1988), pp 259-268.

2] P.M. Sarro, A.W. van Herwaarden and W. van der Vlist, "A silicon-silicon nitride membrane fabrication process for smart thermal sensors", Sensors and Actuators, A41-42, (1994), pp 666-671.

3] S.K. Clark and K.D. Wise, "Pressure sensitivity in anisotropically etched thin diaphragm pressure sensors", IEEE Trans. Electron Dev., ED-26, (1979), pp 1857-1896.

4] H.V. Allen, S.C. Terry and D.W. de Bruin, "Accelerometers systems with self-testable features", Sensors and Actuators, 20, (1989), pp 153-161.

5] R.P. van Kampen, M.J. Vellekoop, P.M. Sarro and R.F. Wolffenbuttel, "Application of electrostatic feedback to critical damping of an integrated silicon capacitive accelerometer", Sensors and Actuators, A43, (1994), pp 100-106.

6] V. Gass, B.H. van der Schoot, S. Jeanneret and N.F. de Rooij, "Integrated flow regulated silicon micropump", Sensors and Actuators, A43, (1994), pp 335-338.

7] H.T.G. van Lintel, F.C.M. van de Pol and S. Bouwstra, "A piezoelectric micropump based on micromachining of silicon", Sensors and Actuators, 15, (1988), pp 153-167.

8] T.A. Kwa and R.F. Wolffenbuttel, "Integrated grating/detector array fabricated in silicon using micromachined techniques", Sensors and Actuators, A31, (1992), pp 256-266.

9] K.M. Mahmoud, R.P. van Kampen, M.J. Rutka and R.F. Wolffenbuttel, "A silicon integrated smart pressure sensor", Proceeding Transducers'93, Yokohama, Japan, pp 217-220.

10] H.C. Nathanson and R.A. Wickstrom, "A resonant-gate silicon surface transistor with high-Q band pass properties", Appl. Phys. Lett., 7, (1965), p 84.

11] H.C. Nathanson, W.E. Newell, R.A. Wickstrom and J.R. Davis jr., "The resonant gate transistor", IEEE Electron Dev., ED-14, (1967), pp 117-133.

12] R.N. Thomas, J. Guldberg, H.C. Nathanson and P.R. Malmberg, "The mirror matrix tube: a novel light valve fro projection displays", IEEE Electron Dev., ED-22, (1975, p 765.

13] R.T. Howe and R.S. Muller, "Polycrystalline and amorphous silicon micromechanical beams: annealing and mechanical properties", Sensors and Actuators, 4, (1983), pp 447-454.

14] L-S. Fan, Y-C. Tai and R.S. Muller, "Pin joints, gears, springs, cranks and other novel micromechanical structures", Proceedings Transducers 87, Tokyo, 1987, pp 849-852.

15] B. Schwartz and H. Robbin, "Chemical etching of silicon; part I", J. Electrochem Soc., 106, (1959), pp 505-508.

16] B. Schwartz and H. Robbin, "Chemical etching of silicon; part II", J. Electrochem Soc., 107, (1960), pp 108-111.

17] B. Schwartz and H. Robbin, "Chemical etching of silicon; part III", J. Electrochem Soc., 108, (1961), pp 365-372.

18] B. Schwartz and H. Robbin, "Chemical etching of silicon; part IV", J. Electrochem Soc., 123, (1976), pp 1903-1909.

19] K.E. Bean, "Anisotropic etching of silicon", IEEE Trans Electron Devices, ED-25, (1978), pp 1185-1193.

20] K.E. Petersen, "Silicon as a mechanical material", Proc. IEEE, 70, (1982), pp 420-457.

21] L. Ristić, "CMOS technology: a base for micromachining", Microelectronics J., 20, (1989), pp 153.

22] H. Baltes, "CMOS as a sensor technology", Sensors & Actuators, A37-38, (1993), pp 51-56.

23] L.K. Nanver, E.J.G. Goudena and H. W. van Zeijl, DIMES-01 baseline BIFET process for smart sensor experimentation, Sensors and Actuators A, 36 (1993), pp 139-147.

24] P.M. Sarro, "Sensor technology strategy in silicon", Sensors and Actuators, A31, (1992), pp 138-143.

25] P.Ashburn, D.J. Roulson and C.R. Selvakumar, "Comparison of experimental and computed results on arsenic- and phosphorus-doped polysilicon emitter bipolar transistors", IEEE Electron dev., ED-34, (1987), pp1346-1353.

26] S. Wolf, "Silicon processing for the VLSI era, vol 2: process integration, Lattice Press, CA, USA.

27] Lj. Ristic ed., "Sensor technology and devices", Artec House, London, UK, ISBN 0-89006-532-2.

28] A.D. Khazan, "Transducers and their elements", PTR Prentice Hall, ISBM 0-13-929480-5, 1994.

29] J.B. Price, "Anisotropic etching of silicon with KOH-H_2O-Isopropyl alcohol", in Semiconductor Silicon ed. H.R. Huff and R.R. Burgess, Princeton, NJ, Electrochemical Society Proceedings, 1973, p 339.

30] H. Seidel, L. Csepregi, A. Heuberger and H. Baumgärtel, "Anisotropic etching of crystalline silicon in alkaline solution; Part I Orientation dependence and behavior of passivation layers; Part II Influence of dopants", J. Electrochem Soc., 137, (1990), pp 3612-3626.

31] C. Hakan Özdemir and J.G. Smith, "New phenomena observed in electrochemical micromachining of silicon", Sensors and Actuators, A31, (1992), pp 87-93.

32] S. Tatić-Lučić and Y.-C. Tai, "Novel extra accurate method for two-sided alignment on silicon wafers", Sensors and Actuators, A41-42, (1994), pp 573-577.

33] T.A. Kwa and R.F. Wolffenbuttel, "Potential defined etch mask for silicon micromachining" Proceedings Eurosensors VII, San Sebastian, Spain, p 370.

34] M.J. Vellekoop, C.C.G. Visser, P.M. Sarro and A. Venema, "Compatibility of zinc-oxide with silicon IC processing", Sensors and Actuators, A21-A23, (1990), pp 1027-1030.

35] B. Peurs and W. Sansen, "Compensation structures for convex corner micromachining in silicon", Sensors and Actuators, A21-23, (1990), pp 1036-1041.

36] R.P. van Kampen and R.F. Wolffenbuttel, "Effects of <110> oriented corner compensation structure on membrane quality and convex corner integrity in (100) silicon using aqueous KOH", Proceedings MME 94, Pisa Italy.
37] K.M. Mahmoud and R.F. Wolffenbuttel, "Compatibility between bipolar read-out electronics and microstructure in silicon", Sensors and Actuators, A31, (1992), pp 188-199.
38] M. Ataka, A. Omodaka and H. Fujita, "A biomimetic micro motion system-a ciliary motion system", Proceeding Transducers'93, Yokohama, Japan, pp 38-41.
39] K. Shimaoka, O. Tabata, M. Kimura and S. Sugiyama, "Micro-diaphragm pressure sensor using polysilicon sacrificial layer etch-stop technique", Proceeding Transducers'93, Yokohama, Japan, pp 632-635.
40] O. Tabata, R. Asahi, N. Fujitsuka, M. Kimura and S. Sugiyama, "Electrostatic driven optical chopper using SOI wafer", Proceeding Transducers'93, Yokohama, Japan, pp 124-127.
41] B. Diem, P. Rey, S. Renard, S. Viollet Bosson, H. Bono, F. Michel, M.T. Delaye and G. Delapierre, "SOI 'SIMOX' from bulk to surface micromachining, a new age for silicon sensors and actuators", Sensors and Actuators, A46-A47, (1995), pp 8-16.
42] M.-A. Grétillat, C. Linder and N.F. de Rooij, "Multilayer polysilicon resonators including shielding for excitation and detection", Proceeding Transducers'93, Yokohama, Japan, pp 292-295.
43] D. Poenar, P.J. French, R. Mallée, P.M. Sarro and R.F. Wolffenbuttel, "PSG for surface micromachining", Sensors and Actuators, 41, (1994), pp 304-309.
44] B.P. van Drieënhuizen, J.F.L. Goosen, P.J. French, Y.X. Li, D. Poenar and R.F. Wolffenbuttel, "Surface micromachined module compatible with BiFET electronic processing", Proceedings Eurosensors 94, Toulouse, France, September 1994, p 108.
45] Y. Onuma and K. Sekiya, "Piezoresistive properties of polycrystalline silicon thin films", Jpn. J. Appl. Phys., 11 (1974), pp 420-421.
46] J.Y.W. Seto, "The electrical properties of polycrystalline silicon films", J. Appl. Phys., 46, (1975), pp 5247-5254
47] M.M. Mandurah, K.C. Saraswat, T.I. Kamins, "A model for conduction in polysilicon: Part 1-Theory, Part 2 - Comparison of theory and experiment", IEEE Trans. Electron Dev., ED-28, (1981), pp 1163-1176.
48] H. Mikoshiba, "Stress-sensitive properties of silicon gate MOS devices", Solid State Electronics, 24, (1981), pp 221-232.
49] P.J. French and A.G.R. Evans, "Piezoresistance in polysilicon and its applications to strain gauges", Solid State Electronics, 32, (1989), pp 1-10.
50] S.P. Murarka and T.F. Retajczyk, Jr., "Effect of phosphorus doping on stress in silicon and polycrystalline silicon", J. Appl. Phys., 54, (1983), pp 2069-2072.
51] H. Guckel, D.W. Burns, C.R. Rutigliano, D.K. Showers and J. Uglow, "Fine Grained Polysilicon and its Application to Planar Pressure Transducers", The Fourth International Conference on Solid-State Sensors and Actuators, Tokyo, Japan, 1987, Digest of Technical Papers, p.277.
52] R.S. Hajab and R.S. Muller, "Residual strain effects on large aspect ratio micro-diaphragms", Proc. IEEE Micro Electro mechanical systems conference (MEMS), Salt Lake City, Utah, USA, 20-22 February, 1989, pp 133-138.
53] K. L. Yang, D. Wolcoxen and G. Gimpelson, "The effects of post processing

techniques and sacrificial layer materials on the formation of free standing polysilicon microstructures", Proc. IEEE Micro-mechanical systems, Utah, USA, 20-22 February, 1989, pp 66-70.

54] H. Guckel, J.J. Sniegowski, T.R. Christenson, S. Mohney and T.F. Kelly, "Fabrication of micromechanical devices from polysilicon films with smooth surfaces", Sensors and Actuators, 20, (1990), pp 117-122.

55] P. J. French, B. P. van Drieënhuizen, D. Poenar, J. F. L. Goosen, P. M. Sarro and R. F. Wolffenbuttel, "Low-stress polysilicon process compatible with standard device processing", Sensors VI: Technology, Systems and Applications, Manchester, U.K., 1993, pp 129-133.

56] H. Guckel, T. Randazzo and D.W. Burns, "A simple technique for the determination of mechanical strain in thin films with applications to polysilicon", J. Appl. Phys., 57, (1985), pp 1671-1675.

57] M.G. Allen, M. Mehregany, R.T. Howe and S.D. Senturia, "Microfabrication structures for the *in-situ* measurement of residual stress, Young's modulus, and ultimate strain of thin films", Appl. Phys. Lett., 51, (1987), pp 241-243.

58] B.P. van Drieënhuizen, J.F.L. Goosen, P.J. French and R.F. Wolffenbuttel, "Comparison of techniques for measuring both compressive and tensile stress in thin films", Sensors and Actuators, A37, (1993), pp 756-765.

59] M. Sekimoto, H. Yoshihara and T. Ohkubo, "Silicon nitride single-layer x-ray mask", J. Vac. Sci. Tech., 21, (1982), pp 1017-1021.

50] P.J. French, P.M. Sarro, R. Mallée and R.F. Wolffenbuttel, "Optimization of a low stress silicon nitride process for surface micromachining applications", Proceedings Eurosensors 94, Toulouse, France, 26-28 September 1994, p 205.

61] Y.X. Li, "Plasma etching for silicon sensor applications", PhD thesis, (1995), Delft University of Technology, Delft, The Netherlands.

62] P.J. French and R.F. Wolffenbuttel, "Reflow of BPSG for sensor applications", J. Micromech. Microeng., 3, (1993), pp 1-3.

63] M. Bartek, P.J. French and R.F. Wolffenbuttel, "Planarization in surface micromachining using selective epitaxial growth", Proceedings Eurosensors 94, Toulouse, 26-28 September 1994, p 210.

64] M. Bartek, Y.X. Li, P. Gennissen, P.J. French and R.F. Wolffenbuttel, "Epitaxial layer optimization for smart silicon sensors by placing read-out electronics in SEG silicon", Proceedings Eurosensors 94, Toulouse, 26-28 September 1994, p 213.

65] D.J. Monk, D.S. Soane and R.T Howe,"Sacrificial layer SiO_2 wet etching for micromachining applications", Proceeding Transducers'91, San Francisco, USA, pp 647-650.

66] D.J. Monk, D.S. Soane and R.T Howe,"A diffusion reaction model for HF etching of LPCVD phosphosilicate glass sacrificial layer", Proceeding Solid State Sensor and Actuator, IEEE Hilton Head, USA, (1992), pp 47-51.

67] D.J. Monk, D.S. Soane and R.T Howe,"Determination of the etching kinetics for the hydrofluoric acid/silicon dioxide system", J. Electrochem. Soc, 140, (1993), pp 2339-2345.

68] D.J. Monk, D.S. Soane and R.T Howe,"Enhanced removal of sacrificial layers for silicon surface micromachining", Proceedings Transducers'93, Yokohama, Japan, pp 280-283.

69] J.J. Gajda, "Techniques in failure analysis of MOS devices", Annual Proceedings of reliability physics, 12, (1974), pp 30-37.

70] P.R. Scheeper, J.A. Voorthuijzen, W. Olhuis and P. Bergveld, "Investigation of attractive forces between silicon nitride microstructures and an oxidized substrate", Sensors and Actuators, A30, (1992), pp 231-239.

71] C.H. Mastrangelo and C.H. Hsu, "Mechanical stability and adhesion of microstructures under capillary forces-part I:Basic theory; part II Experimental", J. Micromechanical Systems, 2, (1993), pp 33-55.

72] G.T. Mulhern, D.S. Soane and R.T Howe,"Supercritical carbon dioxide drying of microstructures", Proceedings Transducers'93, Yokohama, Japan, pp 296-299.

73] S. Boustra, R. Legtenberg, H.A.C. Tilmans and M. Elwenspoek, "Resonating microbridge mass flow sensor", Sensors and Actuators, A21-23, (1990), pp 332-335.

74] A.J. Jacobs-Cook and M.E.C. Bowen, "Planar fibre-optic interface for silicon microresonators", Sensors VI: Technology, Systems and Applications, Manchester, U.K., 1993, pp 147-154.

75] L. Rosengren, L. Smith and Y. Bäcklund, "Micromachined optical planes and reflectors in silicon ", Sensors and Actuators, A41-42, (1994), pp 330-333.

76] J. Brugger, N. Blanc, Ph. Renaud and N.F. de Rooij, "Microlever with combined integrated sensors/actuators function for scanning force microscopy", Sensors and Actuators, A43, (1994), pp 339-345.

77] J. Foerster, Y.X. Li,M. Bartek, P.J. French and R.F. Wolffenbuttel, "Fabrication of recessed micro-tips in silicon for sensor applications", Proceedings MME 94, Pisa, Italy, 1994.

78] K.A. Shaw, Z.L. Zhang and N.C. MacDonald, "SCREAM I: a single mask, single-crystal silicon, reactive ion etching process for microelectromechanical structures", Sensors and Actuators, A40, (1994), pp 63-70.

79] Y.X. Li, P.J. French, P.M Sarro and R.F. Wolffenbuttel, "Fabrication of a single crystalline silicon capacitive lateral accelerometer using micromachining based on single step plasma etching", Proceedings MEMS'95, Amsterdam, The Netherlands, 29 January-2 February 1995, p 398-403.

80] P.T.J. Gennissen, P.J. French, D.P.A. De Munter, T.E> Bell, H. Kaneko and P.M. Sarro, "Porous silicon micromachining techniques for acceleration fabrication", Proceedings ESSDERC'95, Den Haag, The Netherlands, Sept. 1995, pp 593-596.

81] H. Jansen, M. de Boer, B. Otter and M. Elwenspoek, "The black silicon method IV: the fabrication of three dimensional structures in silicon with high aspect ratio for scanning probe microscopy and other applications", Proceedings MEMS 95, Amsterdam, The Netherlands, 29 January-2 February, 1995, pp 88-93.

82] K. Ohwada, Y. Negoro, Y. Konaka and T. Oguchi, "Groove depth uniformization in (110) Si anisotropic etching by ultrasonic wave and application to accelerometer fabrication", Proceedings MEMS 95, Amsterdam, The Netherlands, 29 January-2 February, 1995, pp 88-93.

83] W.H. Ko, J.T. Suminto and G.J. Yeh, "Bonding techniques for microsensors", Micromachining and micropackaging of transducers, ed C.D. Fung, P.W. Cheung, W.H. Ko and D.G. Fleming, Elsevier Science Publishers, (1985), pp 41-61.

84] Y. Kanda, K. Matsuda, C. Murayama and J. Sugaya, "The mechanism of field assisted silicon-glass bonding", Sensors & Actuators, A21-23, (1990), pp 939-943.

85] E.F. Cave, "Method of making a composite insulator semiconductor wafer", US Patent No. 3,290,760, 1966.

86] M. Esashi, A. Nakano, S. Shiji and H. Hebiguchi, "Low temperature silicon-to-silicon anodic bonding with intermediate low melting point glass", Sensors & Actuators, A21-23, (1990), pp 931-934.

87] R.F. Wolffenbuttel and K.D. Wise, "Low-temperature silicon wafer-to-wafer bonding using gold at eutectic temperature", Sensors and Actuators, A43, (1994), pp 223-229.

88] T. Hamaguchi, N. Endo, M. Kimura and M. Nakamae, "Novel LSI/SOI wafer fabrication using device layer transfer technique", Proc. IEDM'86, pp 688-691.

89] A.D. Brooks, R.P. Donovan and C.A. Hardesty, "Low-temperature electrostatic silicon to silicon seals using sputtered borosilicate glass", J. Electrochemical Soc., 119, (1972), pp 545-546.

90] H.J. Quenzer and W. Beneke, "Low temperature silicon wafer bonding", Sensors & Actuators, A32, (1992), pp 340-344.

91] S.Schulze, "Silicon bonding in microsystem technology", Proceedings NEXUS workshop on micromachining, Bremen, 22-24 May 1995.

92] V.M. McNeil, M.J. Novack and M.A. Schmidt, "Design and fabrication of thin-film microaccelerometers using wafer bonding", Proceedings Transducer 93, Yokohama, pp 822-825.

93] P.W. Barth, "Silicon fusion bonding for fabrication of sensors, actuators and microstructures", Sensors & Actuators, A21-23, (1990), pp 919-926.

94] M. Wale and M. Goodwin, "Flip-chip bonding optimizing opto-IC's" IEEE Circuits Devices, No. 11, (1992), pp 25-31.

95] P.J. French and A.P. Dorey, "Frequency output piezoresistive pressure sensor", Sensors and Actuators, 4, (1983), pp 77-83.

96] W.J. Lian and S. Middelhoek, "Flip-flop sensors: a new class of silicon sensors", Sensors and Actuators, 9, (1986), pp259-268.

97] P.J. French, W. Lian and S. Middelhoek, "A study of the NMOS Flip-flop sensor and a comparison with the bipolar technique", Sensors and Actuators, A-24, (1990), pp65-73.

98] T.M. Reeder and D.E. Cullen, "Surface acoustic wave, pressure and temperature sensors, "Proc. IEEE, 64, (1978), pp 754-756.

99] Y. Kanda, "A graphical representation of the piezoresistance coefficients in silicon", IEEE Trans. Electron Dev., ED-29, (1982), pp 64-70.

100] C.S. Smith, "Piezoresistance effect in germanium and silicon", Physics Review, 94, (1954), pp 42-49.

101] Y. Kanda, "Effects of stress on germanium and silicon p-n junctions", Jpn. J. Appl. Phys., 6, (1967), pp 475-486

102] W.M.C. Sansen, P. Vandeloo and B. Puers, "A force transducer based on stress effects in bipolar transistors", Sensors and Actuators, 3, (1982-83), pp 343-353.

103] E. Abbaspour-Sani, R.S. Huang and C.Y. Kwok, "A linear electromagnetic accelerometer", Sensors and Actuators, A44, (1995), pp103-109.

104] F. Rudof, "A micromechanical capacitive accelerometer with a two-point inertial mass suspension", Sensors and Actuators, 4, (1983), pp 191-198.

105] W.Kuehlnel and S. Sherman, "A surface micromachined silicon accelerometer with on-chip detection circuitry", Sensors and Actuators, 45, (1994), pp 7-16.

106] U.A. Dauderstädt, P.H.S. de Vries, R. Hiratsuka and P.M. Sarro, "Silicon accelerometer based on thermopiles", Sensors and Actuators, 46, (1995), pp 201-204.

107] C. Burrer and J. Esteve, "A novel resonant silicon accelerometer in bulk-micromachining technology", Sensors and Actuators, A46, (1995), pp 185-189.

108] S.C. Chang, M.W. Putty, D.B. Hicks, C.H. Li and R.T. Howe, "Resonant-bridge two-axis microaccelerometer", Sensors and Actuators, A21-22, (1990), pp 342-345.

109] G. Stemme, "Resonant silicon sensors", J. Micromech. Microeng., 1, (1991), pp 113-125.

110] J. Neumeister, G. Schuster and W. von Münch, "A silicon pressure sensor using MOS ring oscillators", Sensors and Actuators, 7, (1985), pp 167-176.

111] E. Abbaspour-Sani, R.S. Huang and C.Y. Kwok, "A novel accelerometer", IEEE Electron Dev. Lett., 16, (1995), pp 166-168.

112] P.J. French, H. Muro and Y. Hiramoto, "An angle of rotation sensor using flip-flop photodetecting techniques", Sensors and Actuators, A21-23, (1990), pp 414-419.

113] H.K. Rockstad, T.W. Kenny, J.K. Reynolds, W.J. Kaiser and T.B. Gabrielson, "A miniature, high-sensitivity, electron tunneling accelerometer", Proc. Transducers '95, vol2, Stockholm, Sweden, (1995), pp 675-678.

114] K.W. Yeh and R.S. Muller, "Piezoelectric DMOS strain transducers", Appl. Phys. Lett., 29, (1972), pp521-522.

115] F. Seifert, W-E. Bulst and C. Ruppel, "Mechancial sensors based on surface acoustic waves", Sensors and Actuators, A44, (1994), pp 231-239.

116] S. Middelhoek and S.A. Audet, "Silicon Sensors", Academic Press Inc., San Diego, USA, (1989).

117] R.F. Wolffenbuttel and P.P.L. Regtien, "Integrated tactile imager with an intrinsic contour detection option", Sensors and Actuators, 16, (1989), pp 141-153.

118] P. Greiff, B. Boxenhorn, T. King and L. Niles, Silicon monolithic micromechanical gyroscope", Proceedings Transducers 91, San Francisco, USA, June 1991, pp 966-968.

119] K. Maenaka and T. Shiozawa, "A study of silicon angular rate sensor using anisotropic etching technology" Sensors and Actuators, A43, (1994), pp 72-77.

120] K. Tanaka, Y. Mochida and S. Sugimoto, "A micromachined vibrating gyroscope", Proceedings MEMS 95, Amsterdam, The Netherlands, 29 January-2 February 1995, pp 278-281.

121] J. Söderkvist, "Micromechanical gyroscopes", Sensors and Actuators, A43, (1994), pp 65-71.

122] J.M. Gere and S. Timoshenko, "Mechanics of materials", Chapman and Hall, 3rd edition (1991), ISBN 0-412-36880-3.

124] J.W. Bosman, J.M. de Bruijn, F.R. Riedijk, B.W. van Oudheusden and J.H. Huijsing, "Integrated smart two-dimensional thermal flow sensor with Seebeck-voltage-to-frequency conversion", Sensors and Actuators, A31, (1992), pp 9-16.

125] J.W. Gardner, "Microsensors: principles and applications", John Wiley & Sons, (1994), ISBN 0-471-94135-2.

6. Micro-Actuators for Micro-Robots: Electric and Magnetic

Ronald S. Fearing

Department of EE&CS, University of California
Berkeley, CA 94720-7440

1 Introduction

Many varieties of micro-actuators based on electrostatic and electomagnetic principles have been developed. This chapter examines limitations of actuator force, speed and stroke, and compares representative examples of current micro-actuators. Electrostatic actuators have been simpler to fabricate, but have limited energy density compared to magnetic actuators. Piezo-electric actuators have orders of magnitude more force than electrostatic actuators due to the higher dielectric constant, but have limited strain. An important consideration for selecting actuators for micro-robots is the use of mechanical transformers such as wobble motors and inch worm drives which provide larger forces or displacements at lower speed.

For a very good introduction to the size scaling issues of magnetic and electric actuators, the paper by Trimmer and Jebens [1989] is recommended. A good survey of fully or partly IC-processed actuators can be found in Fujita and Gabriel [1991]. For a good general survey including thermal and chemical actuators, see [Dario et al 1992]. This survey considers two groups: electric field driven and magnetic field driven actuators.

For macro-scale applications, magnetic actuators are dominant. It is not clear yet which technology will be dominant at the micro-scale. One scaling law that is helpful for micro-actuators is the increased break down field strength of very small gaps due to the Paschen effect. Thus for small sizes, the obtainable electrostatic forces can be stronger than electromagnetic forces [Trimmer and Jebens, 1989]. Since electrostatic devices can be constructed using only conductors and insulators, they can be made compatible with silicon micromachining techniques. However, recent improvements in materials processing now allow deposition of thin-film magnetic materials, so integrated micro-magnetics are now possible.

* This work was funded in part by: NSF-PYI grant IRI-9157051.

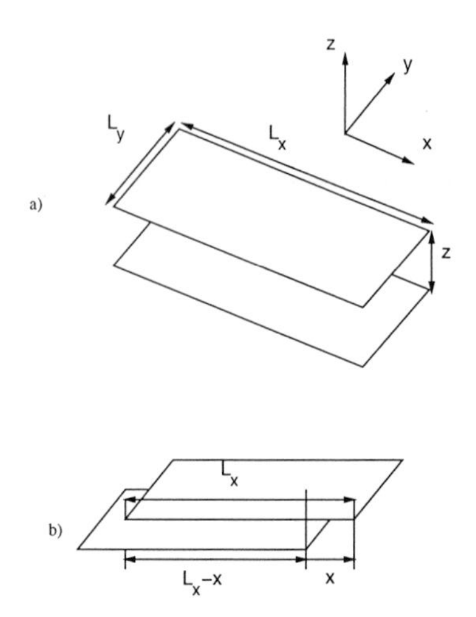

Fig. 1. Parallel plate capacitor geometry. a) normal drive b) tangential drive

For micro-robotics applications, it seems that ideal actuators would have reasonable speed, long strokes, and high force (high torque). However, many of the available actuation principles such as piezoelectric effect and electrostatic actuators tend to have a very short stroke, but high resonant frequency. Methods such as used in the ultrasonic motor and harmonic micromotor provide, in effect, a gear reduction which increases available torque.

2 Electric Field Driven Actuators

There are a variety of electric field driven actuators, based on principles of electrostatic attraction, piezoelectricity, or quasi-static induction. Many types of micro-actuators have been designed, both linear and rotary, using tangential drive and normal drive.

Small tools impractical to fabricate by other methods, for example, an electrostatically driven gripper able to handle 10 μm diameter parts [Kim et al, 1992], can be made by surface and bulk micromachining of silicon. Among the many other novel devices made possible, are planar rotary micro motors [Fan et al, 1989], Trimmer and Gabriel [1987] and dielectric induction motors [Fuhr et al 1992a, 1992b]. The drawback to these devices, even with small gaps, is the low force and torque obtainable. The gripper built by Kim has a grasping force on the order of $1\mu N$, while the electrostatic micromotor has a diameter of 100 μm and torque of about $10pNm$ which may be much too small for micro-robotic applications. Friction is also a problem in the rotary micromotors. Possible ways to build improved rotary actuators with higher torque are to use a vibration-to-rotation conversion [Lee and Pisano, 1991], or a harmonic micromotor [Price, Wood, Jacobsen, 1989], [Furuhata et al 1993], which can achieve torques on the order of $10\mu Nm$ with a 1 mm radius motor.

For many micro-robot applications a linear actuator may be more useful than a rotary one. Recently, integrated force arrays have been described (Bobbio et al [1993], [Kornbluh et al 1991] and Yamaguchi et al [1993]), which take advantage of the very strong normal attractive force between two parallel charged surfaces. Another type of planar mechanism, an electrostatic surface drive is described by Egawa et al [1991] and Niino, Egawa, Higuchi [1993].

Multi DOF actuators can also be driven electrostatically. Fukuda and Tanaka [1990] describe a 3 DOF actuator made using copper foil, and Wood, Jacobsen, and Grace [1987] controlled the position of an optical fiber driven electrostatically. In an important innovation, Shimoyama et al [1991], Suzuki et al [1992] have developed electrostatic actuators fabricated from polysilicon and polyimide which can act out of the plane of the wafer and have multiple degrees of freedom. These actuators are being developed to emulate insect wing joints and muscles.

2.1 Linear Electrostatic Actuators

Figure 1 shows the basic geometry of the parallel plate capacitor which is at the heart of most tangential and normal drive electrostatic actuators. (Even for the induced charge actuators, the parallel plate capacitor model provides an upper bound estimate of the available force). With a potential V applied across the top and bottom plates, the stored energy in the capacitor is given by:

$$W = \frac{1}{2} C V^2, \tag{1}$$

where the capacitance C (neglecting fringing fields) is

$$C(z) = \epsilon \frac{L_x L_y}{z} \tag{2}$$

with z, ϵ, L_x and L_y the plate separation, permittivity, and plate dimensions respectively.

2.1.1 Normal Force Actuators

The force in the normal direction is given by:

$$F_z(z) = -\frac{\partial W}{\partial z} = \frac{1}{2} \epsilon \frac{L_x L_y}{z^2} V^2. \tag{3}$$

The equivalent "electrostatic pressure" is calculated by the normal force per area or:

$$P_{electrostatic} = \frac{F_z}{L_x L_y} = \frac{1}{2} \epsilon |E|^2. \tag{4}$$

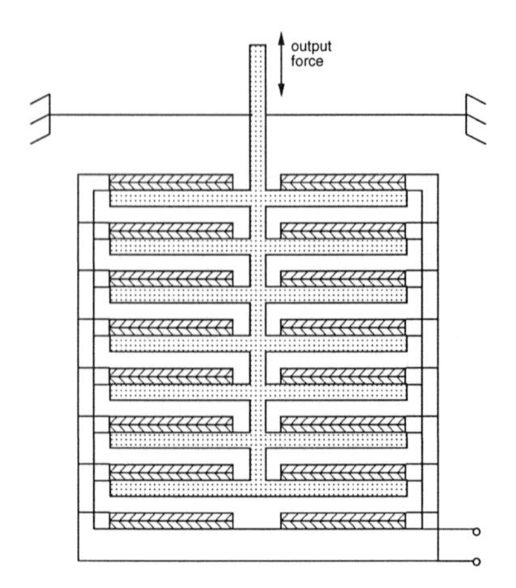

Fig. 2. Parallel plate comb drive actuator.

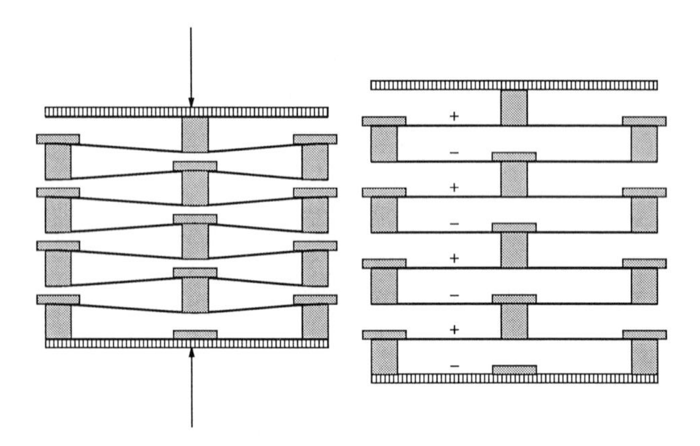

Fig. 3. Integrated Force Array similar to Bobbio et al [1993]. Actuator is unidirectional, with maximum stroke of about 30%.

Figure 2 shows a typical technique for building a normal drive electrostatic actuator on a planar substrate. Multiple sections can be ganged together to increase the effective area, and the actuator is bidirectional. Typically, about two thirds of the actuator is taken up by structural materials. Note that to obtain high fields and large forces, the stroke for the actuator will be very small, typically around $2\mu m$. One approach to getting substantial forces and large stroke from a parallel plate type actuator uses a sandwich of layers with a compliant suspension as shown in Figure 3. Although the peak strain of the actuator is typically less than 30%, the stack could in principle be made substantially thick.

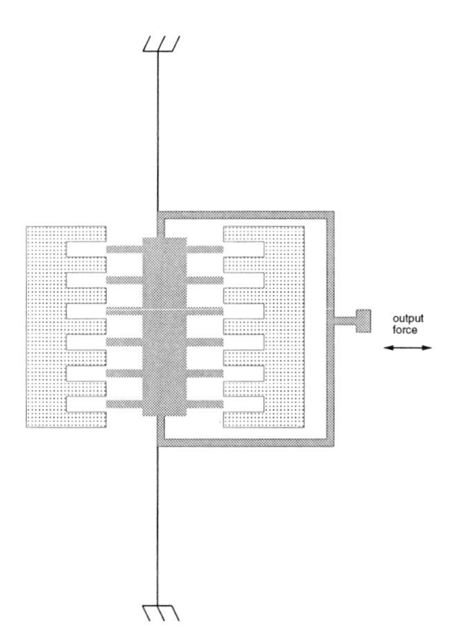

Fig. 4. Side drive comb actuator.

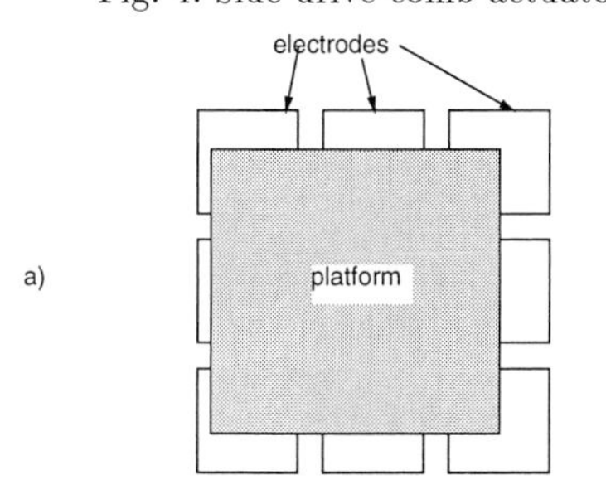

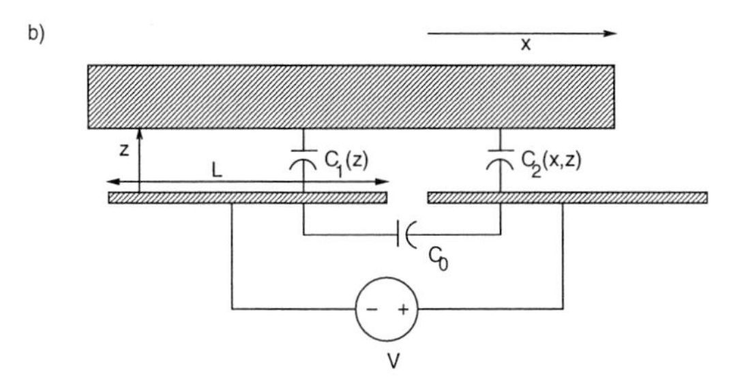

Fig. 5. Planar electrostatic drive with levitated platform. a) Electrode configuration. b) Equivalent circuit.

2.1.2 Tangential Force Actuators

For a tangential drive actuator (which requires a strong bearing to resist the F_z force), the capacitance can be written

$$C(x) = \epsilon \frac{(L_x - x)L_y}{z} \tag{5}$$

and thus the force in the tangential direction is:

$$F_x(x) = -\frac{\partial W}{\partial x} = \frac{1}{2}\epsilon\frac{L_y}{z}V^2. \tag{6}$$

Note that (again neglecting fringing fields) the tangential force is independent of the length of the plate L_x. Thus the tangential force can be increased by having many narrow plates of width L_y connected in parallel. A typical side drive actuator is shown in Figure 4, where the moving member is attached to the substrate using a suspension. A feature of the side drive comb actuator is that the gaps can be made very small (increasing field strength), and the force is relatively constant over the length of the stroke.

A potential limitation of comb drives is the rather short stroke. Figure 5 shows a planar electrostatic actuator, where a mobile platform is levitated on an air bearing [Pister et al 1990]. The total capacitance seen by the power source is

$$C(x,z) = C_o + \frac{C_1(z)C_2(x,z)}{C_1(z)+C_2(x,z)}. \tag{7}$$

The tangential force can be calculated as in part b) of the figure assuming that C_1 is constant:

$$F_x(x) = \frac{1}{2}V^2\frac{\partial C(x,z)}{\partial x} = \frac{1}{2}V^2\epsilon_o\frac{L^2(L+2x)}{z(L+x)^2}, \tag{8}$$

where L is the electrode width. Normalizing by the area orthogonal to the direction of motion, the equivalent pressure is:

$$\frac{F_x(0)}{zL} = \frac{1}{2}\frac{V^2\epsilon_o}{z^2} = \frac{1}{2}\epsilon_o|E|^2 \tag{9}$$

as expected. Acceleration is

$$\ddot{x} = \frac{F_x(0)}{m} = \frac{1}{2}\frac{V^2\epsilon_o}{zL\delta\rho}, \tag{10}$$

where δ is the thickness of the platform and ρ is its density. Note that improved acceleration is possible with gratings on the platform.

2.1.3 Force Limits for Linear Electrostatic Actuators

It is useful to estimate the maximum force which can be obtained by a linear actuator. As we have seen, the force is proportional to the square of the field strength. With smaller gaps, there are fewer ionizable gas molecules available for breakdown, resulting in higher

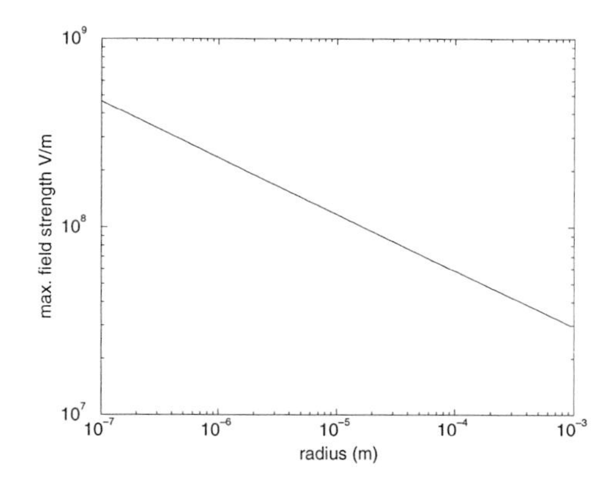

Fig. 6. Surface field at breakdown for spheres [Harper, 1967].

Table 1
Comparison of some representative electrostatic linear actuators.

length	speed	force	stroke	power density	peak field	reference
μm	s^{-1}	N	m	Wm^{-3}	Vm^{-1}	
7×10^4	?	?	3.9×10^{-6}	?	3×10^6	[Fukuda et al 90]
1.2×10^5	100	0.4	4×10^{-4}	200	4×10^5	[Egawa et al 90]
70×10^3	1000	8	240×10^{-6}	2×10^5	2×10^7	[Niino et al 94]
6×10^3	?	70×10^{-6}	9×10^{-6}	?	10^8	[Matsubara et al 91]
400	5000	10^{-7}	6×10^{-6}	200	2×10^7	[Kim et al 92]

sustainable field strengths (the Paschen effect). As an indication of the larger fields which are sustainable at smaller dimensions, Figure 6 shows the maximum local field near a charged sphere before breakdown occurs. Experiments have shown that with sub-micron gaps and very smooth surfaces, air will breakdown at fields of $9 \times 10^8 Vm^{-1}$ [Horn and Smith, 1992]. In a vacuum, or with smaller gaps, the absolute limit is set by field emission at $10^9 Vm^{-1}$ for typical materials [Lang et al, 1987]. In practice, thin films such as SiO_2 withstand $2 \times 10^8 Vm^{-1}$ [Fujita and Omodaka, 1987].

Let's estimate the available force from an actuator such as the parallel plate comb drive shown in Figure 2. New surface micromachining techniques such as HEXSIL allow vertical features approaching $100\mu m$ [Keller and Ferrari, 1994]. Assuming a field of $5 \times 10^7 Vm^{-1}$, with $10\mu m$ per comb, the effective area is $10 cm^2$ giving 10 N force per actual cm^2. This is starting to be enough force to be worthwhile for milli-size robots.

Some representative actuators are summarized in **Table 1.** In addition to clever mechanical design, such as minimizing contrary forces from suspension springs, the most important limit on peak force is the maximum field strength obtained in practice be-

Table 2
Scaling for representative electrostatic linear actuators. (Assuming peak field of $E = 1 \times 10^8 V m^{-1}$ for $2\mu m$ gap).

Actuator Type	Strain	Force	Stroke	pressure	Energy Density
	$\frac{m}{m}$	N	m	Nm^{-2}	Jm^{-3}
parallel plate	$\leq 90\%$	$\frac{\epsilon_o}{2}L^2E^2$	z_o	$\frac{\epsilon_o}{2}E^2$	$\frac{\epsilon_o}{2}E^2$
with air gap					
tangential drive					
Ideal 1 mm cube	30%	44 mN	0.3 mm	$44 \times 10^3 Nm^{-2}$	$44 \times 10^3 Jm^{-3}$

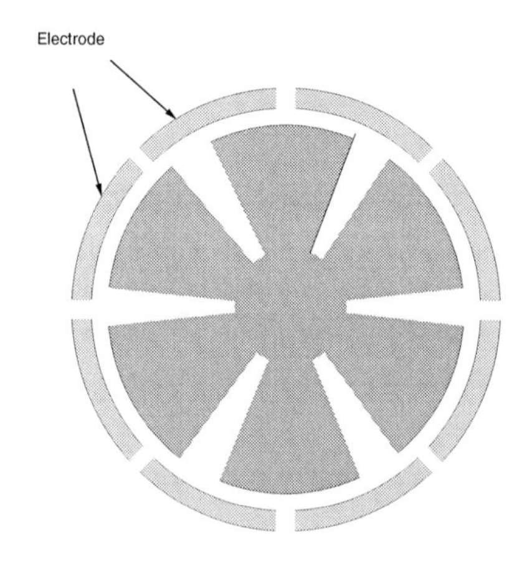

Fig. 7. Side drive rotary electrostatic motor.

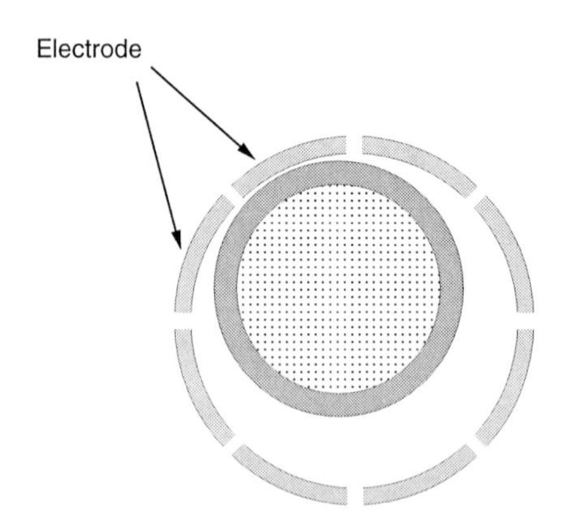

Fig. 8. Wobble motor using insulated rotor.

fore breakdown. The maximum field strength is a function of gap size as well as surface roughness.

Table 3
Comparison of Electrostatic Rotary Actuators.

radius	rot. speed	torque	power density	peak field	reference.
m	$rads^{-1}$	Nm	Wm^{-2}	Vm^{-1}	
65×10^{-6}	1500	10^{-11}	4×10^{-3}	10^{8}	[Mehregany et al 90]
1.6×10^{-3}	?	7×10^{-4}	?	6×10^{7}	[Trimmer and Jebens, 1989]

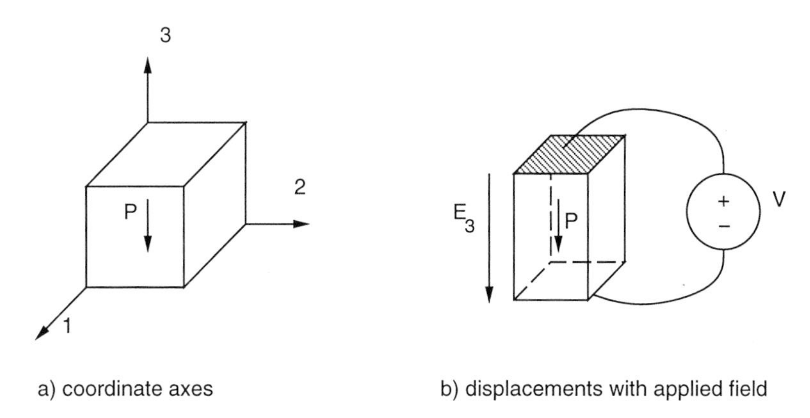

Fig. 9. Piezo electric actuator coordinate system.

2.2 Rotary Actuators

A typical rotary side drive motor is shown in Figure 7. These motors are characterized by high rotational speeds (thousands of RPM) and very small torques. By using an offset rotor in the motor (called a wobble or harmonic drive motor), more torque at lower speeds can be obtained, as seen in Figure 8. The equivalent gear ratio is given by:

$$ratio = \frac{d_{inner}}{d_{inner} - d_{outer}}. \tag{11}$$

With typical control of gap sizes, gear ratios up to several hundred have been obtained. Note that the inner rotor is in rolling contact with the outer stator. High field strengths can be used with thin film insulating materials.

Some representative rotary motors are shown in **Table 3**. Again, the important parameter is the maximum field strength which was actually sustained.

Table 4
Properties of common piezoelectrics. From Linvill [1978] and Dario et al [1983]. Notes: (*) breakdown strength rather than polarization limit.

	PVDF	PZT-5H	thin film PZT [Udayakumar et al 1991]
d_{31} $(CN^{-1} = mV^{-1})$	27×10^{-12}	-274×10^{-12}	-88×10^{-12}
d_{33} $(CN^{-1} = mV^{-1})$	31×10^{-12}	593×10^{-12}	220×10^{-12}
stiffness (Nm^{-2})	3.6×10^9	50×10^9	
E_{max} (Vm^{-1})	30×10^6	0.3×10^6	10^8 (*)
permittivity (constant tension)	$12\epsilon_o$	$3400\epsilon_o$	$700\epsilon_o$
density (kgm^{-3})	1.8×10^3	7.5×10^3	
maximum strain $(E_{max} * d_{33})$	1×10^{-3}	2×10^{-4}	
energy density Jm^{-3}	5×10^4	10^3	6×10^5

3 Piezoelectric Actuators

Piezoelectric actuators in their normal mode of operation have too short a stroke to be useful for micro-robots. However, since the piezoelectric actuator can respond at high frequencies, many small displacements can be added together to give large net motion. In one example, an array of piezoelectrically driven "cilia" can be used to sweep objects along in any direction in the plane (Furuhata et al [1991]). Or transverse surface vibrations can be set up in the surface of a piezo-electric disk which will rotationally displace a disk which is clamped to its top (Raine et al [1993], Udayakumar et al, [1991]). The motor described by Raine et al achieved a torque of 50 nNm at a speed of 20 rad s^{-1}, with a motor diameter of 3.5 mm.

As a piezoelectric element works as both an generator and a motor, there is a coupled set of equations describing charge and strain. In general form [Berlincourt et al 1964], the strain S as a function of applied stress and applied field is:

$$S = s^E T + dE \tag{12}$$

and the charge displacement D is given by:

$$D = dT + \epsilon^T E \ , \tag{13}$$

where S is the strain, s^E is the compliance with electrodes shorted (0 field), T is the applied surface stress, d are the piezo electric constants in strain per applied field, E is the applied field, and ϵ^T is the permittivity at constant stress. For the special case of

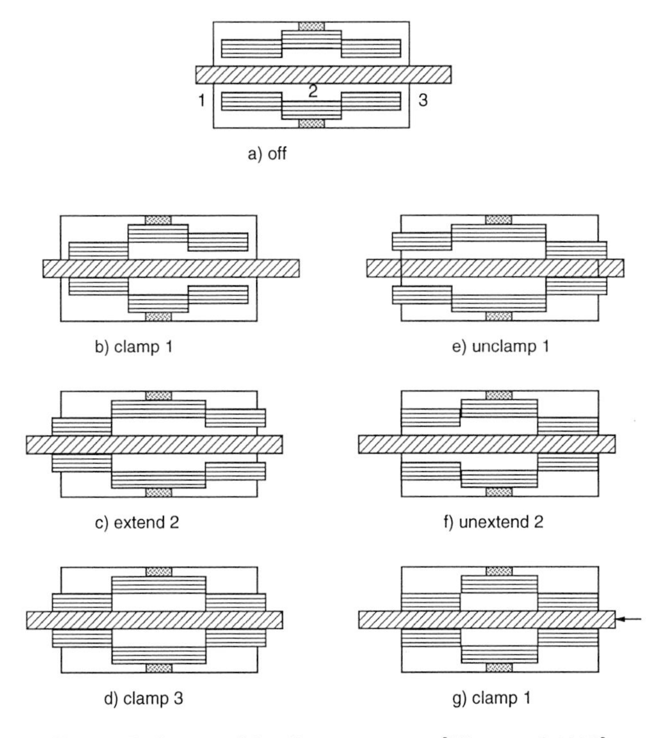

Fig. 10. Inchworm drive linear actuator [King et al 1990].

unclamped sides (stress $T = 0$), and the electric field E_3 applied parallel to the poling direction, the above equations simplify to: $s_3 = E_3 d_{33}$ and $s_1 = E_3 d_{31}$. The maximum field strength which can be applied is limited by depolarization, with a field strength of just $3 \times 10^5 Vm^{-1}$ for PZT. Electrical break down of the piezo electric element is another limitation.

4 Mechanical Transformers

Electrostatic and piezoelectric devices can have large forces with high response frequencies, but small strokes. The small stroke limitations are usually due to the desire to take advantage of greater field strengths at small gaps. Two methods for converting small stroke, high force and high frequency actuation into larger stroke are the inchworm mechanism and traveling mechanical wave mechanisms. The inchworm drive linear actuator of Figure 10 is composed of three linear actuators. Actuators 1 and 3 are simply binary clamps. Actuator 2 is a linear actuator which is controlled longitudinally. By alternately clamping actuators 1 and 3 and extending actuator 2, large effective strokes can be obtained. Note that in principle, the peak force of actuator (assuming no slip) is determined by actuator 2, with no loss in force. The inchworm drive can be used with piezoelectric,

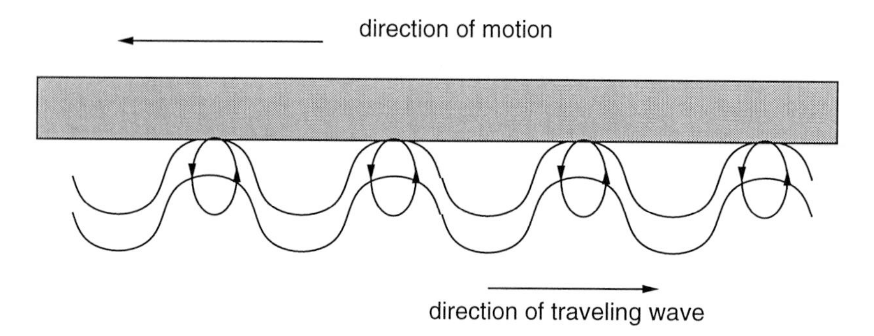

Fig. 11. Traveling wave drive. Elliptical motion of contact points in vibrating body applies tangential force to slider.

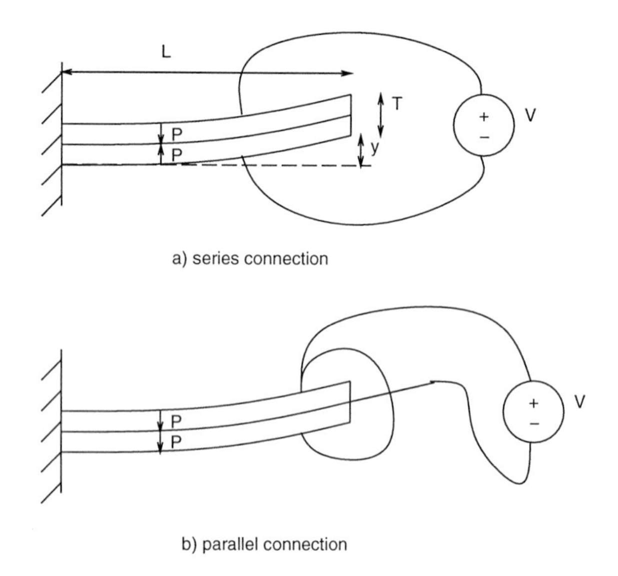

Fig. 12. Piezo electric bender.

electrostatic, or magnetic actuation.

Ultrasonic motors use a piezoelectric element (driven in resonance for higher efficiency) to set up traveling waves (Figure 11). Since points on the surface of the piezo material move elliptically, they drag the slider along with it. While the elliptical motion amplitude is only on the order of nm, the high drive frequencies give relatively high velocities, on the order of cm per second.

One of the classical transformer methods with piezo-electric transducers is the bimorph, made by creating a sandwich of two polarized layers of piezoelectric material (Figure 12).

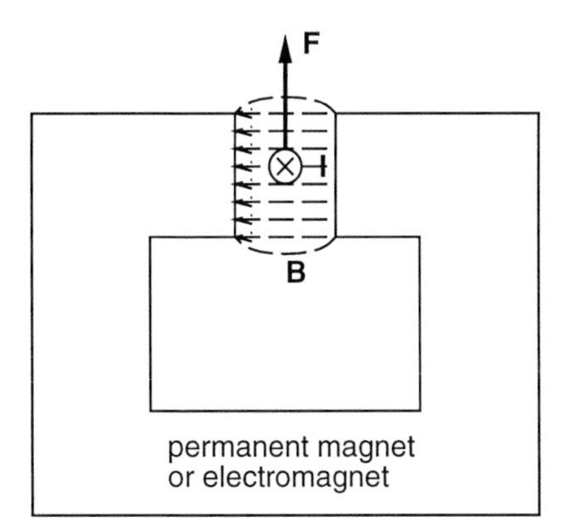

Fig. 13. Lorentz force on a wire in constant magnetic field.

The two types of connection are series [Piezo Systems, 1994]:

$$y = \frac{2L^2}{T} E_3 d_{31} \tag{14}$$

and parallel:

$$y = \frac{4L^2}{T} E_3 d_{31}, \tag{15}$$

with L the length of the cantilever and T the thickness. Significant strokes can be achieved, for example, with a 25 mm long bimorph one can achieve free deflections of several hundred μm and blocked forces of several hundred mN [Piezo Systems, 1994].

5 Magnetic Field Driven Actuators

Electrostatic actuation may have problems with charge accumulation in the dielectric if the two charged surfaces make contact, see for example [Anderson and Colgate, 1991]. Another problem is that electrostatic drive won't work in a conductive fluid medium such as water. Magnetic actuation may thus be an attractive alternative, see for example [Busch-Vishniac, 1991]. Magnetics may also be attractive when using an internal high current, low voltage source like a single cell battery, since the coil will usually be fairly low impedance. One of the most significant drawbacks to (non-superconducting) magnetic actuators is the thermal dissipation in coils while maintaining a constant force; in contrast, electrostatic actuators require no power to maintain a constant force with no displacement.

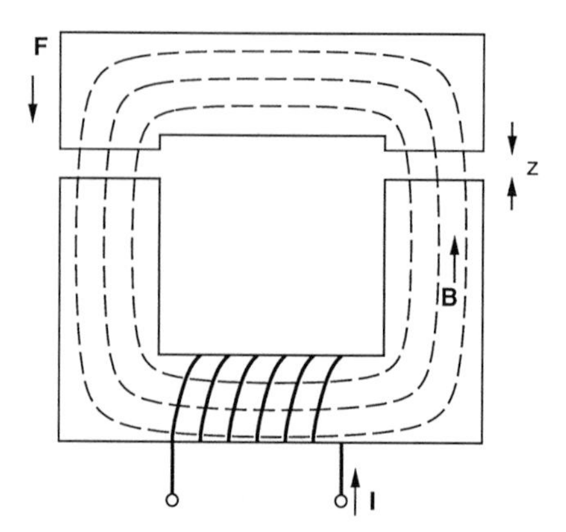

Fig. 14. Electromagnetic actuator.

5.1 Linear Electromagnetic Actuators

The simplest magnetic actuator consists simply of a current carrying wire in a constant magnetic field, as shown in Figure 13, with the force per unit current element:

$$d\vec{F} = I\vec{dl} \times \vec{B}. \tag{16}$$

Since very strong fixed magnetic fields can be easily provided by miniature permanent magnets, this type of actuator could be attractive for the micro scale. Recently Holzer et al [1995] demonstrated a single turn thin film aluminum coil, with 20 μm displacement, and estimated force of $0.1\mu N$ using only 20 mA. This compares quite favorably to electrostatic actuator force and stroke.

A more conventional electromagnetic actuator is shown in Figure 14. This type of device relies more on concentrating the magnetic flux in a narrow gap. The stored magnetic energy in the gap is given by:

$$W = \tfrac{1}{2}\mu_o H^2 = \tfrac{1}{2}\mu_o(\frac{NI}{2})^2(Az), \tag{17}$$

where H is the magnetic field, $\mu_o = 4\pi \times 10^{-7} Hm^{-1}$ is the permeability of free space, N is the number of turns, z is the total gap length, and A is the gap area . The force in the normal direction is given by:

$$F_z(z) \;=\; -\frac{\partial W}{\partial z} \;=\; \tfrac{1}{2}\mu_o \frac{A}{z^2}(NI)^2. \tag{18}$$

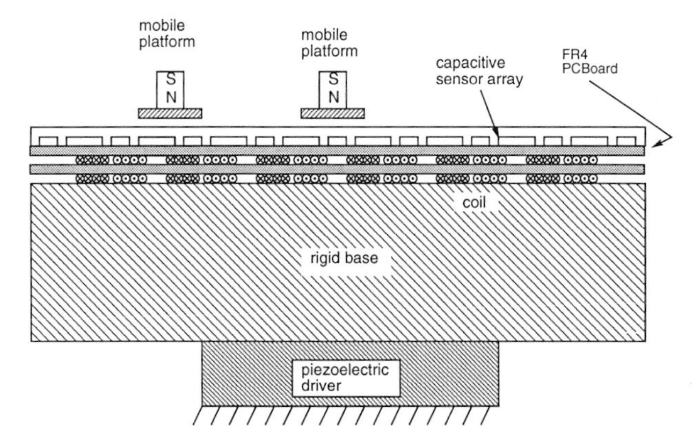

Fig. 15. Air bearing levitated planar magnetic actuator.

The equivalent "magnetic pressure" is calculated by the normal force per area or:

$$P_{magnetic} = \tfrac{1}{2}\mu_o\left(\frac{NI}{z^2}\right)^2. \tag{19}$$

It is interesting to note that in conventional size machines with flux density on the order of 1 Tesla, pressures of $10^6 Nm^{-2}$ are easily obtained.

Recently, electromagnetic motors have been fabricated using lithographic and thick film techniques (Ahn et al [1993], Guckel et al [1993]). The motor of Ahn et al has a predicted torque of $1.2\mu Nm$ with a motor diameter of 1.5 mm. A 3 DOF actuator was built by attaching a permanent magnet to an elastic suspension above a set of coils (Wagner and Benecke, [1991], Wagner, Kreutzer and Benecke, [1991]). They obtained a deflection of $70\mu m$ with a force of $450\mu N$. A linear actuator using superconducting levitation yielded a force of $30\mu N$ and a 5 mm stroke (Kim, Katsurai, Fujita [1990]). A linear actuator using a self-pressurizing air-bearing gave a peak force of about $2mN$ on a 6.3 mm diameter, 5.0 mm length magnet, with a total travel distance of 20 mm [Fearing, 1992]. (See Figure 15).

Conventional electromagnetic motors are also shrinking in size. An electric motor was built into the 4.6 mm car designed by Hisanaga et al [1991].

There are also analogs to the piezoelectric effect for magnetic materials. One advantage of the magnetic field actuation is the comparatively long range which is good for tetherless drive, such as the milli-robot described in Fukuda et al [1991].

6 Summary

While there is currently much research activity in micro-actuators, there are very few off-the-shelf actuators available in the millimeter size region. Hopefully the demand for actuators for micro-robotic applications will stimulate their further development.

Acknowledgement

I would like to thank H. Furuichi, E. Tan, and B. Gray for helpful discussions, and J. Judy for making his bibliography available.

References

[1] C.H. Ahn, Y.J. Kim, and M.G. Allen, "A Planar Variable Reluctance Magnetic Micromotor with Fully Integrated Stator and Wrapped Coils" *Proc. IEEE Micro Electro Mechanical Systems*, pp. 1-6, Fort Lauderdale, FL Feb. 7-10, 1993.

[2] K. M. Anderson and J. Edward Colgate, "A model of the attachment/detachment cycle of electrostatic micro actuators", *ASME Micromechanical Sensors, Actuators, and Systems*, DSC-vol. 32, pp. 255-268, Atlanta, GA Dec. 1-6, 1991.

[3] D.A. Berlincourt, D.R. Curran, and H. Jaffe, "Piezoelectric and Piezomagnetic Materials and Their Function in Transducers', in *Physical Acoustics: Principles and Methods* edited by W.P Mason, New York: Academic Press 1964.

[4] S.M. Bobbio, M.D. Kellam, B.W. Dudley, S. Goodwin-Johansson, S.K. Jones, J.D. Jacobson, F.M. Tranjan, and T.D. DuBois, "Integrated Force Arrays", *Proc. IEEE Micro Electro Mechanical Systems* , pp. 149-154, Fort Lauderdale, FL Feb. 7-10, 1993.

[5] I.J. Busch-Vishniac, "The Case for Magnetically Driven Micro-Actuators", *ASME Micromechanical Sensors, Actuators, and Systems,*DSC-vol. 32, pp. 287-302, Atlanta, GA Dec. 1-6, 1991

[6] P. Dario, R. Valleggi, M.C. Carrozza, M.C. Montesi, and M. Cocco, "Microactuators for microrobots: a critical survey", *J. Micromech. Microeng.* vol. 2, pp. 141-157, 1992.

[7] P. Dario, C. Domenici, R. Bardelli, D. DeRossi, and P.C. Pinotti, "Piezoelectric Polymers: New Sensor Materials for Robotic Applications", *13th Int. Symp. Industrial Robots,*pp. 14:34-49, Chicago IL April 19-20, 1983.

[8] S. Egawa and T. Higuchi, "Multi-layered electrostatic film actuator", *Proc. of the IEEE Workshop on Micro-Electro Mechanical Systems* , pp. 166-171, Napa Valley, CA Feb. 11-14, 1990.

[9] S. Egawa,, T. Niino, and T. Higuchi, "Film Actuators:Planar, Electrostatic Surface-Drive Actuators", *IEEE Micro Electro Mechanical Systems",* pp. 9-14, Nara, Japan, Feb. 1991

[10] R.S. Fearing, "A Miniature Mobile Platform on an Air Bearing", *Third Int. Symp. on Micro Machine and Human Sciences*, Oct. 14-16, 1992, Nagoya, Japan.

[11] G. Fuhr, R. Hagedorn and J. Gimsa, "Analysis of the torque-frequency characteristics of dielectric induction motors", *Sensors and Actuators A*, vol. 33, pp. 237-247, 1992.

[12] G. Fuhr, R. Hagedorn, T. Muller, W. Benecke, U. Schnakenberg and B. Wagner, "Dielectric Induction micromotors: field levitation and torque-frequency characteristics", *Sensors and Actuators A*, vol. 33, pp. 525-530, 1992.

[13] H. Fujita and A. Omodaka, "Electrostatic Actuators for Micromechatronics", *IEEE MicroRobots and Teleoperators Workshop,* Hyannis, MA Nov. 9-11, 1987.

[14] H. Fujita and Kaigham J. Gabriel, "New Opportunities for MicroActuators", 1991 Int. Conf. on Solid-State Sensors and Actuators (Transducers '91) June 1991, San Francisco, CA, pp. 14-20.

[15] T. Fukuda and T. Tanaka, "Micro-Electro Static Actuator with Three Degrees of Freedom", *Proc. of the IEEE Workshop on Micro-Electro Mechanical Systems* , Napa Valley, CA Feb. 11-14, 1990.

[16] T. Fukuda, H. Hosokai, H. Ohyama, H. Hashimoto and F. Arai, "Giant Magnetostrictive Alloy (GMA) Applications to MicroMobile Robot as a Micro Actuator without Power Supply Cables", *IEEE Micro Electro Mechanical Systems"*, pp. 210-215, Nara, Japan, Feb. 1991.

[17] T. Furuhata, T. Hirano, and H. Fujita, "Array-Driven Ultrasonic Microactuators- arrayed microactuator modules that have swing pins", 1991 Int. Conf. on Solid-State Sensors and Actuators (Transducers '91), June 1991, San Francisco, CA, pp. 1056-1059.

[18] T. Furuhata, T. Hirano, L.H. Lane, R.E. Fontana, L.S. Fan, H. Fujita, "Outer Rotor Surface Micromachined Wobble Micromotor", *Proc. IEEE Micro Electro Mechanical Systems*, pp. 161-166. Fort Lauderdale, FL Feb. 7-10, 1993.

[19] H. Guckel, T.R. Christenson, K.J. Skrobis, T.S. Jung, J. Klein, K.V. Hartojo, and I. Widjaja, "A First Functional Current Excited Planar Rotational Magnetic Micromotor", *Proc. IEEE Micro Electro Mechanical Systems*, pp. 7-11, Fort Lauderdale, FL Feb. 7-10, 1993.

[20] M. Hisanaga, T. Kurahashi, M. Kodera, and T. Hattori, "Fabrication of a 4.8 Millimeter Long Microcar", Proc. Second Int. Symp. on Micro Machine and Human Science, pp. 43-46, Nagoya, Japan, Oct. 8-9, 1991.

[21] W.R. Harper, *Contact and frictional electrification*, Oxford: Clarendon Press, 1967.

R. Holzer, I. Shimoyama, and H. Miura, "Lorentz Force Actuation of Flexible Thin-Film Aluminum Microstructures", *Proc. IEEE-RSJ Intelligent Robots and Systems,*pp. 156-161, Pittsburgh, PA August 3-5, 1995.

[22] R.G. Horn, D.T. Smith, "Contact electrification and adhesion between dissimilar materials." *Science*, 17 April 1992, vol.256, (no.5055):362-4.

[23] J.D. Jacobson, S.H. Goodwin-Johansson, S.M. Bobbio, C.A. Bartlett, and L.N. Yadon, "Integrated Force Arrays: Theory and Modelling of Static Operation", *Jnl. of Microelectromechanical Systems,*vol. 4, no. 3, pp. 139-150, Sept. 1995

[24] C. Keller and M. Ferrari, "Milli-Scale Polysilicon Structures", *IEEE Solid-State Sensor and Actuator Workshop* pp. 132-137, 1994.

[25] C-J. Kim, A.P. Pisano, and R.S. Muller, "Silicon-processed overhanging microgripper", *Jnl. of Microelectromechanical Systems*, pp. 31-36, vol. 1, no. 1, March 1992.

[26] Y. Kim, M. Katsurai, and H. Fujita, "Fabrication and Testing of a Micro Superconducting Actuator using the Meissner Effect", *Proc. IEEE Micro Electro Mechanical Systems*, pp. 61-66, Napa Valley, CA, Feb. 1990.

[27] T.G. King, M.E. Preston, B.J.M. Murphy, D.S. Cannell, "Piezoelectric ceramic actuators: a review of machinery applications", *Precision Engineering* , vol. 12, no. 3, pp. 131-136, 1990.

[28] R.D. Kornbluh, G.B. Andeen, and J.S. Eckerle, "Artificial Muscle: The next generation of robotic actuators", *4th World Conf. on Robotics Research,* Sept. 17-19, 1991, Pittsburgh, PA.

[29] J.H. Lang, M.A. Schlecht and R.T. Howe, "Electric Micromotors: Electromechanical Characteristics", *IEEE MicroRobots and Teleoperators Workshop,* Hyannis, MA Nov. 9-11, 1987.

[30] A.P. Lee and A.P. Pisano, "An Impact-Actuated Micro Angular Oscillator–Design, Testing and Dynamic Analysis", ASME Micromechanical Sensors, Actuators, and Systems, DSC-vol. 32, Atlanta, GA Dec. 1-6, 1991.

[31] J.G. Linvill, "PVF2- Models, Measurements and Device Ideas", Stanford University, Integrated Circuits Lab Technical Report #4843-2.

[32] T. Matsubara, M. Yamaguchi, K. Minami, and M. Esashi, "Stepping ELectrostatic Microactuator", 1991 Int. Conf. on Solid-State Sensors and Actuators (Transducers '91), June 1991, San Francisco, CA, pp. 50-53.

[33] M. Mehregany, P. Nagarkar, S.D. Senturia, and J.H. Lang, "Operation of Microfabricated Harmonic and Ordinary Side-Drive Motors", *Proc. IEEE Micro Electro Mechanical Systems*, pp. 1-8, Napa Valley, CA, Feb. 1990.

[34] M. Nakasuji and, H. Shimizu, "Low voltage and high speed operating electrostatic wafer chuck." *Journal of Vacuum Science & Technology A (Vacuum, Surfaces, and Films),* Nov.-Dec. 1992, vol.10, (no.6):3573-8.

[35] T. Niino, S. Egawa, and T. Higuchi, "High-Power and High-Efficiency Electrostatic Actuator" *Proc. IEEE Micro Electro Mechanical Systems* , pp. 236-241, Fort Lauderdale, FL Feb. 7-10, 1993.

[36] T. Niino, S. Egawa, H. Kimura, and T. Higuchi, "Electrostatic Artificial Muscle: Compact High-Power Linear Actuators with Multiple-Layer Structures", *IEEE Micro Electro Mechanical Systems* , pp. 130-135, Oiso, Japan, Jan. 25-28, 1994.

[37] Piezo Systems, Inc. "Piezoceramic Application Data", Cambridge, MA 1994.

[38] K.S.J. Pister, M.W. Judy, S.R. Burgett, and R.S. Fearing, "Microfabricated Hinges", *Sensors and Actuators A* , vol. 33, pp. 249-256, 1992.

[39] R.H. Price, J.E. Wood, and S.C. Jacobsen, "Modelling Considerations for Electrostatic Forces in Electrostatic Microactuators", *Sensors and Actuators* , vol. 20, pp. 107-114, 1989.

[40] G.-A. Racine, R. Luthier, and N.F. de Rooij, "Hybrid Ultrasonic Micromachined Motors" *Proc. IEEE Micro Electro Mechanical Systems* , pp. 128-132. Fort Lauderdale, FL Feb. 7-10, 1993.

[41] K. Suzuki, H. Miura, I. Shimoyama and Y. Ezura, "Creation of an Insect-based Microrobot with an External Skeleton and Elastic Joints", Proc. IEEE Micro Electro Mechanical Systems Workshop, Travemunde, Germany, February 4-7, 1992, pp. 190-195.

[42] H. Saeki, T. Tanaka, T. Fukuda, K. Kudou, et al "New electrostatic micromanipulator which dislodges adhered dust particles in vacuum". *Journal of Vacuum Science & Technology B (Microelectronics Processing and Phenomena)*, Nov.-Dec. 1992, vol.10, (no.6):2491-2.

[43] I. Shimoyama, "Scaling in Microrobots", *Proc. IEEE/RSJ Intelligent Robots and Systems*, pp. 208-211, Pittsburgh, PA August 3-5, 1995.

[44] W.S.N. Trimmer and K.J. Gabriel, "Design Considerations for a Practical Electrostatic Micro-Motor," *Sensors and Actuators,* vol. 11, pp. 189-206, 1987.

[45] W. Trimmer and R. Jebens, "Actuators for Micro Robots", IEEE Intern. Conf. on Robotics and Automation, Scottsdale, AZ, May 1989, pp. 1547-1552.

[46] W. Trimmer and R. Jebens, "An Operational Harmonic Electrostatic Motor", *IEEE Micro Electro Mechanical Systems Workshop* , pp. 13-16, 1989.

[47] K.R. Udayakumar, S.F. Bart, A.M. Flynn, J. Chen, L.S. Tavrow, L.E. Cross, R.A. Brooks, and D.J. Ehrlich, "Ferroelectric Thin Film Ultrasonic Micromotors", *IEEE Micro Electro Mechanical Systems"* , pp. 109-113, Nara, Japan, Feb. 1991

[48] B. Wagner and W. Benecke, "Microfabricated Actuator with Moving Permanent Magnet", *Proc. IEEE Micro Electro Mechanical Systems* , pp. 27-32, Nara, Japan, Feb. 1991

[49] B. Wagner, M. Kreutzer, and W. Benecke, "Electromagnetic MicroActuators with Multiple Degrees of Freedom" 1991 Int. Conf. on Solid-State Sensors and Actuators (Transducers '91), June 1991, San Francisco, CA, pp. 614-617.

[50] J.E. Wood, S.C. Jacobsen, and K.W. Grace, "SCOFFS: A small Cantilevered Optical Fiber Servo System", IEEE MicroRobots and Teleoperators Workshop, Hyannis, MA Nov. 9-11, 1987.

[51] M. Yamaguchi, S. Kawamura, K. Minami, and M. Esashi, "Distributed Electrostatic Micro Actuator" *Proc. IEEE Micro Electro Mechanical Systems* , pp. 18-23, Fort Lauderdale, FL Feb. 7-10, 1993.

7. Energy Source and Power Supply Method

Fumihito Arai and Toshio Fukuda

Department of Micro System Engineering, Nagoya University
Furo-chyo, Chikusa-ku, Nagoya 464-01, JAPAN
TEL: +81-789-3116, FAX: +81-789-3909
E-mail: arai@mein.nagoya-u.ac.jp, fukuda@mein.nagoya-u.ac.jp

7.1. Classification of energy supply methods

One of the final goal of the microrobotics is to realize an ant like mobile robot which is small and intelligent to perform given tasks[1,2,3]. Most of the present micro robot is supplied energy by the cable. However, as long as the robot becomes small, the cable disturbs its motion with great friction. So, the energy supply method of the micro system becomes important. It is classified as an internal supply method(internal energy sources) and an external supply method(external supply of the energy to the system without cable). Noncontact manipulation methods of the small object are also mentioned here as the examples of an external energy supply method.

7.2. Internal supply methods

In this case, energy source is contained inside the moving body. Electric energy is frequently used as an internal energy. As an electric energy supply method, a buttery and a condenser have been developed. Buttery type is good in terms of the output and durability, however it has difficulty in miniaturization. Recently, micro lithium battery whose thickness is micron order, electricity density is 60 mA/cm^2, and which is rechargeable for 3.6V-1.5V has been developed by thin film technology[4]. As for a condenser type, autonomous mobile robot with in 1 cm3 in its volume size has been developed in 1992 by Seiko Epson based on the conventional watch production technology. It uses high capacity condenser of 6 mm diameter, 2 mm thickness, and 0.33 F electric capacity as an energy source. Electric capacity of the condenser is little compared with that of the secondary battery. However, this microrobot uses two stepping motors with current control of pulse width modulation, it can move about 5 minutes after only 30 seconds charge.

7.3. External supply methods and noncontact manipulation

In this case, energy is given to the body(object) from outside. As an external supply methods or noncontact manipulation methods, the following energies are considered to be used.

(1) Optical energy
(2) Electromagnetic energy
(3) Super sonic energy
(4) The others.

7.3.1. Optical energy

Optical energy can be used to produce external driving force for the micro system. Noncontact energy supply method based on optical energy can be classified as follows.

i) Optical radiation pressure type by the laser beam

ii) Optical energy to strain conversion type using UV ray irradiation and photostrictive phenomena of the element
iii) Optical energy to heat conversion type

Optical radiation pressure type

As an example of i), remote operation of the micro object by focused laser beams as the tweezers have been proposed. Laser has superior properties such as coherency, monochromaticity, foucusability, and short pulse. It has been demonstrated that optical pressure can be used for noncontact and remote manipulation of micrometer-sized particles[5]. This technology has been used in optical tweezers for trapping and transporting such micro particles as bacteria and microcapsules containing chemical reagents and assembling the micro objects[6,7]. The basic principle of the optical trapping is summarized in Figure 1. In this figure, the focused laser beam produces the radiation force to the sphere. The sphere can be trapped by changing the radiation power to balance the external forces acting on it. Masuhara et al. used a 1064 nm TEM00 mode gausian beam from a CW Nd: YAG laser as a trapping laser source and focused (1mm) into a sample solution. Based on this principle, micro object can be rotated. Directional control of optical rotation was experimentally demonstrated for artificial SiO_2 micro objects having an anisotropic geometry, which is not bilaterally symmetric but rotationally symmetric in the horizontal cross section(Figure 2)[8]. Mizuno et al. succeeded in manipulating the T4 DNA molecule. They bounded the DNA molecule and the latex bead based on the specific interaction between biotin and avidin. The latex bead was trapped by the laser trapping and the T4 DNA was stretched inside a capillary[9].

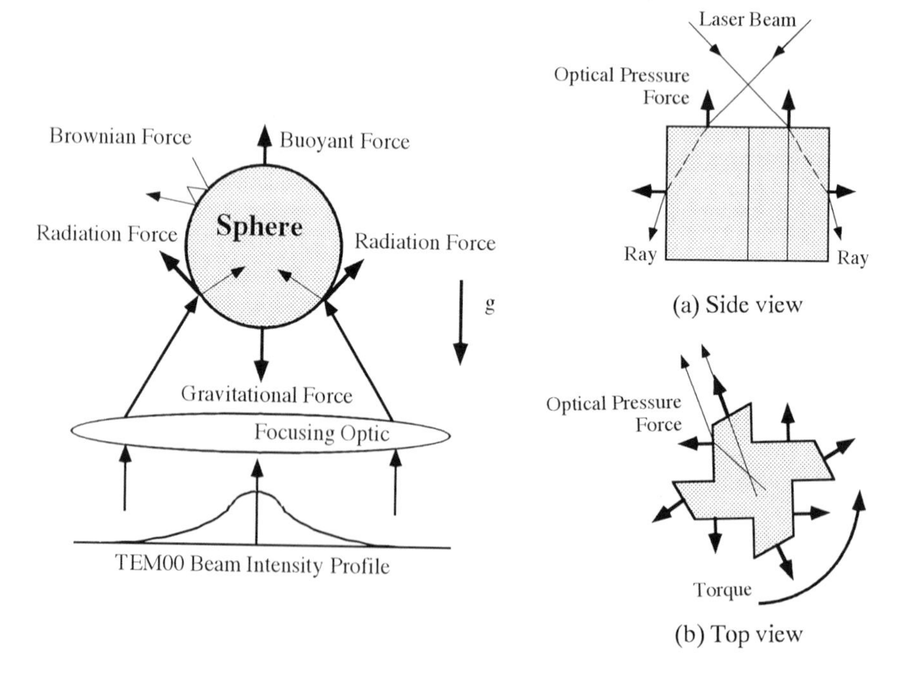

Figure 1. Principle of optical trapping[7]

Figure 2. Geometry of the micro object rotated by the optical pressure force[8]

Optical energy to strain conversion type

As an example of ii), optical piezo electric actuator such as PLZT that employs optical energy for driving is currently focused on intensive study. Merit of this element is not only noncontact energy transmission but also reduction of inductive noise and no need for electrical insulation. Also, an integrated optical system will be expected comprising optical devices and components. For implementation of such an integrated optical system, it is essential to develop advanced optical actuators.

Optical piezo electric element which has photostrictive phenomena has been paid attention as a new actuator [10,11]. This actuator is prospective in the case the cable and noise become nuisance. As the applications, there have been developed a mobile robot and a relay switched on/off by the light beam irradiation. Optical response characteristics of the optical piezo electric element by UV ray irradiation is characterized by the strain, which is combination of the following three different phenomena (Figure 3).

(A) Photostrictive effect of generating the photostrictive voltage, and the strain caused by the piezo electric effect.

(B) Pyroelectric effect of generating pyroelectric current by the temperature difference, and the strain caused by the piezo electric effect.

(C) Thermal deformation caused by the applied heat flux.

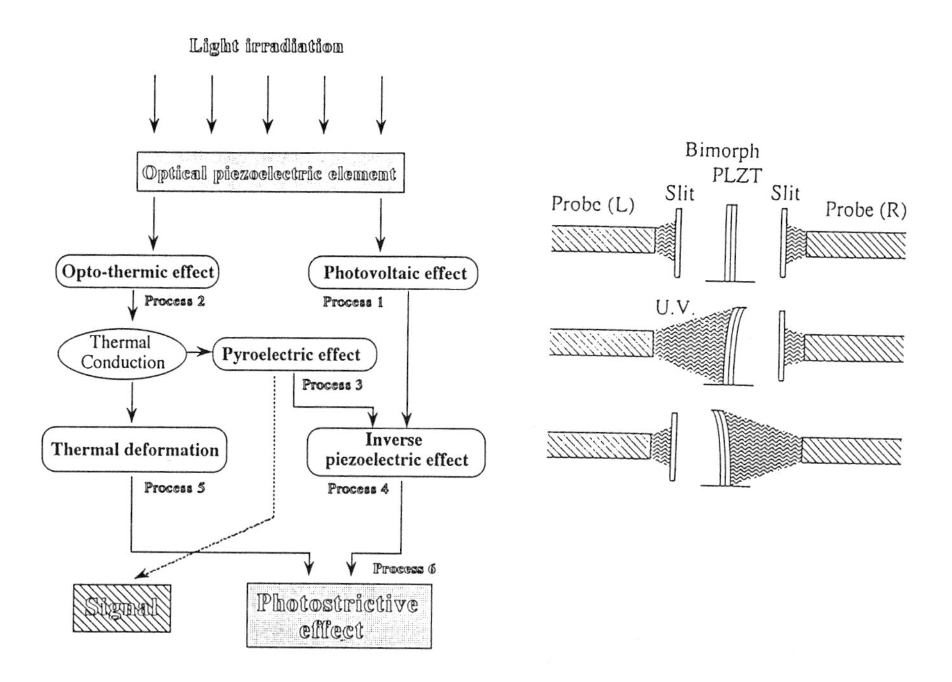

Figure 3. Optical response process of PLZT

Figure 4. UV ray irradiation experiment of PLZT

To improve the response characteristics of the optical piezo electric actuator, bimorph type of PLZT has been developed, and displacement is increased and response time of the strain by UV (ultraviolet) ray irradiation is improved up to couple of ten seconds[11]. Moreover, response characteristics is improved by irradiating the UV rays from both sides of the actuator. Response time of the bending motion of this actuator is extremely improved by this double sides irradiation methods compared with the previous single side irradiation method[12]. Figure 4 shows the experimental device for the UV ray (365 nm wave length peak with narrow spectral band width) irradiation to the bimorph type PLZT ($La/PbZrO_3/PbTiO_3$: 3/52/48). Figure 5 shows displacement at the tip of the PLZT, when left and right sides are irradiated alternately at difference frequencies of 1 Hz, 4 Hz, and 8 Hz by the UV intensity of 170 mW/cm2. Response time of the PLZT is extremely improved.

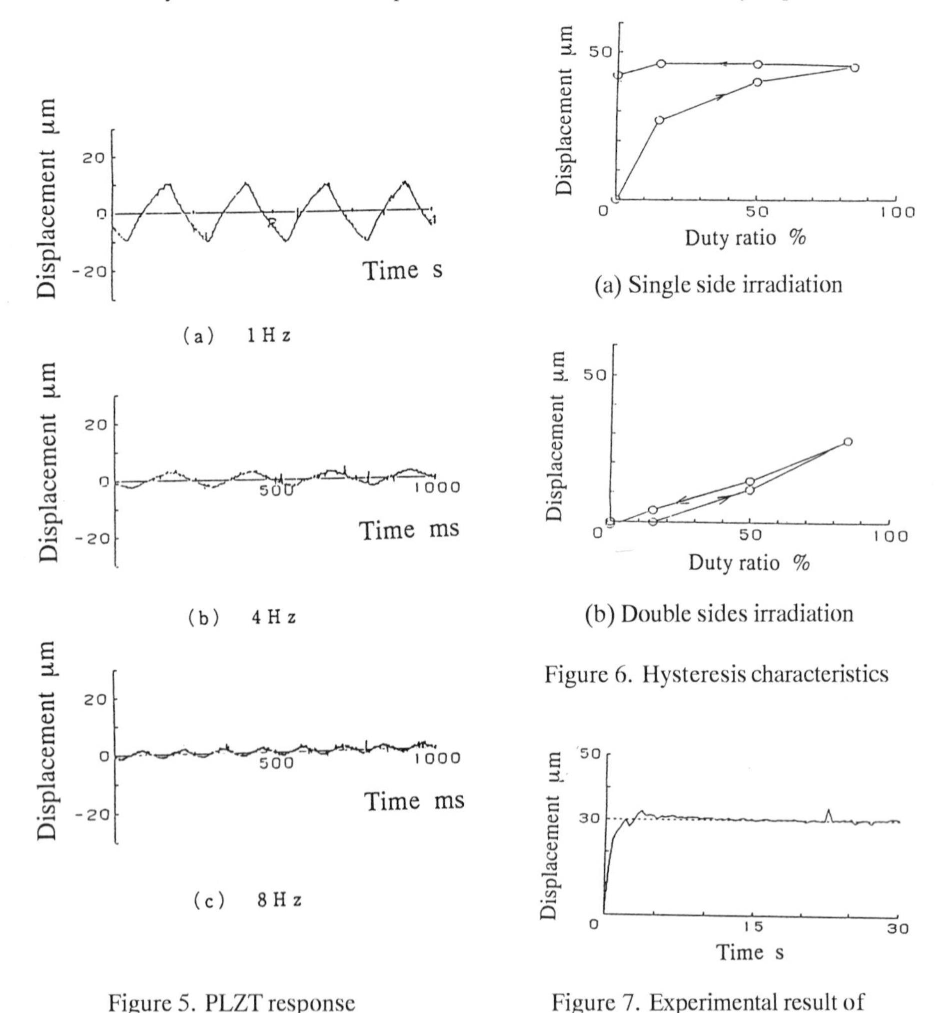

Figure 5. PLZT response to ON/OFF irradiation

Figure 6. Hysteresis characteristics

Figure 7. Experimental result of optical servoing

The hysteresis is a common phenomenon on a piezoelectric devices. We examined the hysteresis caused by accumulation of charges due to photoelectromotive force while the thermal deformation was kept small. For this purpose, intensity of light source is set as low as 20 mW/cm^2. We have proposed the PWM control for this actuator to change strength of the UV rays. UV ray irradiation intensity is almost proportional to the duty ratio. Figure 6 shows the hysteresis characteristics of the bimorph PLZT. We can improve the hesteresis by the double sides irradiation method. We derived the model of the PLZT with this control system, we proposed the optical servo control method. Experimental result of the optical servoing is shown in Figure 7.

As an application example of this actuator, we have developed an optical micro gripper and an optical mobile robots those are driven by the noncontact energy transmission. We consider the integrated optical servo control system with energy and information transmission. PLZT has the different optical response in terms of the three different effect stated before. So, multi-functional use of the PLZT is expected to be developed not only as an actuator but also as an information transmitter.

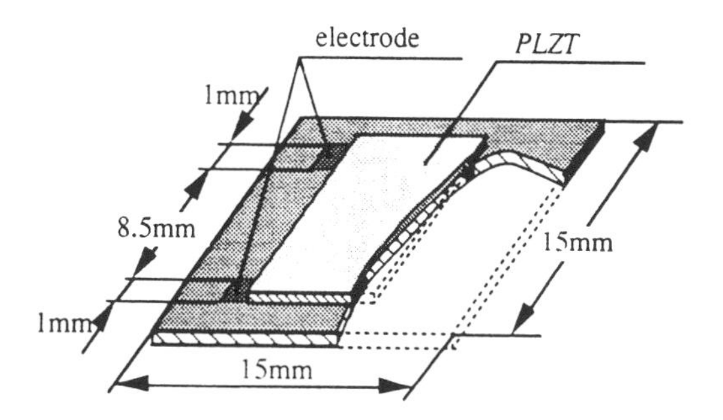

(a) Configuration of the mobile platform

(b) Outline of the mobile platform
on the air table

Figure 8. Optical mobile platform
with noncontact energy transmission
on the air table

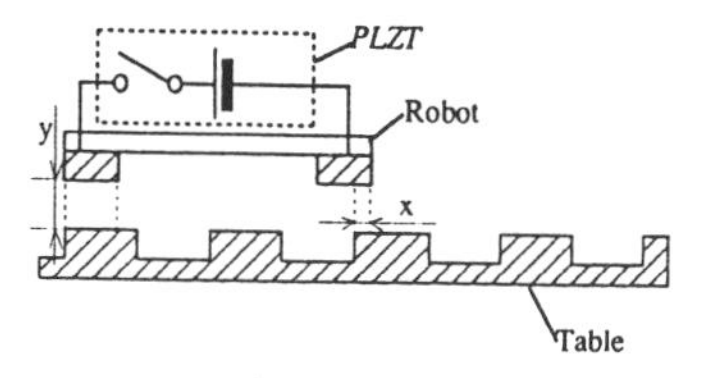

(a) Initial condition

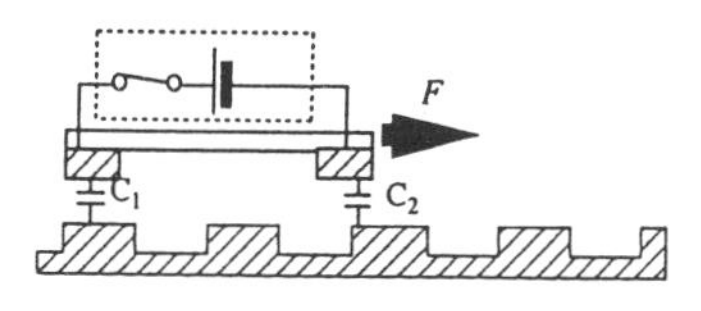

(b) Light is ON

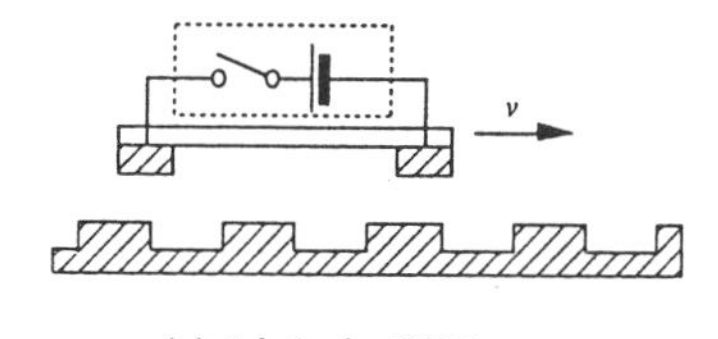

(c) Light is OFF

Figure 9. Moving principle
of the optical mobile robot

Optical energy to heat conversion type

As an example of iii), low boiling point liquid material has been used with an optical heat conversion material. Moreover, utilization of the pyloelectric effect has been proposed to supply energy from outside. Figure 8 shows a moving platform on the air table using pyloelectric current which is generated by the temperature difference by the heat applied by the UV ray irradiation[13]. As an energy transformation method from optical energy to electric energy, PLZT has been employed. Generally, this can be substituted by the other pyloelectric element which can generate pyloelectric current by the temperature change. This system utilizes electrostatic force as a driving force. The moving principle is shown in Figure 9. The field of the moving platform is made by the square shaped electrodes which are arranged squares. Each electrode is 1.5 mm by 1.5 mm width and placed at intervals of 0.5 mm. The field has many halls(diameter 0.18 mm) placed at intervals 1 mm, and the air is brown to float the moving platform. The bottom face of the platform has several electrodes those are 1 mm by 1mm width and are placed at intervals of 4 mm. Each electrodes is connected with the PLZT. By the UV ray irradiation, the thrust force is generated between the bottom face of the platform and the field, and this can be used as driving force of this platform. By the air table, friction is considerably reduced and even weak electrostatic force is enough to move the moving platform fast. By the experiment, the platform moved on the field at the speed of 5 cm/s. The position control of the platform can be attained by controlling the light beam irradiation selectively.

7.3.2. Electromagnetic energy

Microwave transmission

As an example of (2), micro wave, which has been used for the noncontact energy transmission to the airplane and a solar energy generation satellite, has been considered. Sasaki et al. have proposed a wireless system using microwaves to supply energy to micro-robots for performing self-controlled inspections and repairs in thin metal pipes used in heat exchanger, etc. in electric power generation plants[14]. They studied on microwave transmission characteristics for pipes of various discontinuous shapes in which transmission loss might occur. They also established basic structure with a modified monopole for a microwave receiving antenna with a high conversion efficiency(-0.4 dB).

Giant magnetostrictive alloy (GMA)

Giant magnetostrictive alloy (GMA) is actuated by the external magnetic field change and application examples can be considered as the examples of noncontact energy transmission. Magnetic material that generates strain when the magnetic field is applied is called magnetostrictive material.

The magnetostrictive effect was known as Jule effect since old times and its effect was magnitude of 40×10^{-6}. In the 1960's, studied on the magnetic materials were carried out and the magnetostrictive materials (Tb-Dy-Fe alloy) were developed with the giant magnetostrictive effect of a strain magnitude of 10^{-3} level in room temperature. Furthermore, an uniaxial crystal growth technique was developed and consequently a bigger magnetostrictive effect can be obtained in a small magnetic field. The principle of an expansion and contraction of the GMA is shown as follows. In case that the magnetic field does not exist, the magnetic domains disperse in all directions as shown in Figure 10 (a). Once the magnetic field is applied, the magnetic domain flux and the GMA is expanded to the direction and is contracted to the rectangular direction as shown in Figure 10 (b). As the contraction is very small in comparison with expansion, the displacement of the expansion is adopted as the displacement of the actuator. While the piezo-electric elements such as PZT is used as elements for the micro device. The physical properties of the GMA and the piezo-electric elements are shown in Table 1. The GMA shows a bigger output and a larger displacement in comparison with the piezo-electric element. Furthermore, the ratio of a mass per unit stress of the GMA is larger than the piezo-electric element and, therefore, the GMA has the advantage to the piezo-electric element.

Table 1 Comparison between GMA and PZT

	GMA	PZT
Energy density [J/m^2]	2250 ~ 36000	670 ~ 950
$\Delta L/L \times 10^6$ (Room temp.)	1650 ~ 2400	670 ~ 950
Response time	ns ~ μs	μs
Weight/ power [$g/(kgf/mm^2)$]	2.0	78

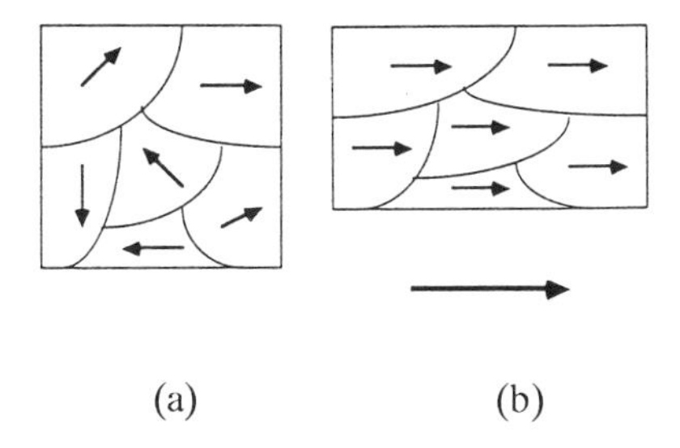

(a) (b)

Figure 10. Deformation model of GMA

The GMA actuator can make the displacement by change of the outer magnetic field. Here we introduce two kinds of magnetic circuits which produce the outer magnetic fields. Generally, the carbon steel is used as a material in a magnetic circuit, while in case of a higher frequency, the eddy current occurs and its efficiency falls down. Therefore, the soft steels making a magnetic circuit must tightly be connected to reduce the amount of the leak magnetic flux. An enamel wire whose diameter is 0.8 mm is coiled thousand times around the soft steel. This magnetic circuit is designed to obtain the magnetic field of 10 kilo-oersted for a current of 7 Amperes.

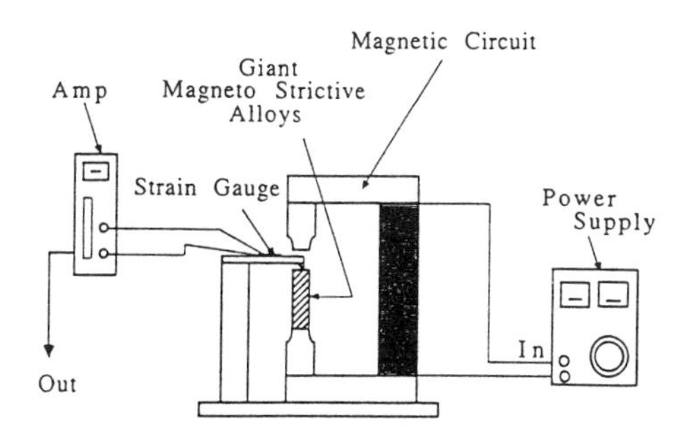

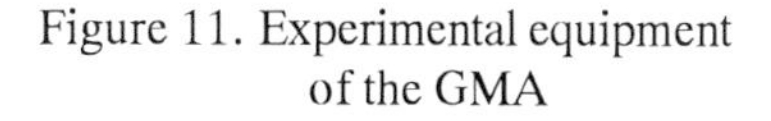

Figure 11. Experimental equipment of the GMA

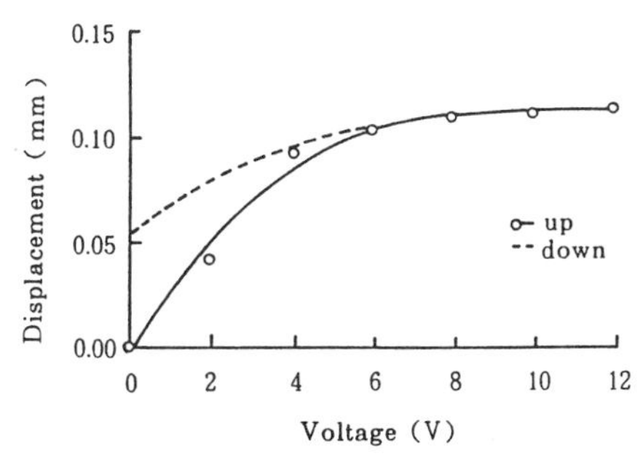

Figure 12. Experimental results of the GAM characteristics

Another magnetic circuit consists of a bobbin whose diameter is 30 mm, and the same enamel wire is coiled around the bobbin. As this magnetic circuit has small self-inductance, it is suitable to obtain alternating magnetic field.

In order to obtain the basic GMA characteristics, an elongation of the GMA to the changes of the magnetic field has been measured by using the experimental equipment shown in Figure 11. This experimental results in Figure 12 show that the GMA is effective as an actuator design for the micro system with consideration of the hysteresis.

The GMA has fragility, so it is difficult to process the GMA mechanically by machining tools. A tensile strength of the GMA is 20-30 N/mm^2 and a compressive strength of the GMA is 100-200 N/mm^2. Therefore, it is desirable to use the GMA under compression. The actuator needs to be used to cause a displacement of the GMA in parallel with the pipe to obtain a larger displacement in restricted space. The macro model of the robot has designed to move by giving magnetic field in parallel with the pipe. The structure of the macro model is shown in Figure 13(diameter 21 mm, length 71 mm, mass 34 g).

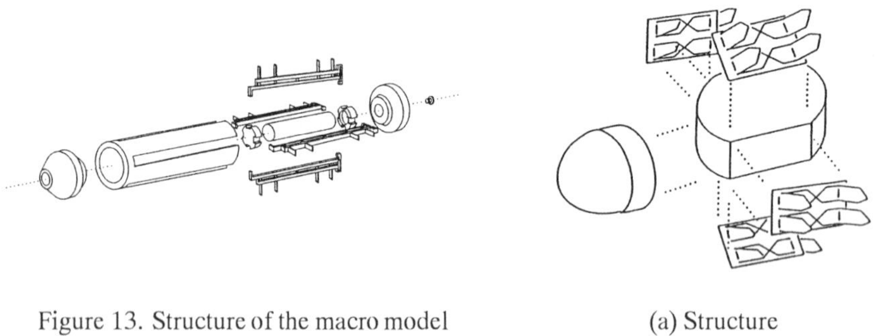

Figure 13. Structure of the macro model

(a) Structure

(a)

(b)

(c)

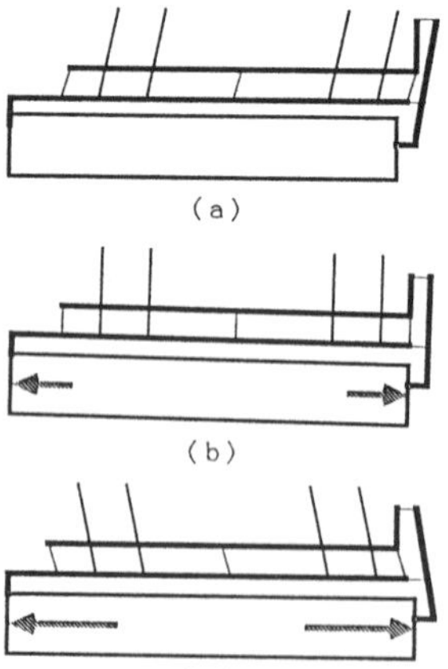

Figure 14. Moving mode of the macro model

(b) Appearance

Figure 15. Micro model

As the displacement of the GMA is very small, the robot can move by legs using a magnifying mechanism of the displacement. The displacement is enlarged by the double enlarging mechanism of a displacement with an inch worm motion. The legs are installed on the body with the opposite inclination to the direction of motion and the tip of the legs are pressed against the inner wall of the pipe. The robot can move by vibrating the legs and the robot can reverse its motion direction by changing the inclination of the legs in the opposite direction. The forward moving mode and the backward moving mode of the macro model are shown in Figure 14. Elongating the GMA, the inclination of the legs decreases. Furthermore, by elongation of the GMA, the inclination of the legs becomes reverse. When the inclination of the legs is reverse to the moving direction, the robot by vibration the legs can move in the left hand direction, while, the robot can do in the opposite direction by vibrating the legs inclined to the moving direction of the robot. These direction changes can easily be made by controlling the applied electric current to the outer coil.

Furthermore, the micro model of the robot has designed to move by locating the GMA actuator in perpendicular with the pipe. The structure and appearance of the micro model is shown in Figure 15(diameter 6 mm, length 4 mm, mass 0.7 g). By locating the GMA actuator in perpendicular with the pipe, the following advantage is obtained. By sacrificing the displacement of the actuator, the micro model is miniaturized and the structure is simplified. The robot has a capability of moving inside a small pipes by the small displacement of the actuator. The macro model can move inside in-pipe diameter of 21 mm, while the micro model can move inside in-pipe diameter of 6mm. Maximum speed of the macro model is 1.5 mm/s and that of the micro model is 0.5 mm/s.

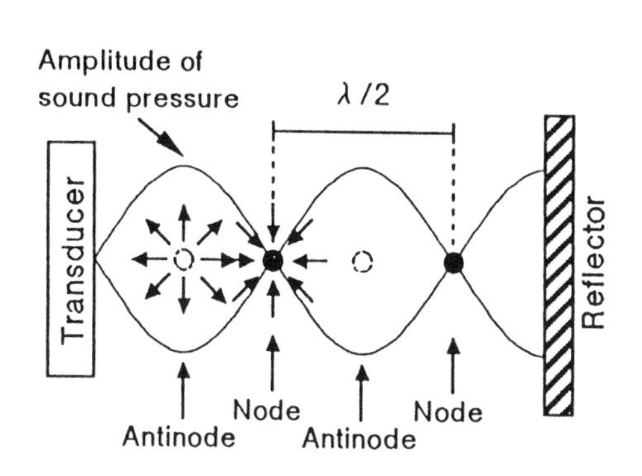

Figure 16. Acoustic radiation pressure[16]

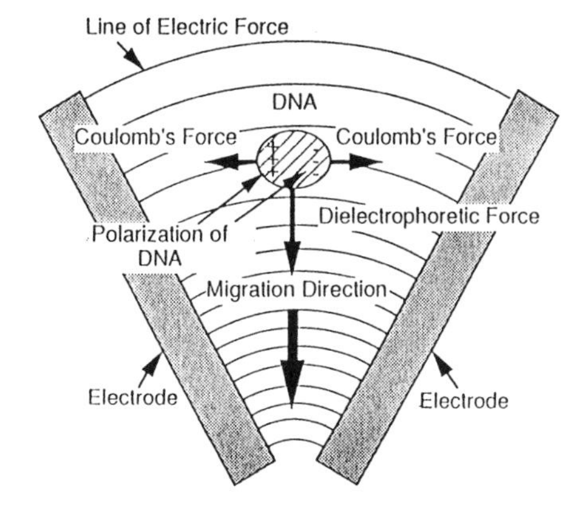

Figure 17. Manipulation of DNA molecule and dielectrophoretic force

7.3.3. Super sonic energy

As an example of (3), radiation pressure of the super sonic wave can be used for noncontact manipulation and driving force of the micro objects. Alumina particles of 16 mm in diameter suspended in water are trapped and formed agglomeration every half wavelength in a standing wave field of 1.75 MHz[16,17]. Acoustic radiation pressure acting at the micro particle is explained in Figure 16. Slight changes in frequency or the distance between transducer and reflector bring lateral shift in the column of agglomerated particles. With the orthogonal standing wave fields, the agglomeration changes its shape depending on the relative force ratio. Applying focused traveling ultrasound on the trapped particles, we can transport only limited clusters of particles. That demonstrates spatially selective manipulation. Since the acoustic radiation force is different depending on the size, shape, density and compressibility of the particle, characteristically selective manipulation is also possible. Concentration and fractionation of small particles in liquid by ultrasound is demonstrated[18].

7.3.4. The others

As the other example, we can consider to get the external force through the external medium such as the maintenance pig robot moves in the pipe line filled with the liquid or the elecromagnetic field.

Moreover, we can control the field to manipulate the micro object[19,20,21]. For example, we manipulated DNA molecule by the electric field control. In a high frequency field, the DNA molecule is drawn in the direction of the high field gradient because the dielectrophoretic force operates in the direction of grad E^2. The DNA molecule moves toward the migration direction in Figure 17[22]. The force acting on the DNA molecule in a nonuniform field is shown in Figure 17. The conformation of a single DNA molecule in an aqueous solution is in the coil state. In generating a high frequency AC field, however, the field around the DNA molecule becomes a nonuniform field. Due to this electric field, the positive and negative charges in the DNA molecule generate toward the electric field by its polarization. As a result, the DNA molecule migrates toward the normal to the line of electric line by dielectrophoretic force.

As the other example, the selective energy transmission to the elastic object on the vibrating plate has been proposed before.

References

1. Fukuda, T. and Arai, F., "Microrobotics-Approach to the Realization," Micro System Technologies 92, Vde-verlag gmbh, 1992, p.15-24.
2. Fukuda, T. and Arai, F., "Microrobotics - On The Highway to Nanotechnology," IEEE Industrial Electronics Society Newsletter, 1993, p.4-5.
3. Fukuda, T. and Ueyama, T., "Cellular Robotics and Micro Robotic Systems," World Scientific Series in Robotics and Automated Systems- Vol.10, World Scientific, 1994
4. Bates, J. B., et al., "Rechargeable Solid State Lithium Microbatteries," Proc. Micro Electro Mechanical Systems, 1993, p.82-86.
5. Ashkin, A., "Acceleration and trapping of particles by radiation pressure," Phys. Rev. Lett., 24, 1970, p.156.
6. Masuhara, H., "Microchemistry: Manipulation, Fabrication, and Spectroscopy in Small Domains," Proc. IEEE Micro Electro Mechanical Systems, 1995, p.1-6.
7. Rambin, C. L. and Warrington, R. O., "Micro-assembly with a Focused Laser Beam," Proc. IEEE Micro Electro Mechanical Systems, 1994, p.285-290.
8. Higurashi, E., and Ukita, H., et al., "Rotational Control of Anisotropic Micro-objects by Optical Pressure," Proc. IEEE Micro Electro Mechanical Systems, 1994, p.291-296.
9. Mizuno, A., Nishioka M., et al., "Opto-Electrostatic Micromanipulation of Single Cell and DNA Molecule," Proc. Sixth Int. Symp. on Micro Machine and Human Science (MHS'95), 1995, p.153-159.
10. Uchino, K. and Aizawa, M., "Photostrictive Actuator Using PLZT Ceramics," Jpn. J. Appl. Phys. 24, Suppl. 42-3, 1985, p. 139-142.

11. Fukuda, T., Hattori, S., Arai. F., et al., "Optical Servo System Using Bimorph Optical Piezo-electric Actuator," Proc. Third Int. Symp. on Micro Machine and Human Science(MHS'92), 1992, p.45-50.
12. Fukuda, T., et al., "Performance Improvement of Optical Actuator by Double Sides Irradiation," Proc. 20th Int. Conf. on Industrial Electronics, Control and Instrumentation (IECON'94), Vol.3, 1994, p. 1472-1477.
13. Ishihara, H. and Fukuda, T., "Micro Optical Robotic System(MORS)," Proc. Fourth Int. Symp. on Micro Machine and Human Science(MHS'93), 1993, p.105-110.
14. Sasaki, K., et al., "Technique of Wireless Energy Service for Micro-robots Using Microwave, " Proc. Fourth Int. Symp. on Micro Machine and Human Science (MHS'93), 1993, p.113-117.
15. Fukuda, T., Hosokai, H., et al., "Giant Magnetostrictive Alloy(GMA) Applications to Micro Mobile Robot as a Micro Actuator without Power Supply Cables," Proc. IEEE Micro Electro Mechanical Systems, 1991, p.210-215.
16. Kozuka, T., et al., "Acoustic Manipulation of Micro Objects Using an Ultrasonic Standing Wave," Proc. Fifth Int. Symp. on Micro Machine and Human Science (MHS'94), 1994, p.83-87.
17. Kozuka, T. et al., "One-Dimensional Transportation of Particles Using an Ultrasonic Standing Wave," Proc. Sixth Int. Symp. on Micro Machine and Human Science (MHS'95), 1995, p.179-185.
18. Yasuda, K., et al., "Concentration and Fractionation of Small Particles in Liquid by Ultrasound, " Jpn. J. Appl. Phys. Vol. 34, Part 1, No. 5B, 1995, pp.2715-2720.
19. Washizu, U., "Electrostatic Manipulation of Biological Objects in Microfablicxated Structures, 3. Toyota Conference, 1989, pp.(25-1)-(25-20).
20. Fuhr, G., et al., "Linear Motion of Dielectric Particles and Living Cells in Microfablicated Structures Induced by Traveling Electric Fields," Proc. IEEE Micro Electro Mechanical Systems, 1991, p.259-264.
21. Miwa, Y., " Automation of Plant Tissue Culture Process," I. Karube (Ed.) Automation in Biothechnology, Proceedings of the 4th Toyota Conference, 1990, pp.217-233.
22. Morishima, K., Fukuda, T., Arai, F., et al. , "Noncontact Transportation of DNA Molecule by Dilectrophoretic Force," Proc. Sixth Int. Symp. on Micro Machine and Human Science (MHS'95), 1995, p. 145-152.

8. Control Method of Micro Mechanical Systems

H. Ishihara and T. Fukuda

8.1 Control Principle

Control of mechanical systems is defined as control of converting system of physical amount, for example, an actuator such as electric motor, a robot integrated with several actuators such as manipulator with multiple links. In this case, we control a position, velocity or posture of control subject. On the other hand, system control is control of integrated system with actuators and robots, for example, group of robots. In this case, control results appear as performance of systems.

The most simple control method of mechanical systems is generally described by the following equation and block diagram as shown in Fig. 8.1:

$$X = G(s) \cdot u \quad (8.1),$$

where, X is output value such as displacement and velocity, $G(s)$ is function modeled control material, and u is input value. In the conventional mechanical systems, input value uses an error between target value and output value to evolve the performance of control. The control using this error is named feedback control or closed-loop control, and a block diagram of feedback control is shown in Fig. 8.2. Besides, another control method shown in Fig. 8.1 is named open-loop control.

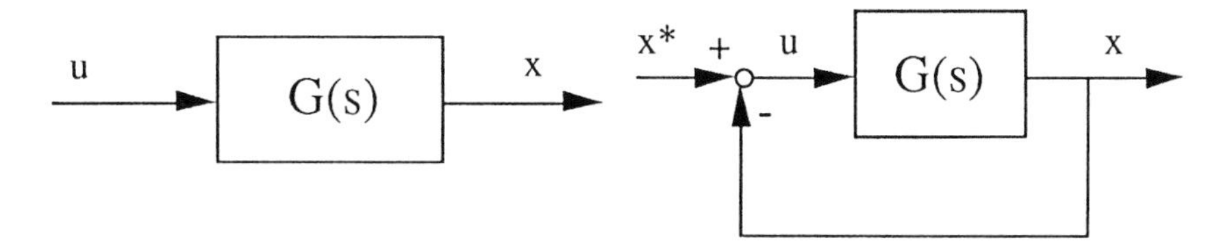

Fig. 8.1 Open-Loop Control System   Fig. 8. 2 Closed-Loop Control System

In the previous research on micro mechanical systems, almost microdevices are control by open-loop control. A reason is difficulty of detecting the output value such as displacement because control subjects are very small. The other reason is difficulty of mounting a sensor in the microdevices. In some examples of controlling using feedback signals, performance of control subjects is detected by the external sensors.

In this section, we show position control of electrostatic microactuator as an example of controlling microdevices.

Figure 8.3 shows the configuration of an electrostatic microactuator with 3 degrees of freedom. In Fig. 8.3, the electrostatic micro actuator consists of movable electrode (parts 3), fixed electrode (parts 2), spacer (parts 4) and cover (parts 1). Figure 8.4 shows prototype of electrostatic microactuator with 3 D.O.F.

Figure 8.5 shows the moving modes of this microactuator. The tip of a probe moves in different direction according to applied voltage at each electrode; No. 1, 2, 3 and 4. Mode 1 is a rotary motion of the probe tip around Y axis. This can attain as follows: At first, voltages are applied to the electrodes No. 1 and No. 2 equally. Then these voltages are switched to be applied at the electrodes No. 3 and No. 4. Then, switching these applied voltages again to the electrode No. 1 and No. 2, and repeating this process, we can produce Mode 1. Mode 2 is a rotary motion around X axis. Likewise, applying voltages to the electrodes No. 2 and No.3 equally, and switching these applied voltages to the electrodes No. 1 and No. 4, Mode 2 can be achieved. In the same 2 way, applying voltage to the all electrodes equally, and switching them to deferent levels, a traveling motion can be generated in Z direction, which is called Mode 3.

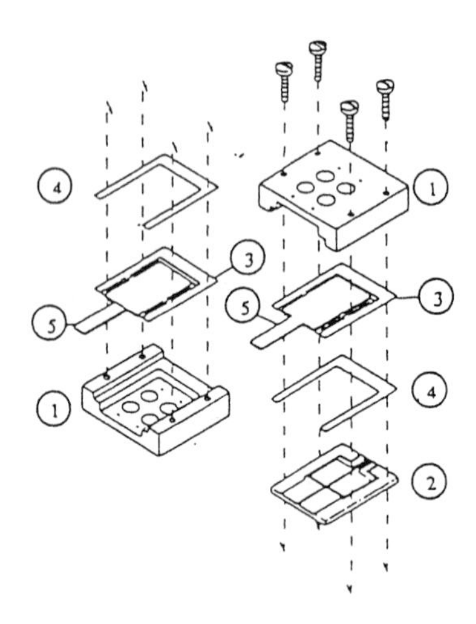

Fig. 8.3 Configuration of Electrostatic Microactuator with 3 D.O.F.

Fig. 8.4 Prototype of Electrostatic Microactuator with 3 D.O.F.

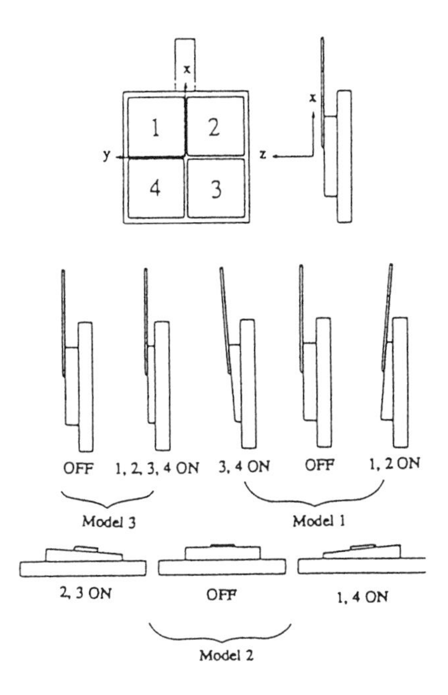

Fig. 8.5 Moving Mode of Electrostatic Microactuator with 3 D.O.F.

The model of the actuator is shown in Fig. 8.6. A base coordinate system is set at the center of a fixed electrode. The coordinate system of a movable electrode is set parallel to the Z direction at the distance of c from the base coordinate system. The relationship between the electrostatic force Fv and the elastic force Fk of the suspension spring that holds the movable electrode. Figure 8.7 shows the relationship between the displacement x and the generated electrostatic force F_v on each supplied electric voltage, and the relationship between the displacement x and the elastic force F_k of the suspension spring.

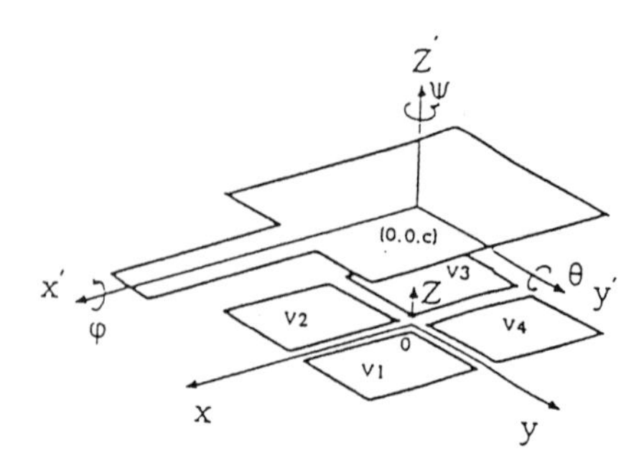

Fig. 8.6 Model of Electrostatic Microactuator

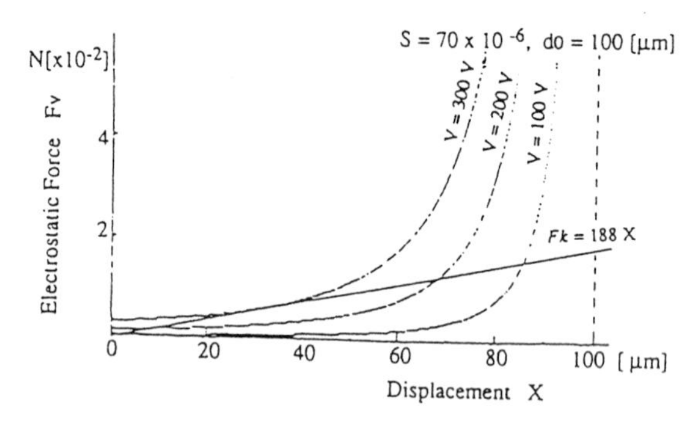

Fig. 8.7 Relationship between Displacement x and Electrostatic Force F_v, Elastic Force F_k

As in Fig. 8.7, the electrode static force curve comes in contact with the elastic force line at the applied voltage of 300 V. In Fig. 8.7, the intersection of the F_v curve and the F_k line means that at the voltage and displacement. "F_v=F_k" is realized, the force of two types is balanced. For example, then the applied voltage is 200 V, there are 2 balanced points. At the points around x=10 mm and 67 mm, that is, the force of two types is balanced. In this case, these points are apart from each other, so they do not affect each other. But, in the case that the applied voltage is around 300 V, the stable points around x = 33 mm are close to each other, and they have possibility to affect each other at that applied voltage. This will degrade the stability of position control. Moreover, if the voltage is applied over 300 V, the performance of systems becomes unstable. We should apply the voltage less than 300 V.

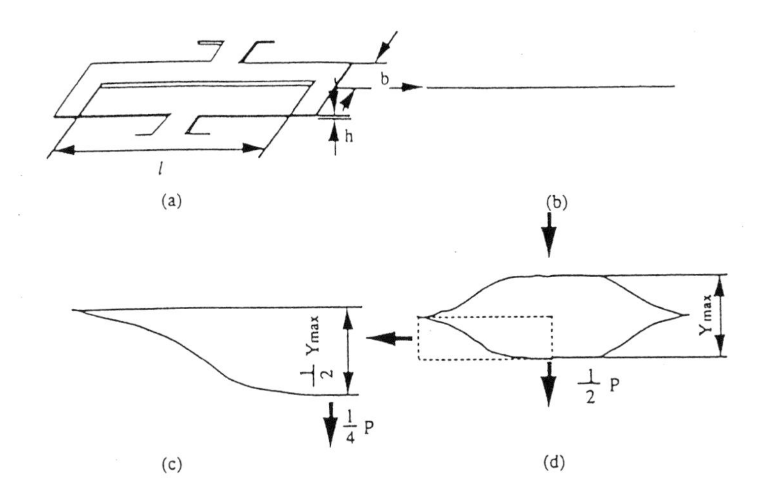

Fig. 8.8 Model of Suspension Spring

Next, we calculate the elastic constants of the spring which supports the movable electrode. In Fig. 8 7, the gradient of the F_k line implies the elastic constant. It is necessary that the stable range for the microactuators must be calculated, since the actuators 300 V. When we set the F_v curve to contact with the F_k line at the possible maximum voltage, we can make the working range of the actuator wide while preserving the conditions denoted, that is, we design that the F_v curve comes in contact with the F_k line at the maximum voltage of 300 V.

Figure 8. 8 (a) shows the configuration of the spring. When the spring is loaded, it is deflected as in Figs. 8.8 (c) and (d). The relationship between the displacement and the load is given by the following equation.

$$y_{\max} = \frac{Pl^3}{16Ebh^3} \qquad (8.2),$$

where P is load, l is length of the suspension spring, E is Young's modules, b is width of the suspension spring, and h is thickness of the spring.

In this case, the elastic constant of the suspension spring is given by the following equation.

$$K = \frac{16Ebh^3}{l^3} \qquad (8.3),$$

By using Eq. 8.3, we determined that the elastic constant is 188.8 N/m. Then, we have design the spring as follows: Their thickness h is 50 μm, the width is 0.49 mm and the length is 8.0 mm.

The relationship between the input voltage and the generated force of electrostatic actuator in nonlinear. Nonlinear effects must be controlled skillfully based on the dynamics. In this section we show a nonlinear feedback control method for the electrostatic actuator in Z direction. The model of motion of the electrostatic actuator in Z direction is shown in Fig. 8.9. The equation of motion of this actuator is written as follows:

$$mz = -k\ (z - z_{std}) - cz - \frac{AV^2}{z^2} \qquad (8.4),$$

where

$$A = \frac{\varepsilon S}{2} \qquad (8.5),$$

e is dielectric constant of air, and S if surface area of the electrode.
Here, let the control rule of the system given as follows:

$$u = k_p\ (z^* - z) - k_v z \qquad (8.6),$$

where kp is position feedback gain, kf is velocity feedback gain.

When the voltage V is given as,

$$V = -\frac{z^2}{A}\{k_p\ (z^* - z) - k_v\ z\} \qquad (8.7),$$

then

$$-\frac{AV^2}{z^2} = k_p\ (z^* - z) - k_v z \qquad (8.8),$$

and the equation of motion is written as follows:

$$mz = -k\ (z - z_{std}) - cz + u \qquad (8.9).$$

The equation of motion can be transformed to a linear equation.

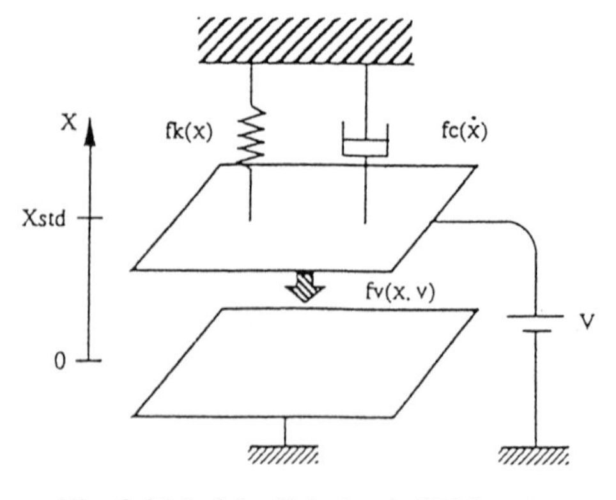

Fig. 8.9 Model of Motion in Z Direction

On the basis of the above equations, we represent the simulation results concerning the position control of the actuator. The simulation of the position control of the electrostatic actuator with 3 D.O.F. are carried out by using the Runge-Kutta method. Parameters of a simulation model are specified as follows:

b: 0.49×10^{-3} m	E : 98×10^{9} Pa	h: 50×10^{-6} m
r: 1700 kg/m	l: 8×10^{-3} m	e: 8.854×10^{-12}
w: 7×10^{-3} m	t: 0.6×10^{-3} m	

These parameters are the same as those of the prototype electrostatic microactuator. The simulation results are shown in Fig. 8.10

Simulation results show that the settling time is rather long. This is because the desired position was set around z=70 (z=30) μm, where stabilizing force is hardly generated. Controllability of the movable electrode around here is bad.

Figure 8.11 shows a system configuration to control the electrostatic microactuator with 3 D.O.F. We carried out the experiments of the position control using the prototype. The results of the position control experiment are shown in Fig. 8.12. The position indicated by a perforated line in the figures of z1 to z4 is the target position of the four electrodes moves to each target position, respectively. In the experiment, because of the lack of the calculation speed of a computer and the accuracy of the displacement sensor, z=90 (z=10) μm, where the stabilizing force is easily generated. The experimental system must be improved in the future.

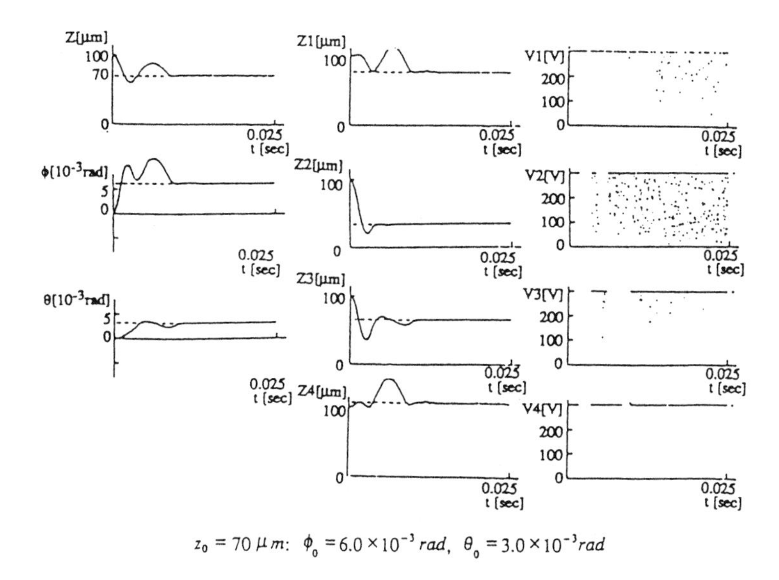

Fig. 8.10 Position Control Simulation of Electrostatic Microactuator with 3 D.O.F.

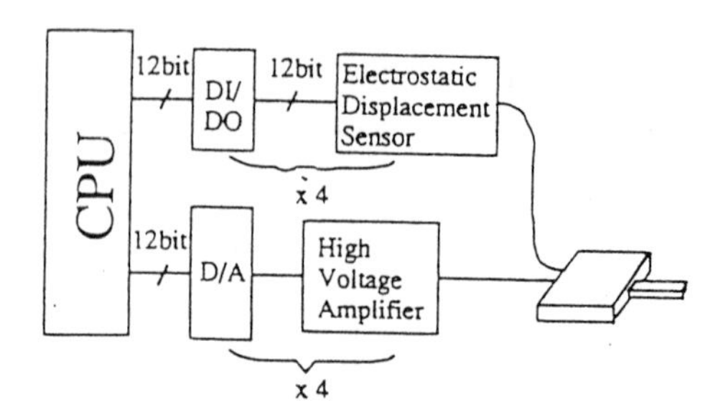

Fig. 8.11 Configuration of Experiment System for Feedback Control

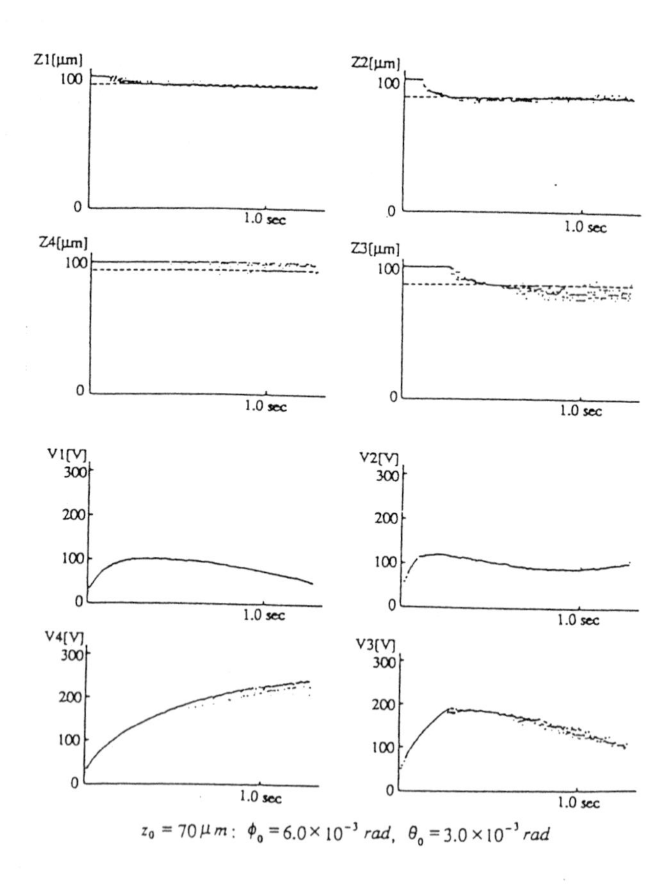

Fig. 8.12 Position Control Experimental Results of Electrostatic Microactuator with 3 D.O.F.

8.2 Scaling Effects

In the case of controlling the microdevices and micro objects, scaling effects plays very important role. Deference of dimensions between the macro and micro world causes a difference in the influential physical phenomena, motion of the objects, and relative change of the system performance between those worlds. These differences are very important in the design , fabrication and control the micro mechanical system because of the miniaturization of the key component. For example, attraction forces are dominant in the micro world compared to gravitational force. Brownian motion is not negligible in a liquid. Since the dominating physical laws are completely different between the macro and micro world, we must consider micro physics in control especially.

Differences of dimensions between the macro and micro world cause a difference in the physical phenomena and the motion of the objects between those worlds. Table 1 shows the dimensions of several forces. As shown in Table1, each force has different dimension, and is influenced by miniaturization. For example, the viscosity forces and friction forces become dominant for the motion of small objects compared with the inertial forces. From this Table, we can evaluate suitable actuation methods for the micro actuators.

In Table 1, the electrostatic force F_{static} is represented by the following equation:

$$F_{static} = \frac{\varepsilon S}{2}\frac{V^2}{d^2} \qquad (8.12),$$

Table 1 Scaling Effects: Dimensions of Forces

Kind of Force	Symbol	Equation	Scaling Effect	
Electromagnetic Force	$F_{magnetic}$	$\frac{B}{2\mu} Sm$	L^2	μ : permeability B : magnetic field density Sm : area of cross section of coil
Electrostatic Force	F_{static}	$\frac{\varepsilon S}{2} \frac{V^2}{d^2}$	L^0	ε : permittivity S: surface area V : applied voltage d: gap between electrodes
Thermal Expansional Force	$F_{thermal}$	$eS \frac{\Delta L(T)}{L}$	L^2	e : Young's Modules L : length T : temperature ΔL: strain
Piezoelectric Force	F_{piezo}	$eS \frac{\Delta L(E)}{L}$	L^2	e : Young's Modules L : length T : temperature ΔL: strain
Inertial Force	F_i	$m \frac{\partial^2 x}{\partial t^2}$	L^4	m : mass t : time x : displacement
Viscosity Force	F_v	$c \frac{S}{L} \frac{\partial x}{\partial t}$	L^2	c : viscosity coefficient S: surface area x : displacement L : length t : time
Elastic Force	F_e	$eS \frac{\Delta L}{L}$	L^2	e : Young's Modules S: cross section area ΔL: strain L : length

where ε is dielectric constant. S os surface area of electrode, V is applied voltage, and d is distance of gap between electrodes. Dimension of Eq. 1 is represented by the following equation:

$$[F_{static}] = [\frac{\varepsilon S}{2}\frac{V^2}{d^2}] = \alpha\frac{[L^2]}{[L^2]} = \alpha\,[L^0] \qquad (\alpha = \frac{\varepsilon V^2}{2}) \qquad (8.13),$$

Equation 8.13 shows that electrostatic force is generated in proportion to S/d^2, $[L^0]$. That is, id S/d^2 is constant in miniaturizing the electrostatic actuator, the generated electrostatic force isn't changed. Therefore, the electrostatic force is said to be suitable for the driving force of the micro actuator.

Dimension of electromagnetic force generated in a solenoid coil is shown as following equation:

$$[F_{magnetic}] = [\frac{S_m B}{2\mu_0}] = [\frac{B}{2\mu_0}S_m] = \beta\,[L^2] \qquad (\beta = \frac{B}{2\mu_0}) \qquad (8.14),$$

where S_m is surface area of cross section of the solenoid coil, B is magnetic field density, and μ_0 is permeability. Equation 8.14 shows that the solenoid coil generates the electromagnetic force in proportion to L^2. Because of that, the electromagnetic force decreases in miniaturizing the structure. Since the electromagnetic force also depends o the power of the electromagnetic field, it isn't said that the electromagnetic force is not suitable for the micro actuator. However, it is true that the efficiency of the electromagnetic force generated in the micro structure is questionable, because the electromagnetic field depends on the size of magnetic elements.

8.3 Intelligent Control

In this section, we introduce some examples of the intelligent control. One of the most popular characteristics of micro mechanical system is mass-production. On the other hands, since a microdevices generates only slight force and small displacement, it is necessary to use many microdevices. That is, the control of multiple microdevices becomes important for micro mechanical systems. One of the methods to control multiple subject is a distributed control. Distributed control is often used as the method of controlling a group robotic system.

The other characteristics of micro mechanical system is tribological problem. Since friction and viscosity influences performance of micro objects more than conventional mechanical systems, performance of prototype is often deferent from respected performance at design. Because of this, measurement of the real parameter of friction, viscosity and so on is needed, but it is very difficult to measure them since amount of them is slight. Therefore, it becomes necessary to infer them from performance of the microdevices. Inference and acquisition of parameters of microdevices become key technologies, and are named "Learning." Learning is generally one of the methods to acquire some parameter from the motion and behavior of devices or robots.

In this section, we introduce distributed control as an example of intelligent control for group of micro mechanical systems and machine learning as an example of acquisition of parameter in the micro mechanical systems.

8.3.1 Distributed Control

The distributed control is defined as that each subject exists independently and constructs the system with corporation. Figure 8.13 shows the conceptual illustration of distributed micro system. In Fig. 8.13, each microrobot controls itself autonomously and has only simple function such as communication, moving and sensing. A microrobot can't execute given task such as carrying the heavy load bigger than the microrobot. In this system, they can achieve their task by connection of them.

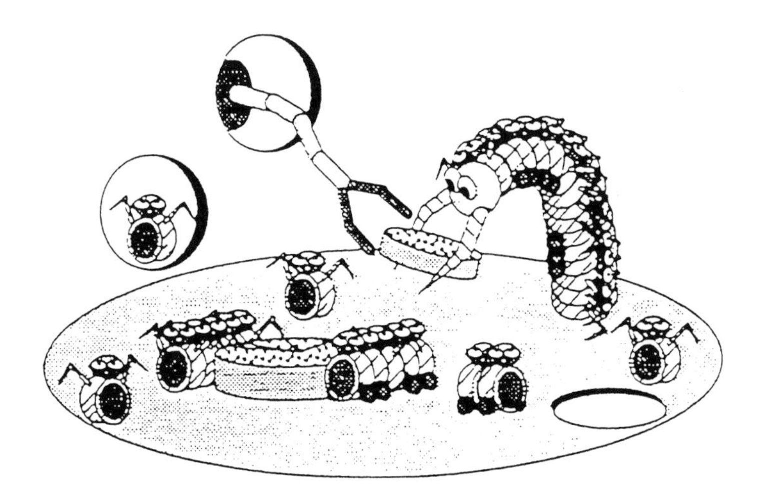

Fig. 8.13 Concept of Distributed Micro System:
Concept of Micro Cellular Robotic System (μ-CEBOT)

The system using distributed control is respected to achieve the following function:
(1) adaptability,
(2) response ability,
(3) reliability,
(4) extendibility,
(5) fault tolerance.

These characteristics directly influence the abilities of the multiple systems.
(1) The adaptability is demanded to correspond to dynamic reformalization or reconfiguration of the robotic system, since the configuration of the robotic system depends on given tasks and environments. This ability is based on the architecture of communication system, that is, the communication architecture should be designed to be able to provide the adaptability.
(2) The response ability is also related to the performance ability of the robotic system, since the communication response or delay influences the behavior of the multiple robots. In the multiple robots, to carry out given tasks it is necessary to cooperate or coordinate, therefore the communication response or delay changes the formation of task execution.
(3) The reliability is an important factor not only for the multiple system, but also for the conventional systems. Especially, for the multiple robots, communication error occurs collision not only in communication but also physically. Therefore, high reliability is demanded.
(4) The extendibility is common ability to construct to communication systems. The communication network spreads and connects with other networks increasing the computer terminal or robots. In addition to the extendibility, the communication network must be designed to adapt itself to the reduction of the dimension of the communication network. The reduction of the robotic system will be occurred by damage of the robots. This is a considerable matter for the design of the communication system in the multiple robots.
(5) The fault tolerance is also considerable ability for the common communication systems. To enhance the fault tolerance, a large number of research has been done. The multiple robots will provide another approach to the fault tolerance concerning the communication. That is, the communication will be based on local communication more than global communication , and the local communication is superior in the fault tolerance to the global communication.

To achieve the above performance, there are several problems about hardware and software. Especially, it is necessary to resolve software problems. The software problems, that is, intelligence is considered to be consists of architecture, decision making, recognition, representation, and learning as shown in Fig. 8.14.

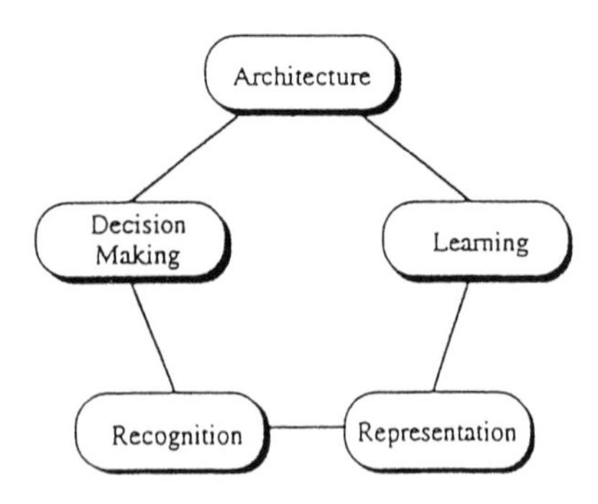

Fig. 8.14 Relation Issue of Intelligence

(1) Architecture: In the case of the group intelligence, the architecture must be optimized depending on the configuration of the intelligence and task execution. In the conventional intelligence, the architecture or the configuration of the intelligence is designed to carry out predicated tasks and knowledge representation. ON the other hand, in the group intelligence, the architecture represents the emergent structure of the group intelligence. The higher intelligence a group has, the higher order the group organizes. That is, in case of swarm intelligence as lower group intelligence, individuals will organize two-layer architecture consisting of macro order and micro dynamics.
(2) Decision Making: Deferent two types of the decision making can be considered on the group intelligence. The first one is carried out;by the interaction among the individuals with non-intelligence. The decision making is regarded as an emergence process of intention of the system. The other has been discussed in the research field of distributed artificial intelligence, where multiple individuals cooperative or coordinate to solve problems, and negotiate to avoid the conflict of solutions that are provided by the individuals. In the distributed artificial intelligence, each individual that refers to a process behaves autonomously and distributively to solve problems with restricted intelligence and knowledge.
(3) Recognition: The word of recognition includes the function of sensing and inference. As a sensing function, active sensing is required for intelligent robots more than passive sensing. Especially, in the distributed systems, distributed sensing has been discussed. The distributed sensing has been treated in many fields, such as traffic control, plant surveillance, autonomous robot control with multiple sensors. Sensor fusion and sensor integration have been discussed to fuse and integrate the sensing information to infer real data.
(4) Representation: The idea of the group intelligence indicates a distributed representation of the intelligence, since the intelligence of the whole system including multiple subjects is redefined by the integration of the intelligence in each individual. Therefore, the intelligence in a robot can be regarded as a part of the intelligence of the system
(5) Learning: The word of learning refers to knowledge acquisition and thinking of entities. The knowledge acquisition is classified into two learning categories, an objective learning and behavioral learning. The former learning function is related to the research on artificial intelligence. In a conventional learning, the knowledge acquisition concerning the objective learning refers to the stores of input data or data about recognized objects. The later learning function closely is related to the research on robotics, since a robot carries out the behavioral learning heuristically. That is, the robot acquires the relationship between external condition, such as environment or goal of the robot, and how to behave in the situation.

These intelligence is quite important for the distributed micro systems, which consist of many agents with intelligence. In the micro mechanical systems, many objects such as actuators, sensors and combination of them are often produced on a substrate by micromachining, especially, surface-micromachining and bulk-micromachining. Because of this, the distributed control becomes key technology of control of micro mechanical systems.

Here, Fig. 8.15 shows the conveyance system produced by Prof. H. Fujita and Dr. S. Konishi. As shown in Fig. 8.15, many actuators, which dimension is 500 x 500 μm^2, are integrated on a wafer, and each actuator is controlled independently. In this system, object which is bigger than actuator is transported by cooperation of actuators, and their cooperative mode achieve movement of object with three degrees of freedom.

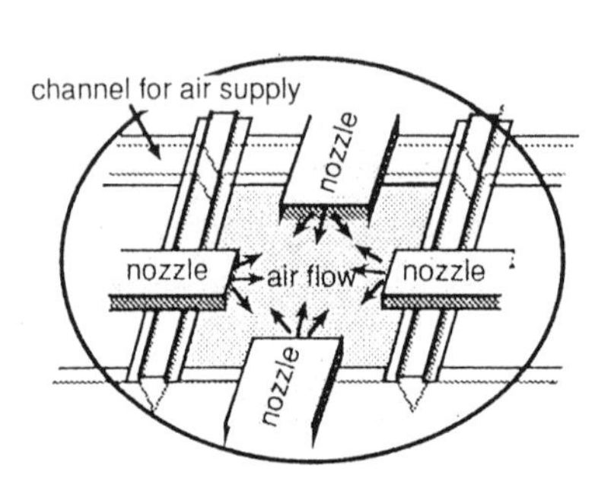

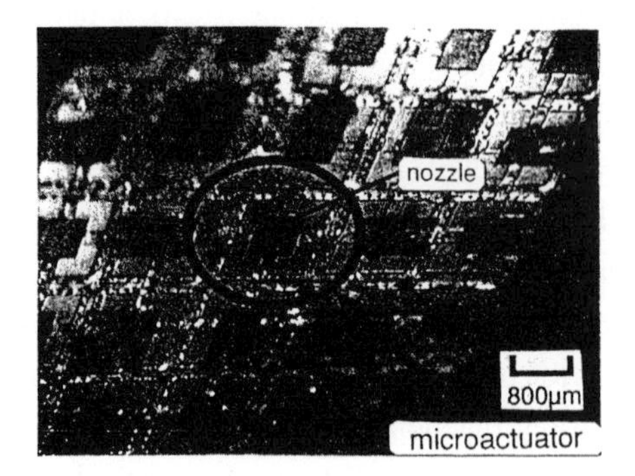

Fig. 8.15 Converance system

8.3.2 Machine Learning

Acquisition of parameters of micro mechanical systems such as friction and size is very important for designing and controlling of them. However, since dimension and different physical phenomena according to it employs performance of micro object, it is difficult to calculate and inference these parameters by the theory on conventional mechanical system in the macro world. Therefore, it is required to estimate them based on observing and measuring them by its performance.

Machine learning is taken account into one of way to estimate them by performance. Machine learning is defined as a method to optimize the parameters such as control gains and system parameters according to motion and output signal of system. One of the most famous machine learning method is the optimization of control signal using Neural Network (NN). Except the method using NN, there are another method using FUZZY, Genetic Algorithm (GA), and combination with them. In the researches on conventional systems and robots, several applied systems have been reported. In the micro mechanical systems, since measurement of their performance, there are a few researches on machine learning for them. In this section, we would show the wave generator using FUZZY controller as a sample of machine learning for the micro mechanical systems.

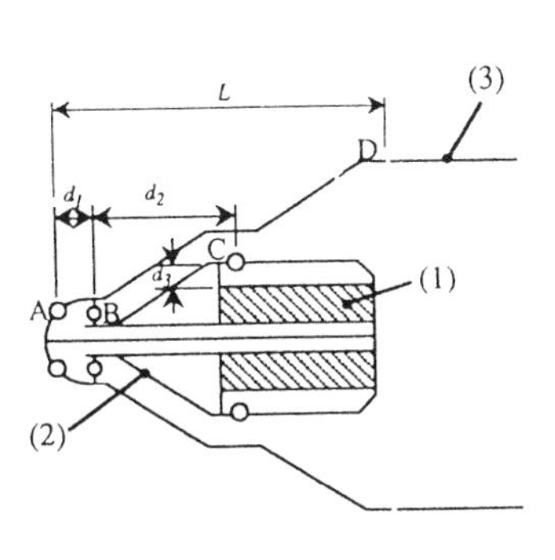

Fig. 8.16 Configuration of Micro Fish

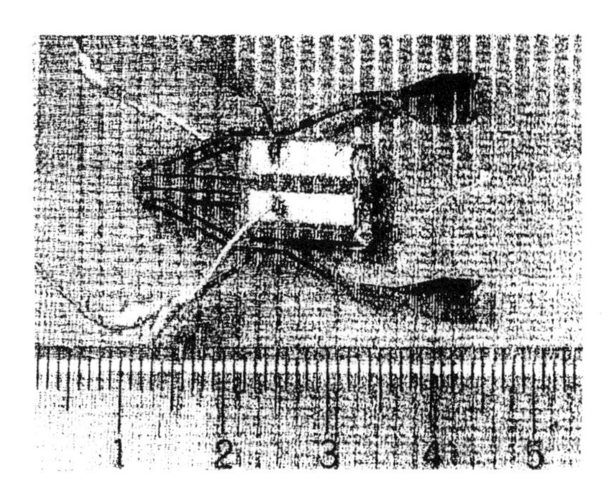

Fig. 8.17 Prototype of Micro Fish

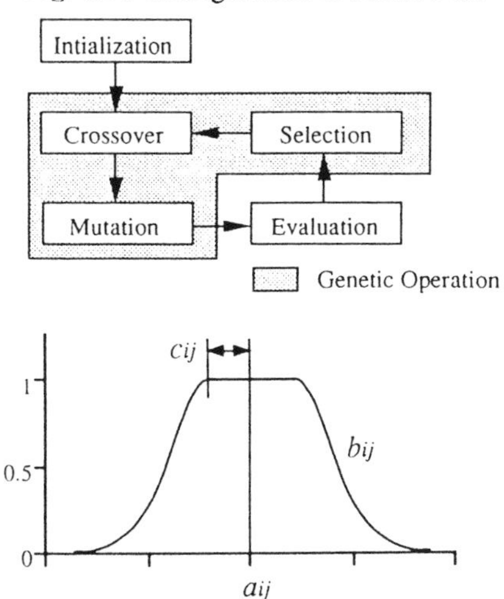

Fig. 8.18 Learning Architecture

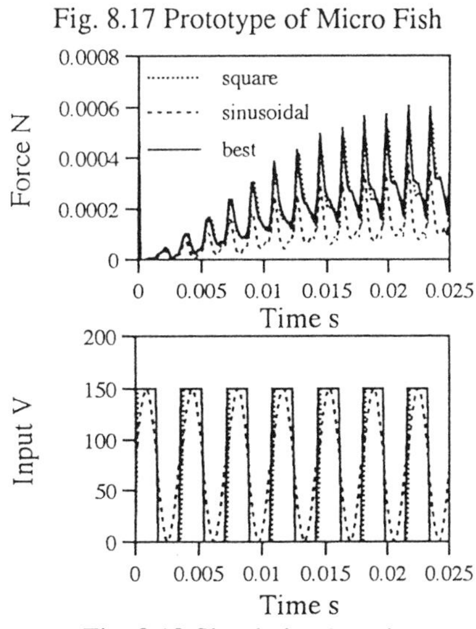

Fig. 8.19 Simulation Results

Figures 8.16 and 8.17 show the configuration and photograph of micro fish, and it consists of piezoelectric element, two fins and body. The body is not transmitting the displacement from piezoelectric elements to fins, but magnifying it. Experimental results shows that the micro fish moves at specific frequency in water. In this experiment, regular square wave is only applied to piezoelectric element, but swinging mode of fin changes as changing applied frequency. If we can control swinging mode of fin, performance of micro fish can be improved. However, it is difficult to analysis and calculate it. So, it is effective to acquire them according to learning. Figure 8.18 shows the architecture of learning system for micro fish. This learning system uses FUZZY controllers to evaluate performance of fins. Figure 8.19 shows the numerical simulation results of learning. As shown in Fig. 8.19, micro fish gets the driving force at the results on learning.

This example shows efficiency of machine learning for micro mechanical systems. However, there are several problems such as difficulty of measuring the behavior of systems.

8.4 Man-Machine Interface

In the micro mechanical systems, since control subject is very small or amount of change by control are very slight, it is difficult to detect them. Because of this, it becomes very important to detect the change of control target and to give information to the operator. As ways of giving the condition of the control target, some methods such as tele-communication and virtual reality have been proposed.

Figure 8.20 shows the concept of tele-operation using virtual reality. In Fig. 8.20, given task is manipulation of small object in the scanning electro microscope (SEM). In conventional system, the operator can handle a micro object using only visual information. However, it is difficult to judge if manipulator grasps the objects completely. Therefore, in the proposed system as shown in Fig. 8.15, visual information from the SEM and force information from the force sensor on the tip of manipulator construct the virtual object, and give it to operator. Then, the operator feels the object in the virtual world, and can handle it completely.

As shown in Fig. 8.20, we can't observe the phenomena in micro world directly. Moreover, the motion and force of micro objects are smaller than the precision of human motion and force. Because of this, magnification of motion and force are also considered to become key technologies of man-machine interface connecting macro-micro world. That is, the force feedback with magnification of motion and force plays very important role. Figure 8.21 shows the simulator system of micro catheter using force feedback. In such system as shown in Fig. 8.21, the operator controls the active catheter which bends its tip optionally , and when the tip of catheter contacts with wall of blood vessel, this information returns to operator as force feedback with magnification of force. According to this force feedback system, the operator can operate the micro surgery using micro catheter without breaking blood vessels.

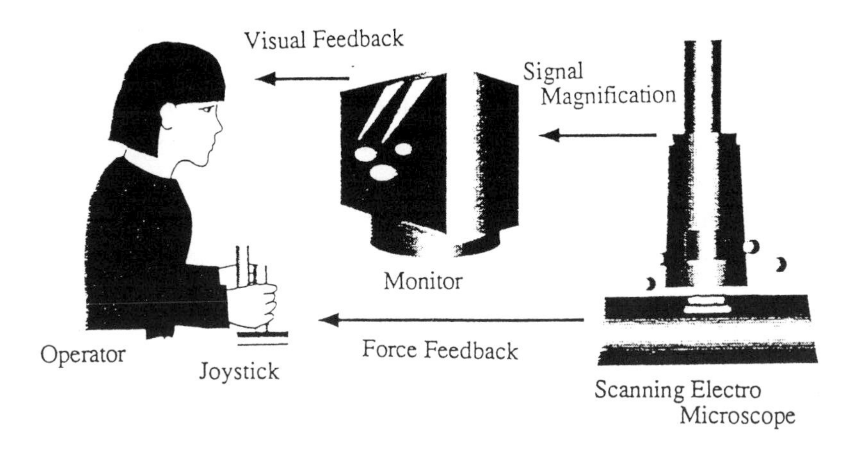

Fig. 8.20 Tele-Operational Micro Manipulation System

Fig. 8.21 Micro-Surgery Simulator System with Micro Active Catheter

8.5 Summary

In this chapter, we introduce some methods of controlling the micro mechanical systems such as microdevices and microrobots. The performance of micro mechanical system is employed by the different physical phenomena from macro world, and so the conventional control methods aren't sometimes used for controlling micro mechanical systems. To achieve control system of micro mechanical systems, we need the observation of these phenomena and make a model from them. In this chapter, we show the distributed control and tele-operation with force feedback and magnification. In such systems, theoretical background is made by some research in the conventional mechanical systems. One of reasons why such systems are completed is considered to be hardware such as sensors and actuators. Recently, microactuator and microsensors are developed and reported from several groups. In future, integration with microactuators and microsensors make it possible to control micro mechanical systems.

References

[1] S. Egawa et al., " Multi-layered electrostatic film actuator," in Proc. IEEE Micro Electro Mechanical Systems, pp.166-171, 1990.
[2] H. Fujita et al., " An Integrated Micro Servosystem," in IEEE Int. Workshop on intelligent Robot and Systems pp.15-20, 1988.
[3] T. Fukuda et al., "Design and Dexterous control of micro manipulator with 6 D.O.F.," in Prof. IEEE Advanced Robotics, pp.343-348, 1991.
[4] F. Arai et al., " Micro Manipulation Based on Micro Physics (Strategy Based on Attractive Force Reduction and Stress Measurement)," in Proc. IEEE/RSJ Intelligent Robots and Systems (IROS'95), p.236-241, 1995.
[5] J.N.Israelachvili, Intermolecular and Surface Force, New York Academic, 1985.
[6] R.S.Fearing, "Grasping of Micro-Parts in Air," IEEE Robotics and Automation Tutorial Text, Micro Manipulation, Micro Motion Control, pp.1-7, 1994.
[7] T. Fukuda et al., "Cellular Robotics and Micro Robotic Systems," World Science, 1994.
[8] H. Ishihara et al., "Micro Optical Robotic System (MORS)," in Proc. Fourth Int. Symp. Micro Machine and Human Science (MHS'93), pp.105-110, 1993.
[9] T. Fukuda et al., "Design and Experiments of Micro Mobile Robot Using Electromagnetic Actuator," in Proc. Third Int. Symp. Micro Machine and Human Science (MHS'92), pp.77-81, 1992.
[10]H. Ishihara et al., " Approach to Autonomous Micro Robot (Micro Line Trace Robot with Reflex Algorithm)," in Proc. IEEE/SEIKEN Symp. Emerging Technology and Factory Automation (ETFA'94), pp. 78-83, 1994.
[11] N. Mitsumoto et al., "Self-organizing Micro Robotic System (Biologically Inspired Immune Network Architecture and Micro Autonomous Robotic System)," in Proc. IEEE Micro Machine and Human Science (MHS'95), pp.261-270.
[12] S. Konishi, et al., "System Design for Cooperative Control of Arrayed Microactuators," in Proc. IEEE Workshop on Micro Elctro Mechanical Systes (MEMS'95), pp. 322-327, 1995.
[13] S. Konishi, et al., "A Conveyance System using Air Flow based on the cnocept of Distributed Micro Motion System," Journal of MIcroelectromechanical Systems, Vol.3, No.2, pp. 54-58, 1994.

[14] T. Fukuda, et al., "Steering Mechanism and Swimming Experiment of Micro Mobile Robot In Water," in Proc. of IEEE Workshop on Micro Electro Mechanical Systems (MEMS'95), pp. 300-305, 1995.

[15] T. Fukuda. et al., "Acquisition of Swimming Motion by RBF Fuzzy Neuro with Unsupervised Learning," in Proc. of Int. Workshop on Biologically Inspired Evolutionary Systems (BIES'95), pp.118-123, 1995.

[16] H. Kobayashi et al., "Micro-Macro Manipulator with Hepatic Interface - 2nd Report:Control by Using Virtual Model," in Proc. IEEE Robot, Human Communication (RO-MAN'94), pp.130-133, 1994.

[17] F.Arai, et al., "Intelligent Assistance for Intravascular Telesurgery and Experiments on Virtual Simulatior," in Proc. on Virtual Reality Annual Int. Symp. '95 (VRAIS '95), pp. 101-107, 1995

9. Examples of Microsystems

P. Dario and M.C. Carrozza

MiTech Lab, Scuola Superiore Sant'Anna, via Carducci, 40, 56127 Pisa, Italy

1. INTRODUCTION

A very large - and constantly growing - number of microsystems have been developed and presented in the scientific and technical literature by many authors in last few years. Those microsystems differ substantially for conceptual design, fabrication technology and targeted application. Although a selection of representative examples of microsystems is certainly difficult and ultimately somewhat arbitrary, we have identified and shall present in this Chapter some cases of microsystems which are particularly interesting either because they implement well the concept of a microsystem or of a micromachine, or because they incorporate important design or technological solutions, or because they illustrate the potential for application, or because of a combination of these factors.

The examples of microsystems that we present can be considered as falling into two broad domains, the one of "*MEMS or microsystems*" and the one of "*micromachines*", terms which reflect a different conceptual approach to the field and thus deserve some explanation.

In the current literature and also in many workshops and conferences, different terms like MEMS (Micro Electro Mechanical Systems), microsystems, micromachines, micromechanics, micromechatronics, and so forth, are often used as synonyms to indicate miniature devices. However the use of a different terminology may not indicate just a personal preference or reveal a "geographical" origin, but it rather implies a different philosophy and reflects a different technical background. This concept is depicted in Figure 1, which illustrates how the field defined as "MEMS or microsystems" (two terms that, according to the authors, can be considered as *real* synonyms) represents the current state of a technical evolution which has its roots in *microelectronics*, whereas the field of micromachines has a different origin and evolution.

Microsystems have evolved from solid state integrated circuits technology, through the development of microfabricated sensors and then of smart sensors, to the invention of microfabricated movable mechanical parts and micromotors, and finally to the concept of the microsystem, viewed essentially as a silicon wafer incorporating the processing circuit and either a microsensor, or a microactuator, or one or more mechanical, or fluidic, or optical components, or their combination [1][2]. The main conceptual difference between an integrated circuit and a microsystem is that the integrated circuit processes information only, whereas the microsystem can also process "matter". But the similarities between the two classes of devices are many: in most cases both are silicon-based, solid state planar devices (although there are now many examples of 3D, non silicon devices), and they are ultimately

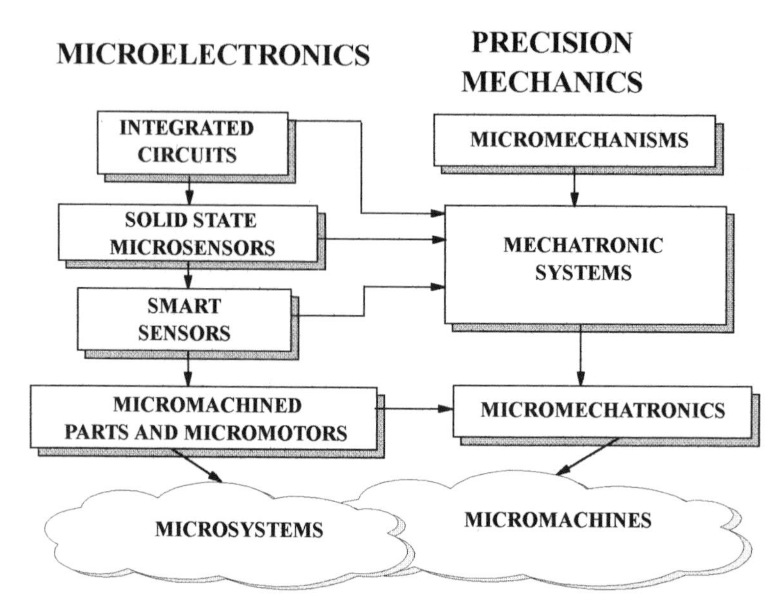

Figure 1. Concept of technology evolution: from integrated circuits to microsensors and *microsystems*, and from precision mechanics to micromechatronic systems and *micromachines*.

intended to be manufactured in large or even very large scale. The expected advantages of a microsystem are virtually the same as those of an integrated circuit: small size, high reliability and high performance (all obtained by the integration of different functions on the same chip or substrate), and low cost (obtained through mass production, possibly by an integrated circuits factory). The concept of MEMS or microsystems as a more complex configuration of a solid state device is particularly popular in the United States and in Europe.

Substantially different are the evolution and the expected market of micromachines. In fact, the concept of micromachine originates primarily from the mechanical engineering community rather than from the microelectronics community, and stems from the idea that there is a need for miniaturizing complete and real (meaning "essentially 3D" machines) [3]. According to this approach, the technical evolution of micromachines started from precision mechanical systems, moved towards mechatronic systems (that is, devices that integrate precision mechanisms, sensors, actuators and embedded controllers), and ultimately - through further miniaturization and integration - to micromechatronic systems. (In fact, a micromachine can be regarded as an implementation of micromechatronics concepts and technologies). Obviously there are many crosslinks between the evolution phases of microsystems and the evolution of micromachines: the very same miniaturization of mechatronic systems would just not be possible without the development of microelectronic circuits, sensors and actuators. In fact the main advantages of mechatronics and micromechatronics machines over traditional mechanical machines derive from the improvement of performance (smaller size, lower complexity, higher reliability and controllability, lower cost) allowed by the use of microelectronic components. Furthermore, just like many mechatronic systems (for example, photo and video cameras, appliances, automobiles, medical instrumentation) already incorporate integrated circuits, microsensors and even some microsystems, the future micromachine will rely heavily on the use of solid state microsystems.

However, we wish to underline again the main aspects that, in the authors' opinion, make a difference between microsystems and micromachines: first, micromachines are inherently three-dimensional "real" machines which are designed and built to work as ordinary machines do in the macro world, but are just smaller (this, of course, requires not only a reduction of size, but a completely different analysis of the machine and of its functioning in the micro world); and, secondly, micromachines are not necessarily intended to be mass fabricated. They can be even manufactured in small quantities and maybe sold at high price per unit. (Consider, for example, miniature endoscopes for medical diagnostics or industrial inspection). The concept of micromachines as micromechatronic systems is very popular in Japan, where it has been proposed first and where its practical applications are actively pursued [4][5].

In this Chapter, rather than classifying different microsystems and micromachines using a taxonomy based on the type of application, or of technology, we shall discuss many examples of devices according to the concepts and definition given above. In particular, examples of devices corresponding to the concept of microsystem are described in Section 2, and devices corresponding to the concept of micromachine are described in Section 3.

2. EXAMPLES OF MICROSYSTEMS

Many papers and reports quote silicon micromachined sensors as examples of the potential impact of microsystems [6]. The reason for this can be understood based on the concept and evolution of microsystems: microsensors are a fundamental part of microsystems and a preliminary step towards their implementation, and it is likely that the potential market of microsystems could be extrapolated from considerations on the market of microsensors. Pressure and acceleration silicon microsensors are now a capable, manufacturable, production-tested technology and a high-volume market for automotive, medical and consumer electronics products. However, in this Chapter we shall try to adhere more strictly to the ideal notion of a microsystem as a more complex microdevice incorporating different components and/or functions.

2.1 Digital micromirror array

The field of microoptics is considered as most promising for microsystems applications [7]. An important potential market for microoptics could be the field of displays. The competition on the development of high performance and low cost displays for standard and high definition projection television systems is very strong worldwide. A light modulator for high-definition projection television systems has been developed by Texas Instruments Inc., Dallas, Texas, using an array of micromirrors [8].

The reflective, spatial, light modulator is composed of an array of rotatable aluminum mirrors, fabricated by thin film and surface micromachining technology. The array is built up over conventional CMOS SRAM address circuitry. The micromirror element is an aluminum mirror suspended over an air gap by two thin, post-supported, mechanically compliant torsion hinges that permit a mirror rotation of ± 10°, as depicted in Figure 2.

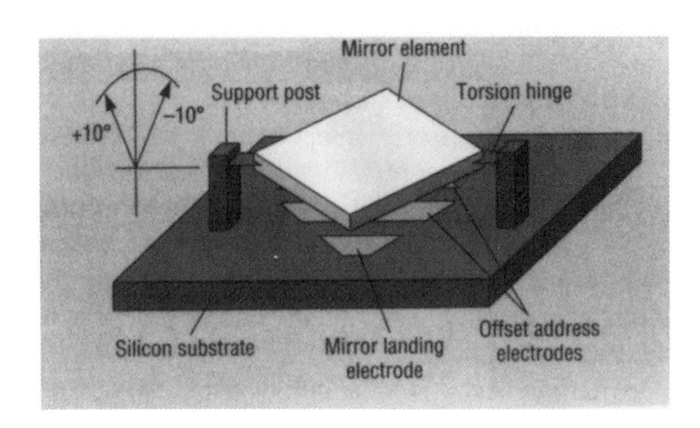

Figure 2. Digital micromirror element with conventional torsion hinge suspension. (From [8]).

The posts are electrically connected to an underlying bias/reset bus that connects all the mirrors of the array directly to a bond pad, as shown in Figure 3.

Each micromirror is 16 μm wide on a pitch of 17 μm. A portion of the digital micromirror device (DMD) is illustrated in the photograph of Figure 4.

The operation of a single chip, DMD-based projection television system is the following: a condenser lens collects light from a white light source; the illumination is directed at a ± 20° angle from the normal to the DMD surface and orthogonal to the rotational axes of the individual mirrors of the array. A lens located above the chip projects an enlarged image of the DMD on a screen. When rotated at + 10°, the individual mirror elements reflect incoming light into the projection lens pupil to produce a bright image on the screen. When rotated - 10°, the mirror elements appear dark. Light reflected, scattered, or diffracted into the optical path by the hinge or mirror support posts increases the illumination level during a DMD pixel-off condition and reduces the projection system's contrast ratio.

The projection television system can utilize either a single- or a three-chip DMD arrangement. Both arrangements can be used for color projection. In particular, even the single-chip system is practical because of the DMD's high light efficiency. In this case a color wheel may be used to sequentially illuminate the DMD with the three primary colors.

Four photolithographically defined layers are surface micromachined to form the electrode, sacrificial layer, hinge, and mirror. The sacrificial layer is an organic material that is plasma-ashed to form the air gap between the address electrodes and mirror. The other three layers are formed from dry plasma etched, sputter-deposited aluminum. The most recent DMD device is a 768 x 576 pixel array with 442,368 mirrors, and has been demonstrated with projection screen diagonals ranging from 42 in. to 13 ft. In conclusion, the DMD offers performance equal to or better than a cathode ray tube. Fundamentally the DMD is an advanced microsystem that utilizes a mature address circuit technology. It offers potential as a display device that can be manufactured with high yield at marketable cost [9].

2.2 Microsystems for gas and liquid analysis

Research on microfluidic devices fabricated by micromachining technology, had its origin about 20 years ago. At Stanford University research was aimed at the development of a gas chromatography system [10], and at IBM at the development of ink jet printer nozzles [11]. The follow up on these early devices has been modest for a long time, but within the last few years a dramatic increase in research activities has taken place, and microfluidics is now developing into a "hot" research topic. Comprehensive discussions of the needs, perspectives, fabrication technologies and results achieved in the field of microsystems for fluidics have been presented by some authors [12-14].

Many different devices are under development, ranging from single components such as flow sensors and valves for gas pressure regulation, to complex fluidic handling microsystems for chemical analysis, consisting of pumps, valves, flow sensors, separation capillaries, chemical detectors, etc., all integrated on a single substrate or as sandwiched modules.

Fluid control will almost inevitably involve valves, and the design of valves has consequently attracted much attention. Micromachined valves have a number of advantages over traditional valves, since small size is beneficial in terms of response time, power consumption, small dead volume and fatigue resistance. Additional advantages include the ability to be batch fabricated with high reproducibility, and the ability to be integrated with other functions. The majority of micromechanical valves have been designed for gas flow control, whereas only a few are intended for liquid applications. The trend in microvalve development is expected to evolve in three main directions: pilot valves for macroscopic valves, pressure or flow regulation, and integrated system applications. Several micromechanical valves have already entered the production phase, including microvalves from Redwood Microsystems, IC Sensors, and Microflow Analytical, all in the U.S.A.

A scheme of the microvalve marketed by Redwood Microsystems, CA, is shown in Figure 5.

The valve is part of an electro-fluidic multi-chip module that also incorporates a piezoresistive pressure sensor and feedback electronics to create a complete electronically programmable pressure regulator. The valve is actuated thermopneumatically and its key component is a silicon diaphragm made by etching [15].

An additional example of microvalve is the three-way microvalve system manufactured by LIGA process at the Forschungszentrum Karlsruhe, Germany [16], which is shown in Figure 6.

A number of different membrane micropumps have been presented using different actuation principles [17-24]. The highest yield, obtained with a piezoelectrically actuated membrane micropump with ball valves and pump body fabricated by stereolithography in the authors' laboratory is 2.4 ml/min of water at 70 Hz pump frequency and zero delivery head [25]. This pump is depicted in Figure 7.

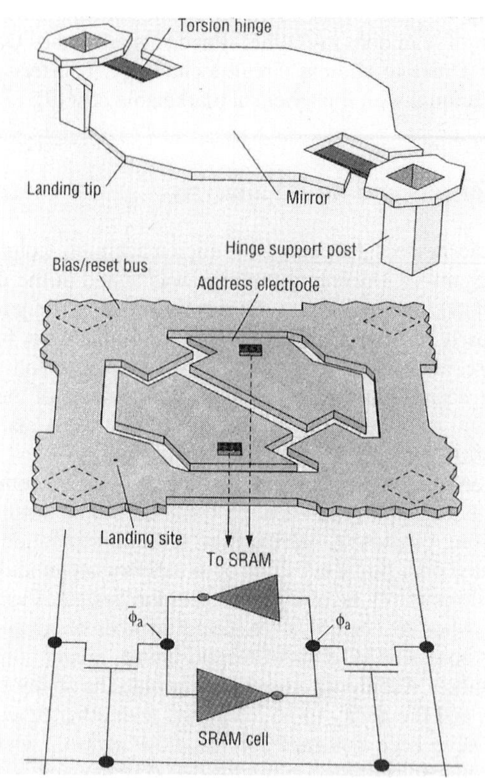

Figure 3. Mirror configuration with schematic of the SRAM cell. (From [8]).

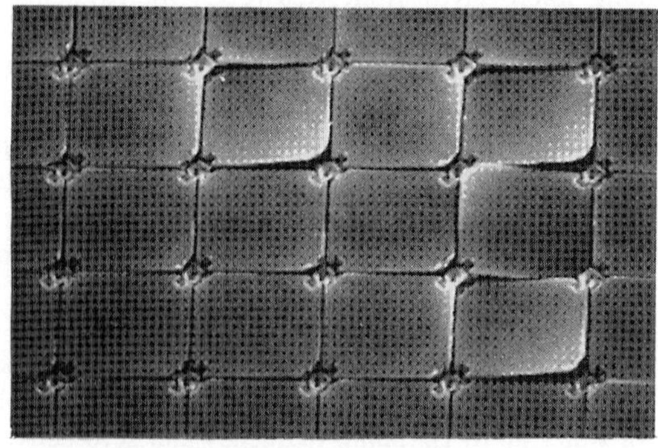

Figure 4. A portion of a DMD array showing selected micromirror elements deflected. (From [8]).

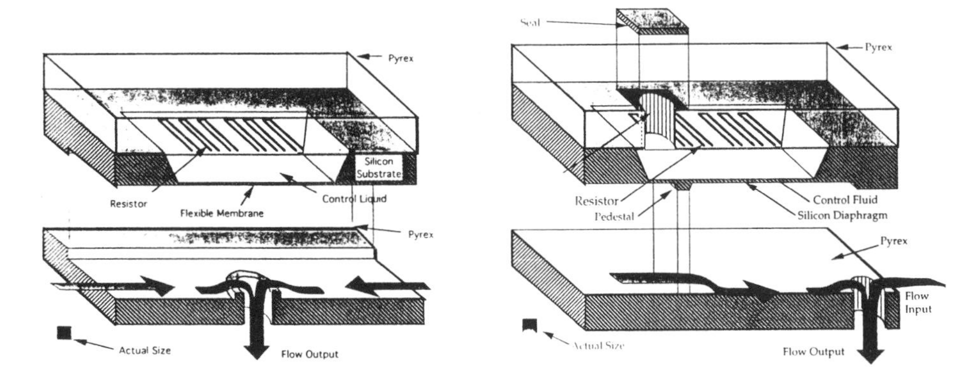

Figure 5. Scheme of the Redwood Microsystems microvalves ("Fluistors"). Shown are the normally open configuration (left) and normally closed configuration (right) of the valve. (From [15]).

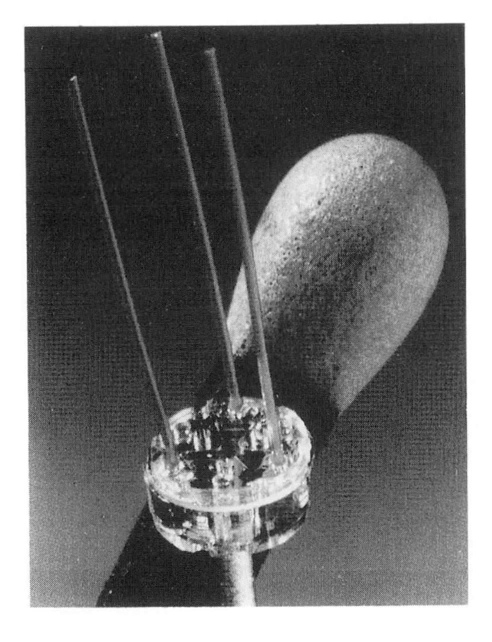

Figure 6. Microvalve system consisting of three independently controlled memebrane valves with one common inlet and three outlet hoses (from [16]).

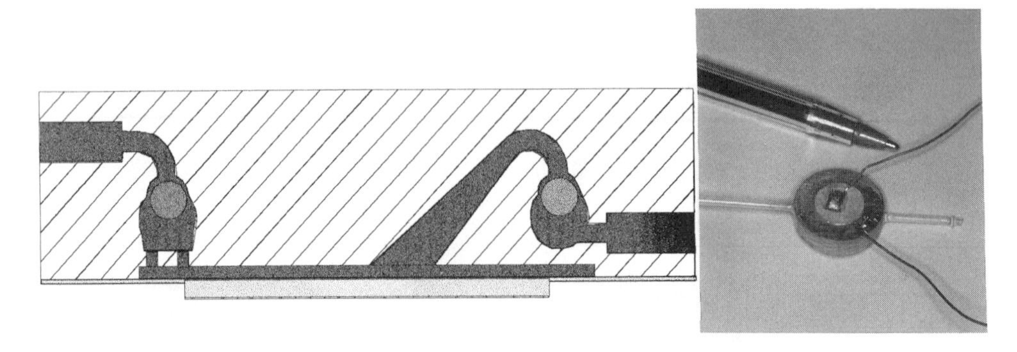

Figure 7. Piezoelectrically-actuated, stereolitography-fabricated micropump fabricated in the authors' laboratory. Its yield is extremely high for pumps of similar size. Cross section of the pump, showing the ball valves (left), and photograph of the micropump (right).

A quite high yield (550 μl/min) has also been reported for a piezoelectrically actuated membrane micropump, whose most interesting and promising feature is flow regulation through the integration of a flow sensor with the pump [26]. The flow sensor consists of a cantilever beam, with integrated piezo-resistive strain gauges, situated in the flow stream. As a liquid flow is passed through the sensor a drag force act on the beam. A scheme of this fluidic microsystem, which allows constant flow-rate dosing, is illustrated in Figure 8.

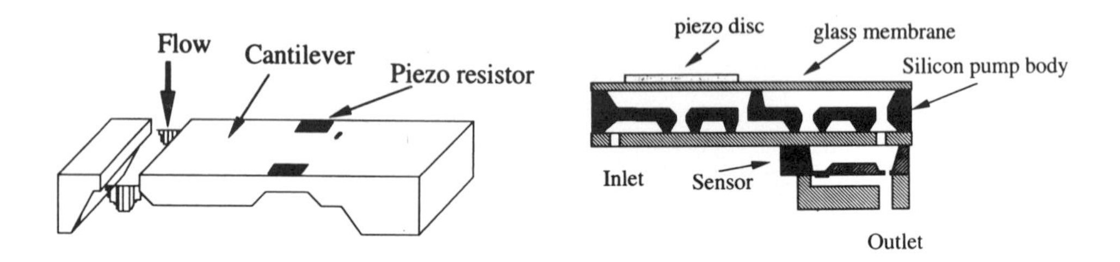

Figure 8. A liquid handling microsystem incorporating a flow-regulated silicon micropump. Shown are a schematic representation of the flow sensor (left) and of the complete microsystem (right). (From [26]).

Another example of micromembrane pump manufactured exclusively by batch processes, such as plastic molding and thin film techniques, and actuated by an electro-thermo-pneumatic actuator integrated in the pump [27], is shown in Figure 9.

Figure 9. A micromemebrane pump manufactured by molding and actuated electro-thermo-pneumatically (From [27]).

Three application fields show clear promise for fluidic microsystems: ink jet printer heads with nozzle arrays, micro dosing systems and micro chemical total analysis systems (also known as μTAS). Chemical analysis systems combine different fluidic components, like valves, pumps, flow-sensors, mixing units, channels and chemical detectors built together to perform the explicit function of chemical analysis. The chemical sensor is often a critical component in chemical analysis instrumentation.

An interesting example of a chemical sensor (that is actually a microsystem on its own) has been presented recently [28]. The device is a microfabricated surface plasmon sensor that integrates optical, mechanical and electronic components. The sensor contains elements that are common to numerous electro-optical systems, thus creating a sort of miniature optical bench. Surface plasmon sensors are used to detect adsorption or absorption of a species of interest in chemical, gas and bio-sensing applications [29]. Detection is based on changes in the permittivity of thin interfacial layers. A surface plasmon is excited in a thin metal film by the transfer of energy from an incident light beam. The plasmon resonance occurs when the exciting light is at a specific incident angle on the metal film. Surface plasmon sensors have four basic elements: a collimated transverse magnetic polarized beam, a method of coupling this beam to a thin metal film so as to achieve incidence angles above the critical angle, a means of varying the incidence angles, and a detector to measure the reflectance of the metal film as a function of the incidence angle. In contrast to most surface plasmon detection configurations which require expensive optical equipment and a large stable optical bench, and are thus confined for the most part in laboratory environments, the proposed microfabricated sensor is very small, self-contained, and batch fabricated, which makes it potentially portable and inexpensive. The microsystem includes a single-mode optical fiber, a graded index collimating lens, a scanning rotating micromirror, a cavity filled with index matching fluid, and a position sensitive photodiode. Figure 10 is a cross-sectional diagram of the sensing microsystem that identifies individual components and shows the path of the incident light beam.

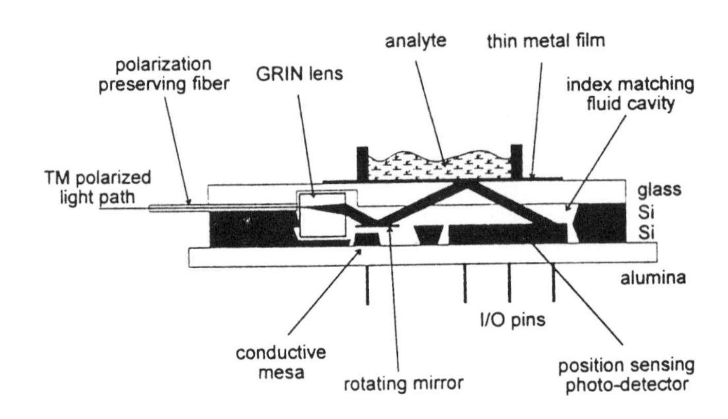

Figure 10. Cross-sectional diagram of a surface plasmon sensing microsystem, labeling individual components and showing the light path. (From [28]).

Just like in the case of the Digital Micromirror Device (DMD) described in Paragraph 2.1, a critical component of the microsystem is the micromirror. In the plasmon sensing microsystem the micromirror is a square silicon plate 2.5 mm on edge, suspended by two center torsional silicon tethers. The mirror plate is 20 μm thick and the tethers are typically 500 μm long x 30 μm wide x 10 μm thick.

The microsystem is contained in a hybrid package, which integrates optical, mechanical and electronic components in a small volume: 10 x 25 x 2 mm^3. Tested for plasmon sensing and chemical response, the microsystem recorded shifts of the resonance angle when drops of different solutions (including BSA, or bovine serum albumen) were placed on the sensing film.

An interesting example of microsystem usable for chemical sensing is the microspectrometer developed in Germany by microParts GmbH using LIGA technology [30]. The complete microspectrometer, illustrated in Figure 11, is fully equipped with input fiber and read-out electronics and compatible software.

The LIGA microspectrometer is placed on a photodiode array. Its dimensions are 18 x 7 x 6 mm^3. The system is sensible to energy in wavelengths from 400 to 1100 nm; transmission efficiency is 25%; spectral resolution is 7 nm; dynamic range up to 20,000. The device can be used for color measurement and analytical applications, in particular for gas detection.

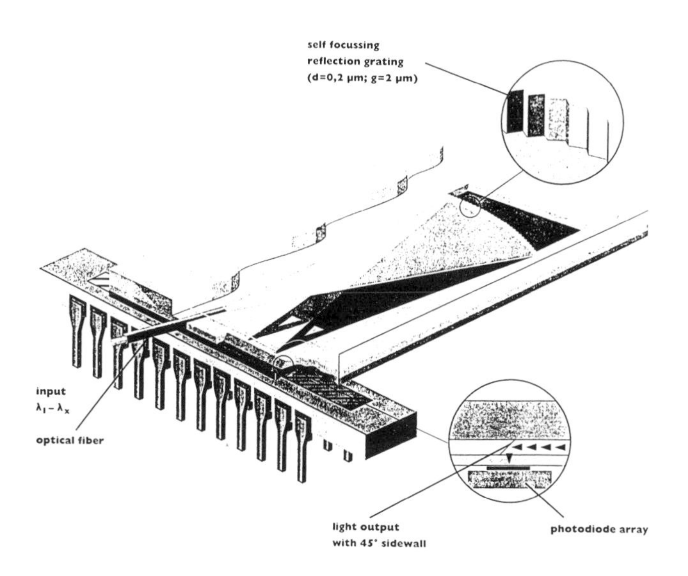

Figure 11. Complete microspectrometer fabricated by LIGA technology.

An important example of microsystem for chemical analysis is the silicon-micromachined gas chromatography system. Gas chromatography (GC) is an important analytical tool for a variety of disciplines, including environmental analysis and pollution management. However, most modern GC systems are too bulky and fragile to be operated in remote and hostile locations. This problem has motivated research to develop portable and robust GC systems. Very well known is the already mentioned pioneer work in this field carried on at Stanford by S.C. Terry et al. [10].

There are five major components in a conventional GC system: carrier gas, sample injector, column, detector, and data processing system. The carrier gas (known as the "mobile phase") is a high-purity gas (e.g. hydrogen or helium) that transports the gas sample of interest through the column. The sample injector introduces a precise and finite gas sample volume into the carrier gas that continuously flows through the column. The column is coated with a "stationary phase" that chemically interacts with the injected gas sample to produce a propagation delay for each of its components based upon their heat of adsorption and vapor pressure. The sample components will emerge from the column at different times, and the detector located at the column's output can be used with the data processing system to identify and quantify the concentration of each species.

A photograph of the Stanford's micromachined GC system is shown in Figure 12.

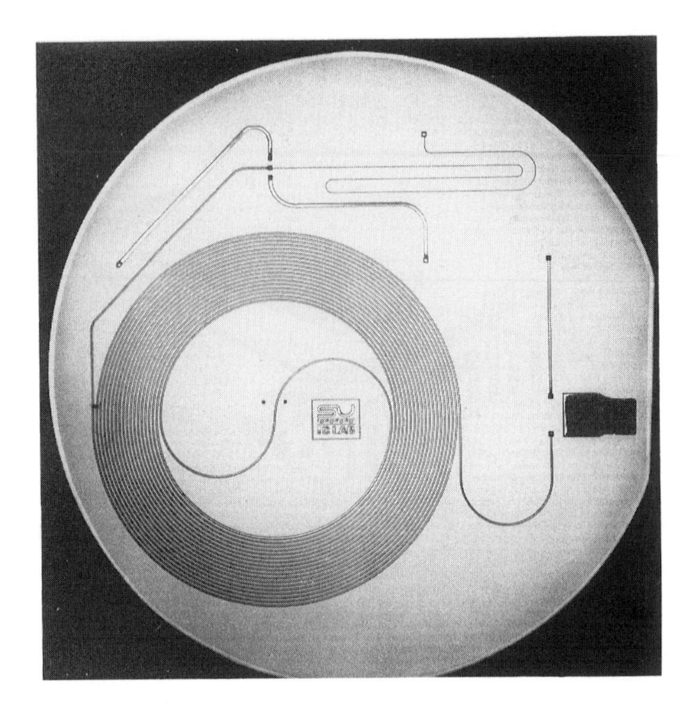

Figure 12. Prototype of a silicon GC developed at Stanford University with funding from the U.S. National Institute for Occupational Safety and Health. The GC is built on a 5 cm - diameter silicon wafer. The capillary column is a 1.5 m - long spiral obtained by etching the silicon wafer and bonding a sealing glass to the wafer (From [10]).

The most critical limitation of the initial design by Terry et al. involved the deposition of the stationary phase within the micromachined column. A liquid medium was injected into the column after the GC system was assembled: this resulted in a stationary phase's inhomogeneous surface coverage, which compromised its overall performance.

A new silicon micromachined gas chromatography (MMGC) system that addresses Terry's fundamental limitation has been presented recently by R.R. Reston and E.S. Kolesar [31]. The MMGC system is based upon conventional analytical chemistry laboratory instrumentation and it incorporates the same five major components as an ordinary GC system. A scheme of the MMGC is shown in Figure 13.

The critical component in a GC system is the column, and central to its operation is the stationary phase. Therefore particular attention was devoted to the design of the column, that also influenced many of the other design choices of the MMGC. To effectively utilize the micromachinable silicon wafer's surface area, the miniature GC column was configured as two

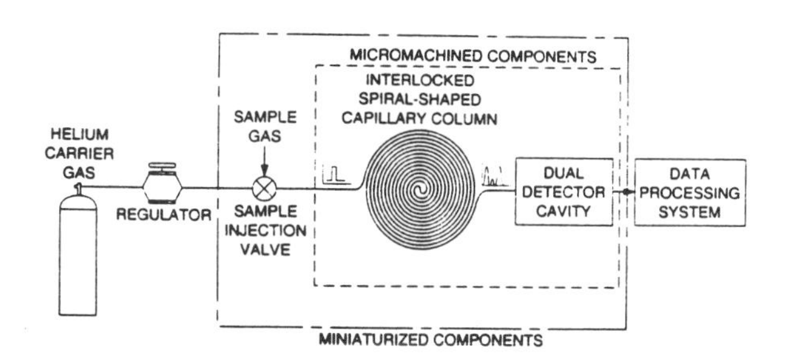

Figure 13. Block diagram of a micromachined gas chromatography system. (From [31]).

interlocking spirals. This also facilitated the location of the gas inlet and outlet far away from the complex spiral structure. The miniature sample injector incorporates a 10-μm-long sample loop; a 0.9-m-long, rectangular-shaped (300 μm width and 10 μm height) capillary column coated with a 0.2-μm-thick copper phthalocyanine (CuPc) stationary phase, and a dual-detector scheme based upon a CuPc-coated chemiresistor and a 125-μm-diameter thermal conductivity detector bead. Silicon micromachining was employed to fabricate the interface between the sample injector and the GC column, the GC column itself, and the dual-detector

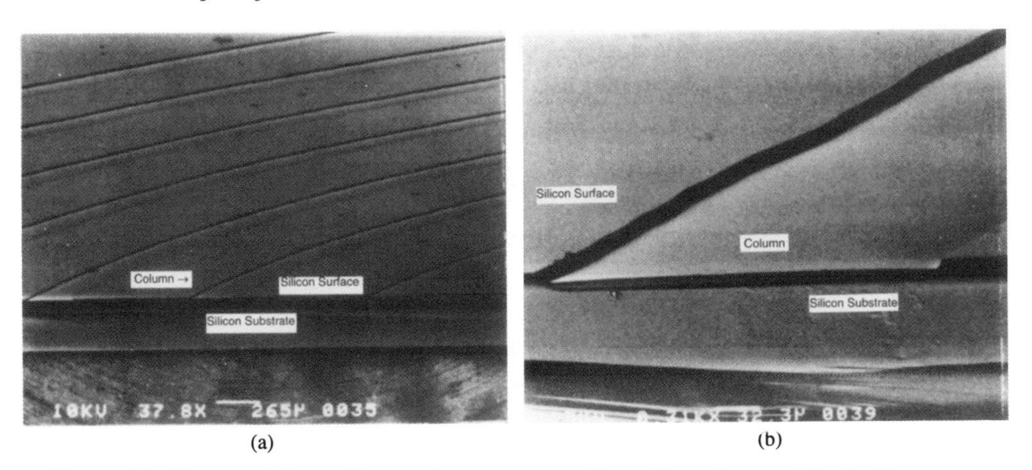

Figure 14. Representative silicon column cross-section of a micromachined silicon gas chromatography system, after the isotropic column etch process (a). Expanded view of the same column cross-section (b). (From [31])

cavity. A novel processing technique was developed to sublime a homogeneous CuPc stationary-phase coating on the GC column walls. The complete MMGC system package is approximately 4 in. square and 2.5 mm thick. The miniature GC column was etched into two different substrates: a silicon wafer and a borosilicate glass cover plate. Two different isothermal etch processes were implemented for each substrate. The results of the silicon wafer etch process are shown in Figure 14.

The actual performance of the MMGC system was evaluated. The MMGC system can separate parts-per-million ammonia and nitrogen dioxide concentrations in less than 30 minutes when isothermally operated. The heat of adsorption of nitrogen dioxide (0.38 eV) on a CuPc thin film (0.2 μm thick) was also independently established. The authors of the project foresee future improvements by increasing the MMGC column's length by decreasing the inter-column spacing without significantly sacrificing yield, resulting in a proportional increase in the separation factor, that is in the theoretical efficiency of the column. Also, the sensitivity of the chemiresistor could improve by incorporating a chemiresistor IC directly within a detector cell. An integral heater (with controller) could also be incorporated as part of the MMGC system using standard IC fabrication techniques, further reducing the requirement for external equipment.

An interesting example of complete "fluidic" microsystem is the so-called "Space Bioreactor", designed and fabricated for experiments in space [32]. Flight models of the bioreactor, mounted on the base of the experiment container are shown in Figure 15.

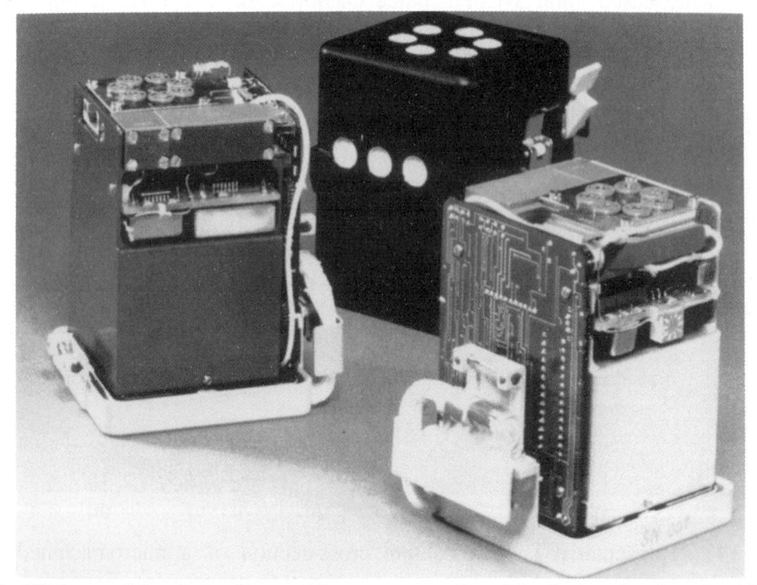

Figure 15. Flight models of the Space Bioreactor produced in Switzerland under European Space Agency contract. The Type II experiment container is the black box at the back. Dimensions are: 84 x 60 x 60 mm^3. Four units of the Space Bioreactor flew on STS-65 in 1994.

Produced by Mecanex S.A., Switzerland, in collaboration with the Institute of Microtechnology at Neuchatel and the Space Biology Group, ETH, Zurich, the Space Bioreactor contains a micropump for fluid circulation and microsensors to monitor vital parameters in the reactor chamber. Four units of the Space Bioreactor flew in space in 1994.

2.3 Microsystems as neural prostheses

Medical applications of microsystems are particularly promising but also rather challenging: in fact biological systems are a very delicate and, at the same time, very aggressive environment, thus requiring great care to interface them [33].

An attractive but extremely difficult field of application of medical microsystems are neural prostheses, that is those devices designed to substitute some sensory functions, or to interface the peripheral or central nervous system, for example for nervous signal recording or for functional nervous stimulation [34]. The next generations of high performance external and internal miniature hearing aids are probably the neural microsystems which will have the wider application and largest market impact in the near future. In this paragraph, however, we shall discuss two different applications of neural microsystems that, although of more remote applicability, are particularly innovative and intriguing. The first relates to visual neuroprosthetics, and the second to regeneration-type neural interfaces.

Some researchers are considering the prospect of providing a functionally useful visual sense to the profoundly blind by electrically stimulating their visual system at the brain cortical level via an array of implanted microelectrodes [35].

The basic components of a cortically based visual neuroprosthetic system will include a video encoder (for example a miniaturized video camera mounted on a pair of eyeglasses, or even a "silicon retina" that performs some of the image preprocessing functions of the human retina and will make the signals compatible with the neurons they are intended to stimulate); and a dense array of implanted stimulating microelectrodes, each able to excite only a small population of neurons in the vicinity of the electrode. The signals generated and preprocessed by the video encoder will be delivered to the implanted electrodes by either a hard-wired, percutaneous connection, or a telemetry link. A scheme of the possible visual prosthetic system, in which microsystem technologies could be exploited both at the "silicon retina" and at the microelectrode array levels to increase miniaturization and functional integration, is illustrated in Figure 16.

The functioning of the proposed visual neuroprosthesis is based on the combination of a number of well documented physiological observations, according to which: a) most forms of blindness are of retinal origin and leave the higher visual centers unaffected; b) the visual pathways are organized in a rational scheme; c) electrical stimulation of neurons in the visual pathway evokes the perception of points of light (called *phosphenes*); the visual system is a highly adaptive neural network with a great capacity to make appropriate correlations between the sensed visual world and the physical world around the observer.

Functionally useful artificial vision could not be possible without stimulating electrodes about the same size as the neurons they are intended to stimulate [36]. New generation electrodes are microsystems built from silicon and containing integral signal processing circuitry.

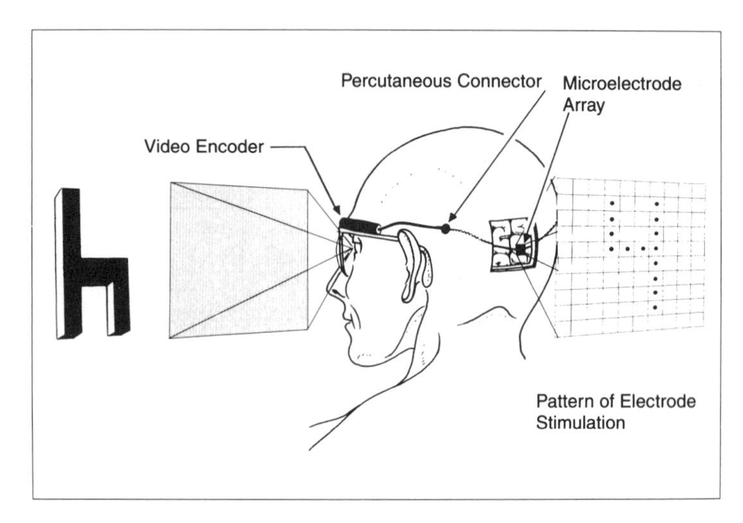

Figure 16. A visual prosthetic system consisting of a video encoder to transform optical images into electrical images, and a microelectrode array implanted in the visual cortex for focal stimulation.(From [35])

The electrode arrays can be inserted into neural tissue, and the active electrode regions can be directly apposed to the neurons they are intended to stimulate to achieve very focal stimulation. However, other important issues must be addressed in the design of a microsystem incorporating a microelectrode array: the integrated circuitry on the array must be hermetically sealed from the corrosive environment of the brain; any parts extending outside of the neural tissues must be as unobtrusive as possible; electrodes must penetrate the visual cortex down to about 1.5-2.0 mm below the cortical surface. The so called "Utah Intracortical Electrode Array" is probably, at the moment, the most advanced example of microsystem for visual neuroprosthetics. The array has a three-dimensional architecture described as a "hair brush", which is illustrated in Figure 17.

The electrode array, designed for use in animal experiments, incorporates 100 needles, 1.5 mm long and projecting from a 200 μm thick silicon substrate. Electrode fabrication has been described in detail in [37]. The electrodes are 80 μm at their base and taper to a sharpened and metalized tip. Each electrode is electrically isolated from its neighboring electrodes by a glass dielectric that surrounds its base. A demultiplexing unit integrated onto the back surface of the array allows electrical access to each electrode. The entire structure, with the exception of metalized tips, is insulated with 1 μm thick dielectric material. A dedicated pneumatically actuated inserter system allows complete insertion of the array with little tissue insult. The biocompatibility of the Utah electrode array has been evaluated in a series of acute and chronic histological and electrophysiological experiments, with most encouraging results.

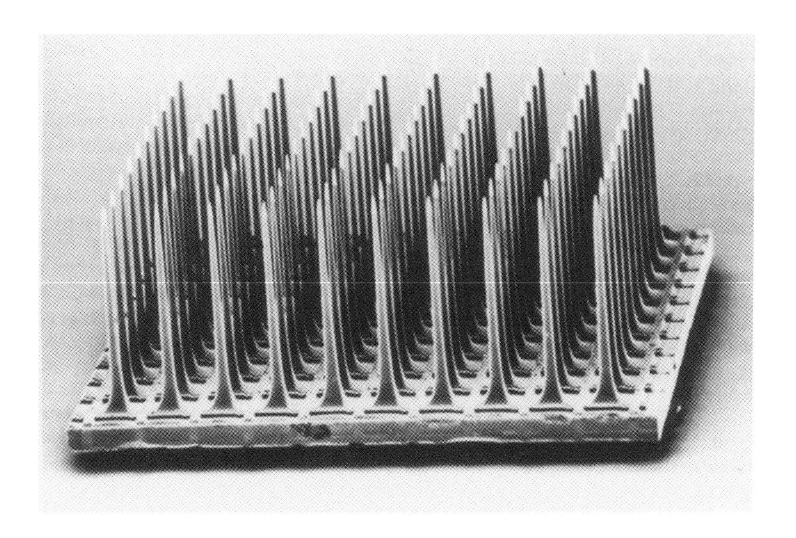

Figure 17. Scanning electron microscopy of an intracortical electrode array, a key component of future microsystems for visual neuroprostheses. (From [35]).

An even more challenging concept of microsystem for visual neuroprosthetics involves the implant of the prosthesis directly on the surface of the retina of a blind person [35]. Electrical stimulation of the retinal ganglion cells by implanted microelectrodes has been demonstrated to evoke phosphenes in human volunteers. The retinal implant of a microsystem integrating photosensors, preprocessing circuitry and stimulating electrodes could eliminate the need for an external video camera, provide a retinal map more rational than a cortical map, and be less traumatic than a cortical implant.

In conclusion, the perspectives of visual neuroprosthetics based on microsystem technologies are very attractive, so that it is not unreasonable to expect that a visual microsystem able to provide a useful visual sense to a blind volunteer could be implanted within the next five years, and commercial visual prosthetic systems could be available after the turn of the century.

A second class of neural microsystems aims at a goal that may be even more challenging than realizing a visual neuroprosthesis: intimately interfacing the nervous system of a living being by means of a chronically implanted regeneration-type device allowing direct bilateral communication between the nervous system and the external world.

So far Information Technology (IT) has dealt mostly with processors and networks totally external to the human operator. In those artificial systems, information is exchanged by wires and cables or, in a few cases, by radio links. The need for developing interfaces that could *intimately* connect these two "worlds", by preserving the wealth of information that can be processed by each separate system, is widely recognized as strategic for the development of the next generation of IT systems [38].

Particularly interesting are *neural interfaces*, that promise to become a most powerful tool for such applications as the control of motor/sensory limb prostheses for amputees, the direct stimulation of limbs in spinal cord injury cases, as an input/output device to directly control machines and computers, and for many other applications.

At present, interfacing with the central nervous system (CNS) or with the peripheral nervous system (PNS) is commonly accomplished via indirect techniques (speech recognition, eye tracking, hand gesture recognition). Direct access to the nervous system should provide sizeable increases in performance over these interfaces and allow for replacement or bypass of defective nervous tissue.

The basic idea of regeneration-type neural interfaces is to integrate a microelectrode array into the neural tissue by allowing the axons to grow through a plurality of via holes through the substrate of the array. Thus the regenerative capacity of the PNS is used to spatially fix the locations of the microelectrodes with respect to the axonal population with which they communicate. A conceptual view of a regeneration-type neural interface is proposed in Figure 18.

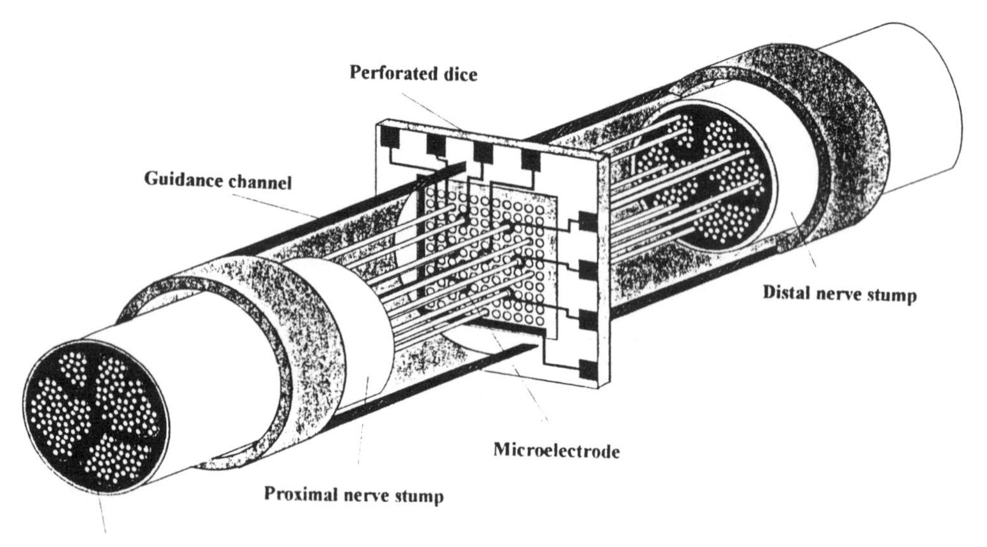

Figure 18. A conceptual scheme of a regeneration-type neural interface.

The first recordings obtained with a silicon substrate regeneration array were reported in the doctoral dissertation of Edell, published in 1980 using potassium hydroxide etching of <110> orientation silicon wafers [39]. Edell etched large (120 μm) slots in a substrate with final thickness of 140 μm. Platinum/tantalum interconnect metalization was applied and contacts in a passivation layer were selectively opened to expose ten 650 μm^2 microelectrode sites, where gold was deposited. Recordings with amplitudes on the order of 150 μV P-P were obtained from such devices implanted in the sciatic nerves of rabbits and represented the action potentials from fairly large numbers of axons in proximity to each microelectrode [40].

The group lead by Gregory Kovacs at the Integrated Circuits Laboratory of Stanford, CA, USA, has carried out what is probably the most complete work described so far in the field of neural interfaces, covering almost all of the technological and biological aspects of this research area. Kovac's work has been aimed at the development of a potentially high resolution, regeneration-type neural interface capable of stimulating and of recording, and fabricated with technologies that are compatible with the inclusion of on-chip active microcircuits. Significant advances were made in the areas of via hole fabrication (using plasma etching), passivation, the design, fabrication and testing of microelectrodes. Different microelectrode geometries were investigated and some of them tested *in vitro* and *in vivo*, in the frog sciatic nerve. Passive (no electrodes) and active electrodes (with no microelectronic circuits on-chip) were tested *in vivo* (in the rat peroneal nerve), leading to a demonstration of recording from, and stimulation of, the nerve at more than one year post-operatively [41]. Preliminary considerations on the use of artificial neural networks to form an adaptive link to the natural network of the patient's nervous system were also proposed. A photograph of a passive interface developed at Stanford by Kovacs and co-workers [6] is shown in Figure 19.

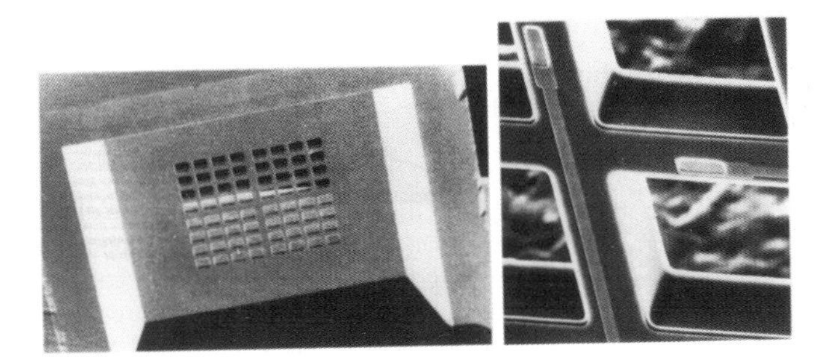

Figure 19. A photograph of the neural interface chip for long-term implantation in nerves developed by Kovacs et al. The chip has been fabricated using anisotropic wet etching to form a thin diaphragm and plasma etching to create holes in it. Size is about 3mm x 3 mm x 0.4 mm. When the chip is implanted, the nerve cells grow through the holes (close-up, right) and establish basic electrical communication with external electronic circuits. (From [6]).

Akin et al. at the University of Michigan have also fabricated sieve electrodes for nerve regeneration applications [42]. A SEM view of one of these sieve electrodes along with its integrated silicon ribbon cable is shown in Figure 20.

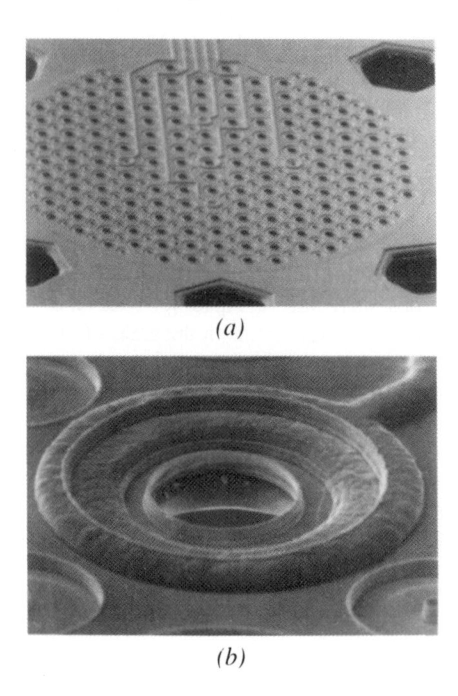

(a)

(b)

Figure 20. A SEM view of a silicon sieve electrode, showing the electrode substrate, the silicon ribbon cables, and the thin diaphragm in the middle with a diameter of 400 μm. The diaphragm contains many through via holes (a). SEM view of one of these holes with a recording site around it (b). (From [36]).

The electrode consists of a thin diaphragm of diameter 400 μm in the middle. The diaphragm contains many holes, through which a nerve can regenerate, and also supports recording sites around the holes for electrical signals traveling through the nerve. Each hole is about 5 μm in diameter and the recording site is made from iridium and has an area of 100 μm^2. An important challenge in the eventual application of implantable neural microsystems is the development of a reliable and flexible interconnect technology to transfer signals from and to the outside world. Many investigators have used different techniques to interconnect to neural microsystems, including insulated wires (typically platinum or gold wires insulated with Teflon or various other polymers), but results have been often poor. The fabrication process developed by the group at the University of Michigan also allows the monolithic fabrication of multi-conductor interconnect silicon ribbon cables with the microsystem [43]. These cables are extremely flexible and elastic and do not break easily. They have been tested under electrical voltages in saline solutions for over two years exhibiting sub-picoamp leakage currents, and carry a large number of conductors.

A project supported by the European Union (ESPRIT/INTER #8897) and involving six different groups from European Countries coordinated by the authors' group, is presently pursuing the goal of developing an implantable neural microsystem comprising various designs of perforated dices, multiple electrodes with on-chip integrated preprocessing circuitry, functional guidance channels for support and fixation of regenerating axons, and interconnection assemblies for bi-directional nervous signal transmission [44]. Special emphasis is given on a light-weight design of the device and on the biocompatibile integration and packaging of the chip [45].

These promising results could ultimate lead to the development of microsystems featured by real physical symbiosis with a living biological system. New perspectives would thus open for the application of the so called "cybernetic" prostheses, artificial devices perceived by the disabled person as the lost natural ones [46].

3. EXAMPLES OF MICROMACHINES

As anticipated in the introduction to this Chapter, the differences between "microsystems" and "micromachines", although conceptually clear, are sometimes rather "subtle" and the two classes of devices partly overlap. In this Section the authors will present and discuss several examples of devices that correspond to the concept of "micromachine" more closely than to the one of "microsystems", along with some cases in which the distinction is not too sharp.

3.1 Silicon micromachined three-dimensional structures

A demonstration of partial overlap between the proposed definitions of microsystem and micromachine is the clever solution to the problem of obtaining real three-dimensional structures and mechanisms starting from conventional silicon micromachined planar structures, proposed simultaneously by two groups of investigators at the University of California at Berkeley and at the University of Tokyo.

At the University of California K.S.J. Pister et al [47] developed a new surface micromachining process which allowed a vast variety of three-dimensional structures to be fabricated. The three-mask process allowed structures to be hinged to the substrate as well as to each other. By fabricating the structures in the plane of the wafer, conventional lithographic techniques could be used to define features with high resolution. These structures could then be rotated out of the plane of the wafer and assembled into three-dimensional designs with detailed features in three dimensions.The hinge fabrication process used by Pister et al. is illustrated in Figure 21.

Several structures were fabricated and tested, including a hot-wire anemometer, a tissue-growth dynamometer, and a gripper.

A photograph of a three-dimensional microfabricated structure after rotation out of the plane of the wafer, developed by Pister after he moved to the University of California at Los Angeles (UCLA) is shown in Figure 22.

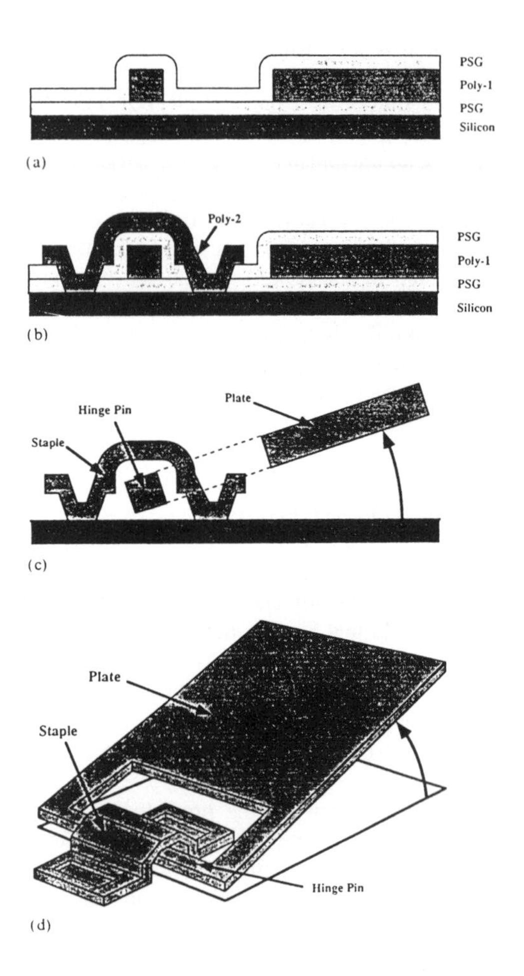

Figure 21. A cross section of a simple hinge during fabrication. (a) A first sacrificial phosphosilicate glass (PSG) layer is deposited, followed by an undoped polysilicon layer (poly-1) that is also patterned. Then the second layer of PSG is deposited. (b) Contacts are etched through both layers of PSG. A second layer of polysilicon (poly-2) is deposited and patterned. (c) The oxide is removed in a sacrificial etch, and the poly-1 layer is free to rotate out of the plane of the wafer. (d) A perspective view of a hinged plate after release. (From [47]).

Figure 22. Three-dimensional structure obtained by fabricating a hinged structure in the plane of the wafer by conventional lithographic techniques, and rotating it out of the plane.

At the University of Tokyo, H. Miura and I. Shimoyama were motivated in their efforts not by a particular interest in new microfabrication technologies, but by a specific application they wanted to investigate, that is to develop real microrobots replicating insects anathomy and behavior. They soon realized that although silicon micromachining technologies enable one to fabricate micromechanical elements and actuators, there are several drawbacks to the structures fabricated by conventional IC process technologies. Specifically, since silicon technology is planar, and the thickness of the material is in the range of only several micrometers, it is difficult to make three-dimensional structures and it is especially difficult to produce large motions out of the plane of the wafer [48]. Friction is also a serious problem in movable microstructures and structures which exhibit surface-to-surface friction like rotational joints are not suitable for microrobotic systems.

The solution devised by Miura and Shimoyama was found examining insects, whose structure, although small and simple, includes sensors, actuators and control systems which work effectively in the microscale. Insects exhibit many interesting features, including external skeletons, elastic hinges, and contracting-relaxing muscles. These characteristics provide clues to designing microrobots. In fact, a 3D mechanism imitating the model of an external skeleton and consisting of rigid plates connected by elastic hinges was devised and implemented with an assembly process based on paper folding. Polysilicon was used to form rigid plates and patches of polyimide to form the elastic hinges. Since the typical values for Young's modulus are 140 Gpa and 3 Gpa, respectively, deformation occurs only at the polyimide hinges. Examples of microcube and prismatic structures which can be obtained by fabricating silicon planar structures and folding them, real 3D structures like those shown in Figure 23 can be obtained.

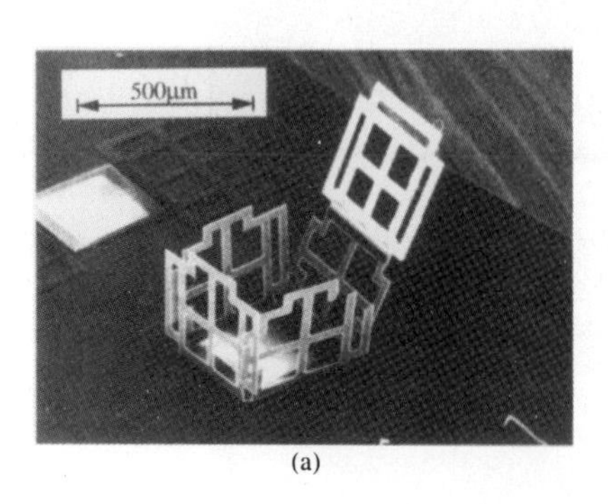

(a)

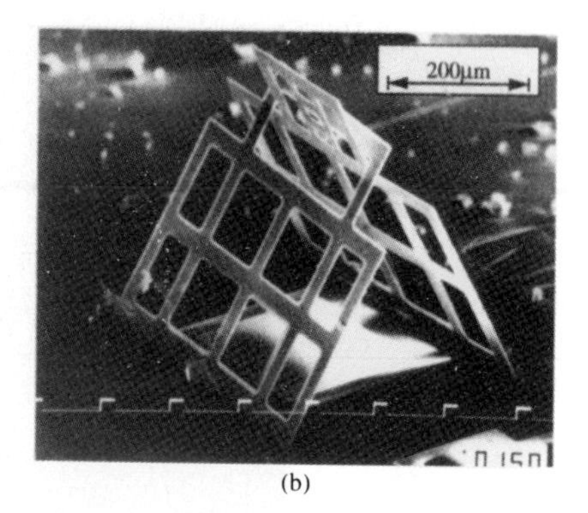

(b)

Figure 23. Examples of three-dimensional structures consisting of polysilicon plates and polyimide hinges: (a) Microcube; (b) Prismatic structure. (From [49]).

The external skeleton model described above is easily actuated by electrostatic force. Microsized models of wings can also be fabricated, and may eventually lead to the development of flying micromachines. A paper model of a flapping mechanism and a silicon-based microsized model of the same mechanism are illustrated in Figure 24.

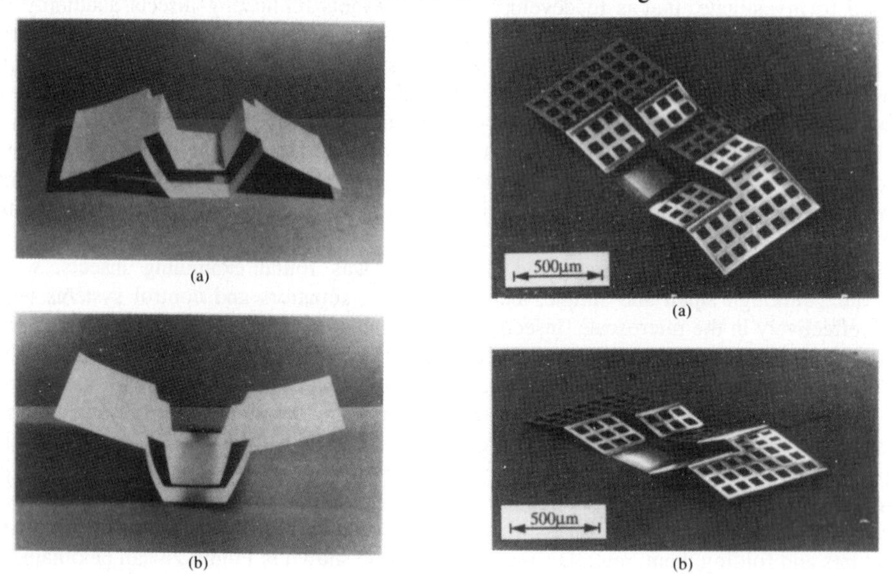

Figure 24. Paper model of a flapping mechanism (left), and microsized model of the same flapping mechanism consisting of polysilicon plates and polyimide hinges (right). (From [49]).

3.2 Self-driven functional microcar

A second example of micromachine, that clearly illustrates a different conceptual approach than the one discussed in the Section on microsystems, is the millimiter-size microcar developed by Nippondenso Co., Ltd., in Japan. The microcar is a 1/1000-scale model of the first Toyota's passenger car, and was designed and fabricated as a conceptual demonstrator of a micromachine and to identify important future technical problems in the field of micromachine technology. The microcar, shown in Figure 25, is a self-driven fully functional model car.

Figure 25. The 4.8 mm long, 1/1000 model of microcar with an embedded electromagnetic motor. The microcar was manufactured by Nippondenso Co., Ltd. as a self-driven fully functional model car and as a conceptual model of a micromachine.

The solid body of the microcar was three-dimensionally machined by a NC machining center; other components were individually produced either by NC machining, by micro electrodischarge machining and by IC processing, and then assembled to the body. As driving source a microstepping motor having a structure similar to those used in quartz watches was selected because of its low applied voltage (more compatible with IC circuits than, for example, electrostatic and piezoelectric micromotors) and relatively large torque. The car propelled by the motor did not move smoothly in the wheel rotation mode, but moved very smoothly at a speed of 0.02 m/s in the wheel vibration mode [49].

3.3 Micromachines for medical applications

Medical applications are considered as an elective field for micromachines. Both for diagnostics, and for therapy and intervention, instruments with improved performance, multi-functionality and miniature size are particularly desirable. In particular, new domains of application are being disclosing for micromechatronics and micromachines in the fields known as Minimal Invasive Therapy (MIT) and Minimal Invasive Surgery (MIS) [articolo nostro Mechatronics]. The potential advantages provided by MIT to the patient, the medical doctor and the society are tremendous in terms of quality of care, shorter hospital stay and reduction of complications [mettere o libro Taylor, o articolo Buess], and in fact such techniques are already well established in current clinical practice in important areas as laparoscopy and arthroscopy.

A conceptual representation of a possible future scenario of an endoluminal manipulator system intended for local tumor diagnosis, therapy [50] and is shown in Figure 26.

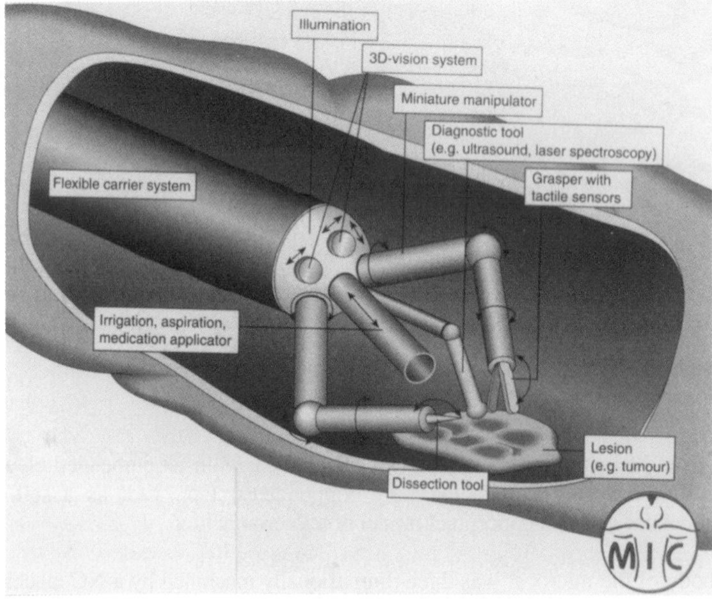

Figure 26. Scenario of an endoluminal manipulator system for local tumor surgery. The multifunctional head includes a manipulator module, a vision module and a diagnostic module. (From [50]).

Planned for detection and eradication of bowel tumors, the endoluminal manipulator system consists of two major parts, a flexible carrier system for reaching the operating area via the anus with minimal trauma, and a multifunctional head with diagnostic and therapeutic modules at the front of the flexible carrier.

The endoluminal manipulator system concept described above, although technically quite difficult to implement, is still an evolution of an existing technique (colonscopy). An even more ambitious scenario has been conceived in Japan, as one of the objectives of the ten year japanese Micro Machine Project, which started in 1991 [51]. The scenario involves the development of microminiature, remotely controlled tools for diagnostics and therapy in such districts very difficult (or impossible, with currently available medical instrumentation) to reach, such as the pancreatic and biliary ducts (average diameter 5 mm) and the brain blood vessels (average diameter 1 mm). This scenario is depicted in Figure 27.

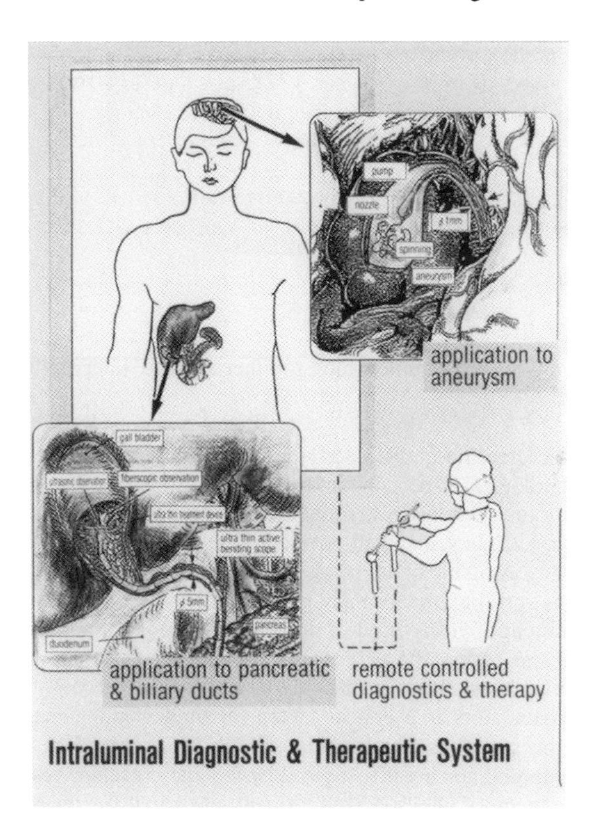

Figure 27. Scheme of the intraluminal diagnostic and therapeutic system based on micromachine technologies and being developed in Japan by the Micro Machine Project.

An even more complex configuration of endoluminal surgical tool includes semi-autonomous micromachines which leave the main carrier system (for example a catheter) and perform various tasks such as the injection of drugs or other substances for diagnosis and therapy, or various local treatments. A scheme of advanced surgical tool for endoluminal operation, comprising two tethered micromachines is depicted in Figure 28.

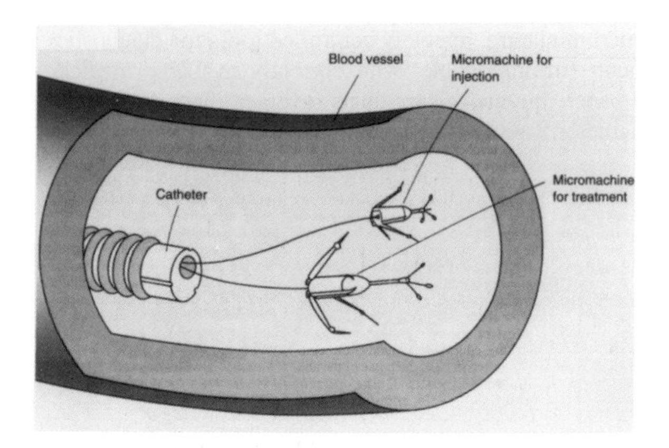

Figure 28. Scheme of an endoluminal tool for therapy and intervention comprising two tethered micromachines.

Implementing the micromachines discussed above requires the availability of many new components and technologies, especially in the fields of new actuators, new mechanical components and new tools. (Microsensors and sensory microsystems like those described in Section 1 will also be necessary. and will be incorporated into many future surgical tools). Some of them are already available or are being developed.

Dexterous behavior is very important in any intraluminal tools. Manually actuated and cable transmitted motion, although widely used in ordinary endoscopes, exhibits severe limitations when the diameter of the endoscope decreases, or when its kinematic degrees of freedom increase to allow reaching sites otherwise difficult or impossible to access.

Designing new microactuators is a critical factor for implementing real microsystems and micromachines such the ones described previously for MIT and MIS. In fact, many new actuators have been proposed and are investigated by the micromachine research community in the recent past. Shape memory alloys (SMA) are very attractive for the actuation of medical devices of different kinds and for the actuation of steerable endoscopes in particular [3]. The scheme of an endoscope incorporating SMA coil springs actuators at the distal part capable of providing steerable motion of the tip in order to facilitate insertion along narrow vessels is depicted in Figure 29 [52].

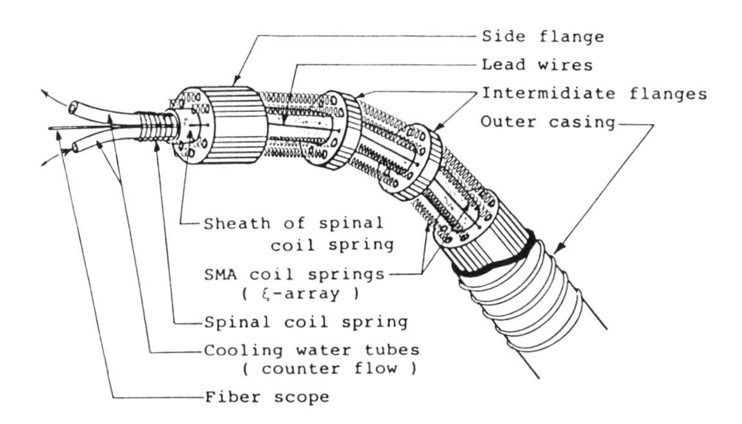

Figure 29. Scheme of the shape memory alloy actuated steerable tip of an endoscope. (From [52]).

SMA actuators are usually heated electrically, and may be cooled by water, as in the example of Figure 29, or by air or other means, to increase their operating frequency if required by the intended application. Water control was used both for heating and cooling tip-mounted SMA microactuators in a steerable active catheter developed in the authors' laboratory [53]. An example of electrically controlled, SMA-actuated steerable tip developed by the authors for incorporation in a new semi-autonomous robotic colonscope [54] is shown in Figure 30

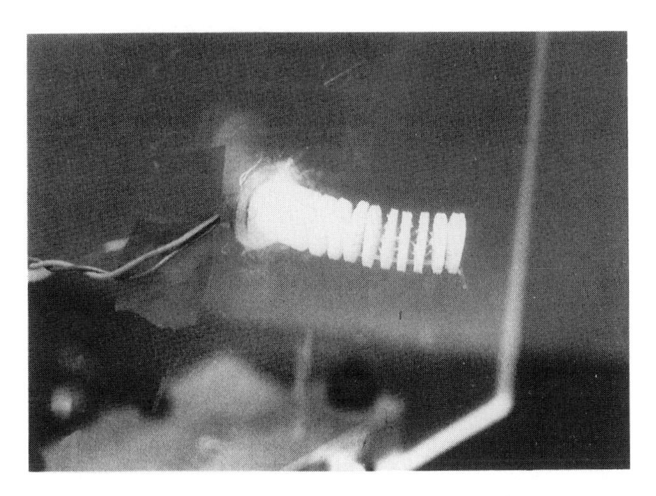

Figure 30 Photograph of a SMA-actuated steerable tip developed in the authors' laboratory.

The actuator is based on 0.15 mm diameter SMA wires, arranged in order to generate the largest possible bending angles for the same tip length. Bending angles of ±90° were obtained for a tip diameter of 8mm and length of 30mm.

An example of how a SMA-actuated catheter can successfully manage difficult situations occurring in medical catheterization or endoscopy is shown in Figure 31, for the case of a 1,8mm diameter catheter developed at Nagoya University [55].

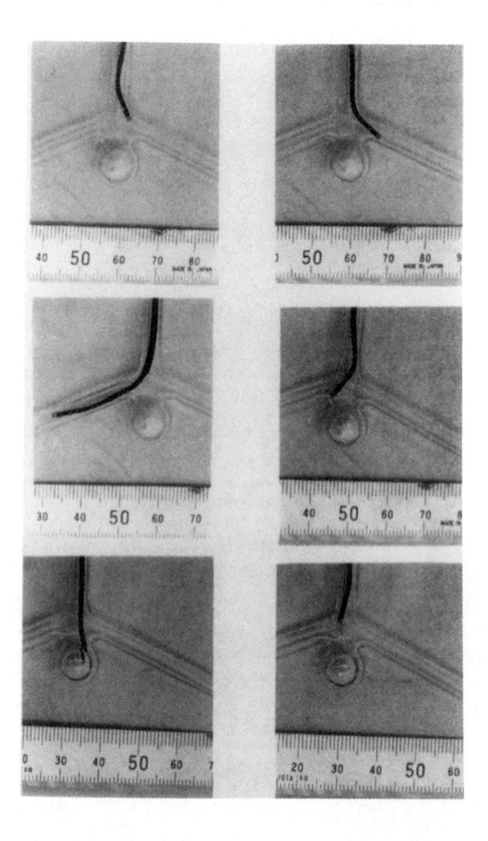

Figure 31. In vitro demonstration of the ability of a SMA-actuated steerable-tip catheter to deal with difficult anatomical configurations such as bifurcations and arrays. (From [56]).

SMA-actuators can be miniaturized to extremely small size and controlled by means of non electrical means as in the case illustrated in Figure 32 for a microcoil actuator developed by Mitsubishi Cable Industries, Ltd. for incorporation in novel catheter [56].

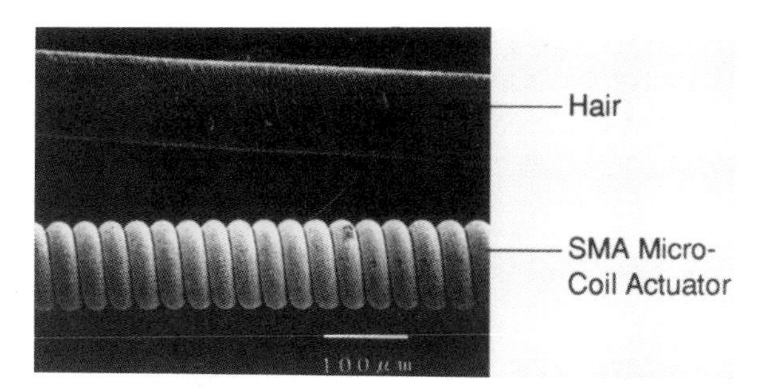

Figure 32. SMA microcoil actuators developed by Mitsubishi Cable Industries Ltd. for the control of steerable microendoscopes. (From [56]).

The diameter of the SMA wire is 30 microns, the coil outer diameter is 30 microns, the shape recovery force is 35 mN at a shearing strain of 3% in the wire. Further size reduction is sought to a coil outer diameter of 75 microns.

Olympus Optical Co., Ltd. is also investigating micro bending mechanisms based on SMA microactuators for use in future microendoscopes. An interesting feature of the Olympus

Figure 33. Magnetic variable reluctance micromotor and microgears fabricated by sacrificial LIGA technique. (From [6])

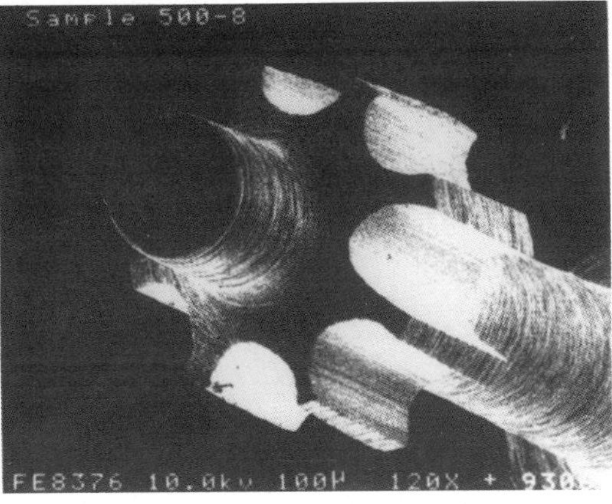

Figure 34. Microgear manufactured by mechanical micromachining (grinding). (From [57]).

approach is the technique used to heat the SMA coil microactuators. Instead of electricity (that may be dangerous in medical endoscopic applications), light carried by optical fibers to the tip is used to illuminate a photo-thermal material that converts light into heat and promotes the phase conversion of the SMA microcoil that is required for generating force and displacement.

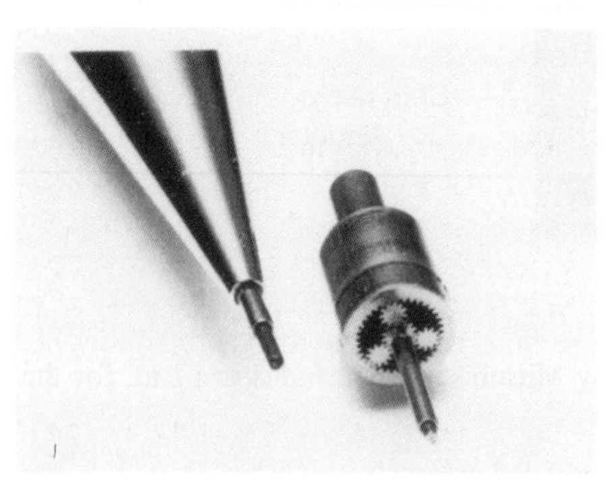

Figure 35. Microgear reductor fabricated by Toshiba Corporation by precision mechanical machining and assembly. (From [58]).

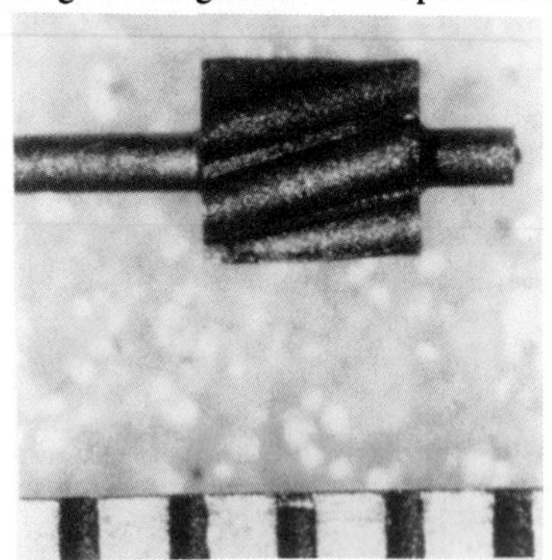

Figure 36. Micro air turbine rotor fabricated by micro electro-discharge machining (μEDM) (From [59]).

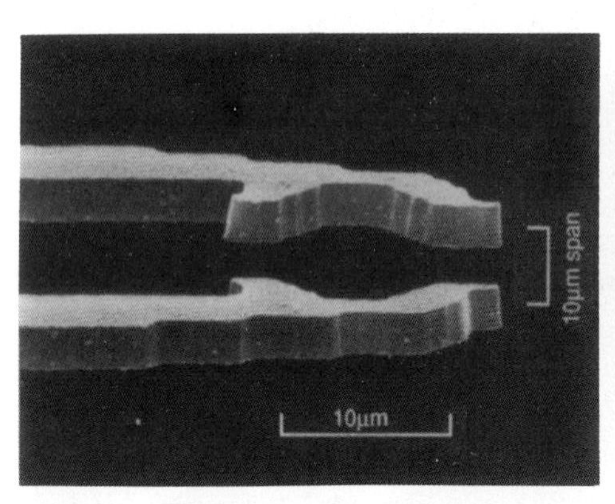

Figure 37. Polysilicon microgripper fabricated using surface micromachining on a silicon wafer. The microgripper is actuated by electrostatic comb-drive actuation ([60]).

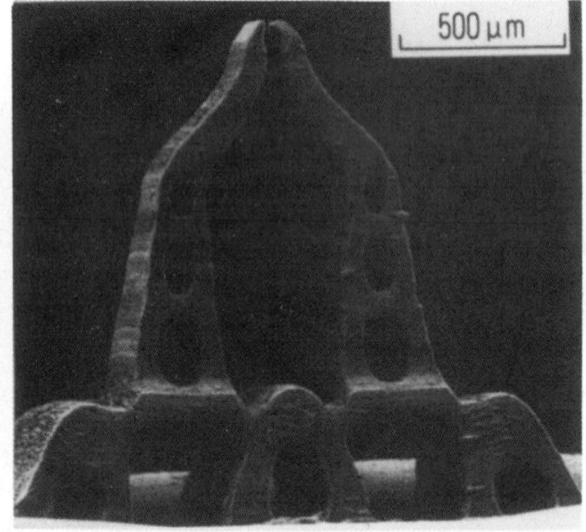

Figure 38. Microgripper fabricated by high-resolution photoforming process using photopolymerizing resin and argon laser (microstereolithography). A piezoelectric actuator mounted in the substrate drives the gripper tips. (From [61]).

The miniaturization of medical intraluminal instrumentation requires additional components and tools, such as the microgears, microgear reductors, microturbines and microgrippers. Examples of such components, whose fabrication technologies range from silicon micromachining to less “conventional’ (in the microsystems domain) technologies such as LIGA, microgrinding, microEDM and micro stereolithography, are illustrated in Figures 33-38.

The current evolution of medical instrumentation for MIS and MIT involves the incorporation of miniature components into traditional carriers, like catheters or endoscopes. In some cases, hower, like in gastroscopy and colonscopy, traditional instruments are somewhat large and stiff, and examination is very unconfortable for the patient. Methods for overcoming the limitations of current endoscopes have been investigated by many groups [62] [63] [55] [64] [65], including the authors’one [53]. K. Ikuta at Nagoya University developed not only the first active endoscope actuated by SMA actuators, but also a “hyper active endoscope” for abdominal surgery for which the degrees of freedom were increased using the so-called cybernetic actuators, based on piezoelectric drive [66].

An interesting example of robot endoscope designed for performing colonscopy by semi-autonomous navigation along the intestine is shown in Figure 39.

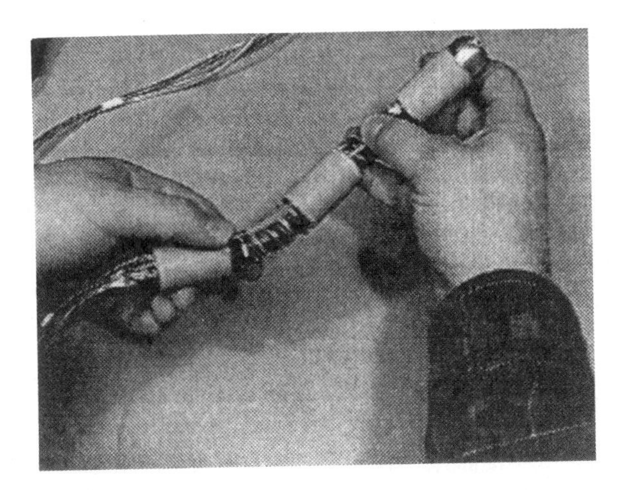

Figure 39. Photograph of a robotic endoscope prototype. (Three gripper/two extensor configuration). (From [65]).

Developed by J. Burdick and co-workers at the California Institute of Technology, the robotic endoscope is designed to directly access and visualize, in a minimally invasive manner, the entire gastrointestinal tract [65]. The long term goal of the project is to produce a surgeon guided robot that can semiautonomously locomote in the small bowel to perform medical diagnostic procedures. The prototype shown in Figure 39 is an endoscopic robot with a trailing cable which consists of: electrical wiring for control signals; tubing to connect the pneumatic actuators to high and low pressure sources; and an optical fiber bundle for illumination and imaging of the area in front of the robot system. Fluid power actuation was chosen because conventional endoscopic procedures require carbon dioxide gas, saline solution, and partial vacuum for insufflation, irrigation, and suction of gastrointestinal lumina. Hence it is convenient and efficient to use these fluids as power sources for locomotion. Solenoid valves are located within the robotic endoscopes which, in order to propel itself, employs mechanisms along its length which can be described as "grippers" and "extensors". The primary purpose of the grippers, or traction devices, is to provide traction against the lumen wall by expanding radially outward. The extensors provide extensibility between the grippers, i.e. they cause the mechanism to locally expand or contract in length.

In the authors' laboratory, a similar robotic endoscope has been developed, also making use of an inchworm-like locomotion principle, but exploiting new SMA-actuated pneumatic microvalves in order to miniaturize the robot, eventually to the size (a few millimeters diameter) of a real micromachine [54].

3.4 Micromachines for inspection and maintenance of industrial plants

A field of application that attracts large interest and poses technical problems not much different than the one encountered in the area of medical endoscopy is the inspection and maintenance of industrial plants.

The japanese Micro Machine Project has identified the development of a maintenance system for power plant as one of its main objectives. The tasks to be performed by the system are to inspect and repair defects within the heat exchangers and piping of electric power generating plants. The proposed concept is very advanced and involves the development of four different modules, each representing extremely well the notion of a fully functional micromachine [66]. The scenario of the maintenance system and the schemes of the four modules are depicted in Figure 40.

Each micromachine has its own functions and, to this aim, it incorporates advanced mechanisms, actuators, sensors and energy generators. The 2 mm diameter microcapsule is a self-powered micromachine which includes a steering mechanism, a microdynamo, a microgyroscope, and a flaw detector. The mother ship is the main module. It has a diameter of about 10 mm and includes a microbattery, an electrostatic driving mechanism, pneumatic clampers, and an optical scanner. The wireless inspection module incorporates a piezoelectric motion drive, an inchworm drive mechanism, a CCD micro visual sensor, an ultrasonic microsonar, a microwave transmitter, a photoelectric converter, and a micro photo spectrometer, all in a 2.5 mm diameter structure

Finally, the operation module with wires contains in a 2.5 mm diameter structure gear train locomotion mechanisms, a multidegree of freedom manipulator, a device for photo stimulation, and a photo electric generator and booster. A scheme of the wireless inspection module is illustrated in Figure 41.

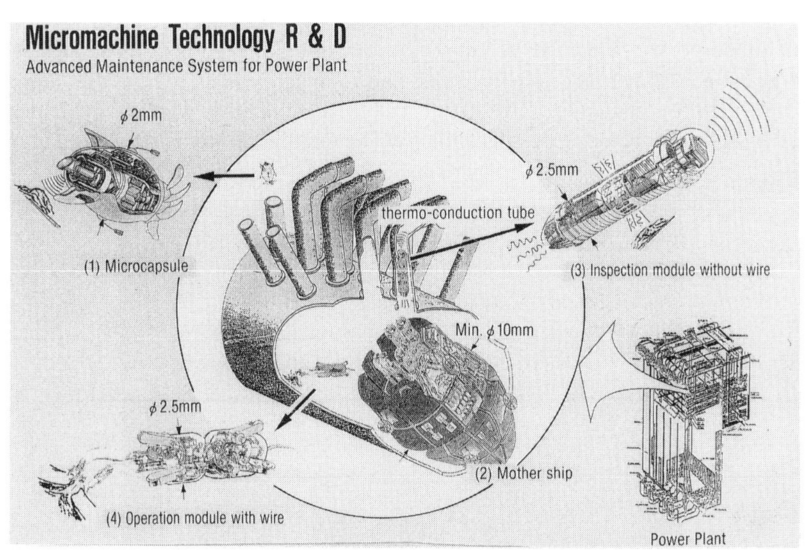

Figure 40. Scheme of an advanced maintenance system for power plants. The development of the system, which comprises four different micromachines: a microcapsule; a mother ship; a wireless inspection module; and an inspection module with wires, is investigated in the framework of the japanese Micro Machine Project. (From [66]).

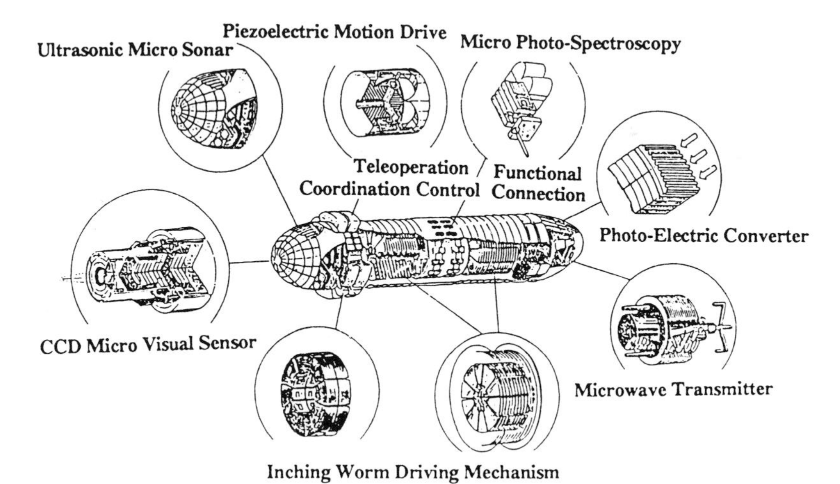

Figure 41. Scheme of the wireless inspection module, which travels from the mother ship to the location of defects, performing precise inspection and analysis of defects. Data on defects are then transmitted through the mother ship to the control center. (From [66]).

One of the components of the microcapsule is a microgenerator with an outer diameter of 1.2 mm, which incorporates a number of innovative technologies. The microgenerator contains a high energy cylindrical permanent magnet thin film rotor fabricated by 2 μm thin film and with a maximum energy product of 210 kJ/m^3; and a high density winding cylindrical stator fabricated by semiconductor process technology. The microgenerator is shown in Figure 42.

Figure 42. 1.2 mm diameter microgenerator for micromachines. (From [67]).

As an intermediate step towards the development of the inspection microcapsule, Nippondenso Co., Ltd. fabricated a micro inspection machine measuring 5.5 mm in diameter and with a mass of 1 g. The micromachine, shown in Figure 43, supports an eddy current sensor of 2.2 mm diameter and a piezoelectric actuator that measures 2 x 3 x 9 mm^3, and exploits a proprietary processing technology to fabricate the body of the extremely thin three-dimensional shell structure.

Figure 43. Micro inspection machine, 5.5 mm diameter, developed by Nippondenso Co., Ltd. (From [68]).

3.5 Micromachines for micromanipulation

In order to fabricate real micromachines (and microsystems as well) it is important not only to develop microcomponents and system integration methods and techniques, but also to devise appropriate assembly systems. In the last 100 years small objects, such as living cells (10 to 100 µm diameter) have typically been manipulated using linear translation stages (often with roller bearing slides) stacked in a serial x-y-z configuration to provide 3-axis motion. Movements of the stages are usually effected by micrometers which in turn are driven either by hand or by some sort of electromagnetic rotatory motor (e.g. stepping motor), or by piezoelectric actuators. A microrobot having two high-performance parallel drive limbs has been developed by I. Hunter et al. for manipulation, mechanical testing and assembly of very small objects such as living cells [69]. The end-points of each limb moved in overlapping spherical workspaces of 1 mm diameter with minimum open- and closed-loop movements of 1 nm and 10 nm, respectively.

An ultra precise manipulation system has also been investigated at the University of Tokyo [70]. A number of tools are necessary for the operations required for micro assembly. Some tools are shown in Figure 44.

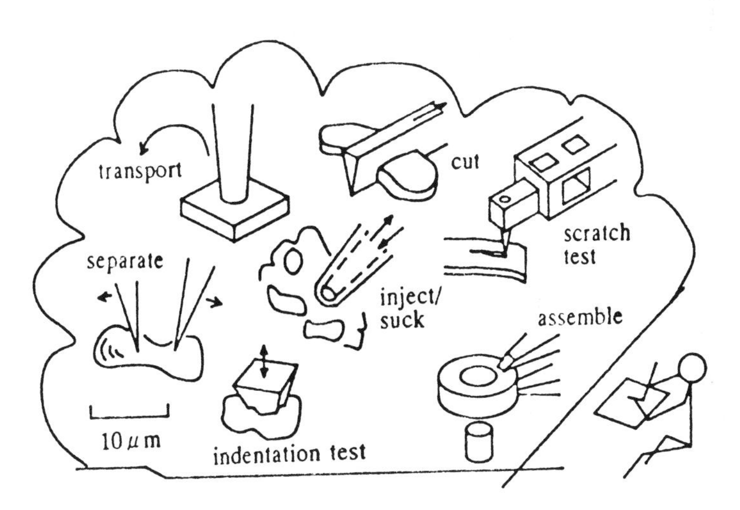

Figure 44. Tools necessary for assembly operations in a future nanorobot system (From [71]).

Three key technologies have been identified as necessary in order to realize a "nano robot" system for micro assembly tasks: a) a method for three-dimensional observation of very small objects; b) techniques for grasping and holding very small objects, and for sensing very small forces; c) a technology for generating nano displacements. Whereas optical, SEM and even

AFM microscopes are investigated for observation, the problem of grasping microparts has attracted much research efforts recently.

A technique for grasping miniature objects is micromechanical manipulation. A micromanipulator making use of two tiny needles like "chopsticks" for holding small mechanical parts is shown in Figure 45.

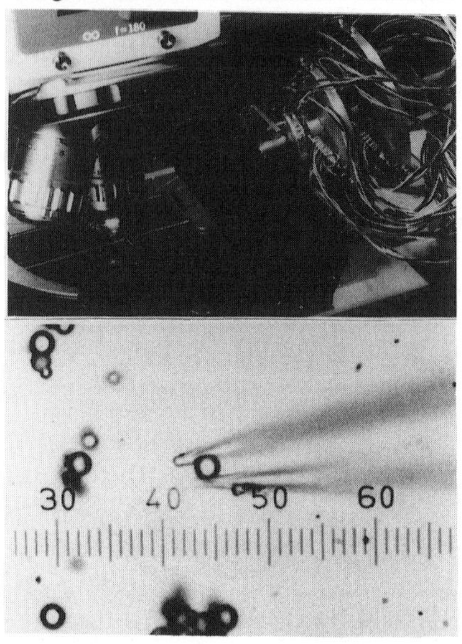

Figure 45. Photographs of a micromanipulator system (upper part) used to manipulate two needles in order to grasp tiny particles (lower part). (From [72]).

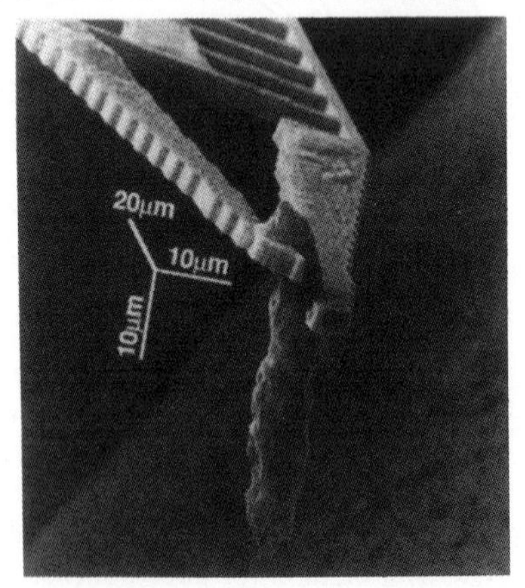

Figure 46. SEM picture of a one-celled protozoa, a euglena, being held by a microgripper. The euglena is 40 μm long and 7 μm in diameter. (From [60]).

Micromanipulation can also be performed using silicon micromachined grippers. The grasping of a biological cell by a microgripper is shown in Figure 46 [60].

Additional manipulation methods using electrical [73][74] or ultrasonic [74] fields have also being proposed. Laser manipulation has also attracted attention as a handling method for small objects [76]. This method utilizes radiation pressure produced when a laser beam is reflected or diffracted by materials to remotely operate fine particles of micrometer dimensions. This technique is easy to use and has excellent space control characteristics. Figure 47 shows a block diagram of a laser system used successfully to hold fine structure objects (polystirene spheres of 3 microns diameter) by laser scanning.

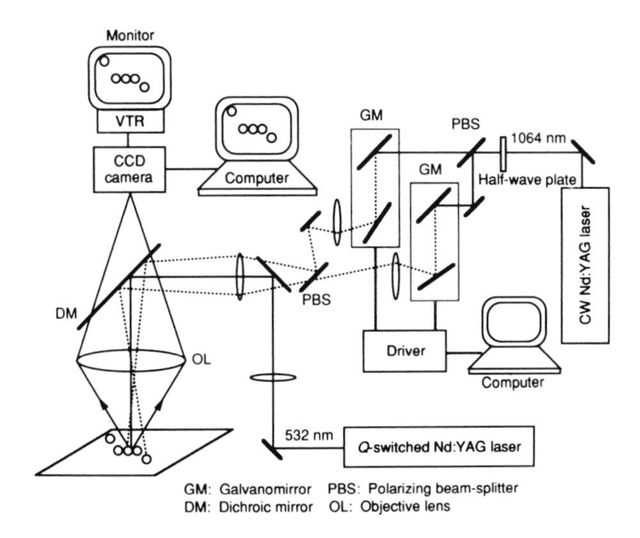

Figure 47. Ultra-precision laser scanning manipulation system for holding micron-size objects. (From [77]).

4. CONCLUSIONS

In this Chapter several examples of microsystems and micromachines developed by many research groups around the world in the last ten years have been presented and discussed. The variety of applications demonstrates the importance and strategic value of the field of micromechatronics, and the deep and growing interest of researchers in academy, research institutes and industries.

The conclusion that can be drawn from the analysis of the results obtained so far and of the work in progress is that the technologies for miniaturization and integration, and the overall field of micromechatronics, are very likely to become a key (or even *the key*) area of the next industrial revolution. However, it is quite clear that the time required before these technologies and applications fully impact the market will be rather long.

Furthermore, an important lesson that can be learnt is the importance of multidisciplinary research for microsystems and micromechatronics. As outlined in the introduction to this Chapter, the strategic value and the implications of the field of micromechatronics can be perceived, in the authors' opinion, only by observing it from a broad perspective (like using a "wide-angle lens"), rather than looking separately at each individual technology and application, (like using a sort of "magnifying lens"). In this sense, the impact of miniaturization and integration, whether achieved by planar or non planar, technologies, by silicon or non-silicon materials, by micromehanical, mechanical or even biological technologies, by microsystems or micromachines, may have a truly revolutionary impact on technology evolution and on industry in the next decades.

REFERENCES

1. Proceedings of the IEEE Workshops on Micro Electro Mechanical Systems (MEMS), 1987-1996.
2. Proceedings of Micro System Technologies (MST), VDI-Verlag GmbH, Berlin, Germany, 1989-1994.
3. Special Issue on "Micro-Machine System", J. Robotics and Mechatronics, 3, 1 (1991) 1.
4. Proceedings of the International Symposia on Micro Machine and Human Science, Nagoya, Japan, 1990 -1995.
5. Micromachines in Japan, Techno Japan, 24, 4 (1991) 8.
6. J. Bryzek. K. Petersen and W. McCulley, IEEE Spectrum (May 1994) 20.
7. S. Valette, Journal of Micromechanics and Microengineering, 5, 2 (June 1992) 74.
8. M.A. Mignardi, Solid State Technology (July 1994) 63.
9. J. Younse, IEEE Spectrum (November 1993) 27.
10. S.C. Terry, J.H. Jerman and J.B. Angell, IEEE Trans. Electron Devices, ED-26, 12 (1979), 1880.
11. K.E. Petersen, IEEE Trans. Electron Devices, ED-26, 12 (1979), 1918.
12. F.C.M. van de Pol and J. Branebjerg, Proceedings of Micro System Technologies '90, VDI-VDE, 10-13 September 1990, Berlin, Germany 799.
13. P. Gravesen, J. Branebjerg and O.S. Jensen, J. Micromech. Microeng., 3 (1993), 168.
14. A. van den Berg and P. Bergveld (eds.), Micro Total Analysis Systems, Kluwer Academic Publishers, Dordrecht, The Netherlands, 1995.
15. M. J. Zdbeblick, R. Anderson, J. Jankowski, B. Kline-Schoder, L. Christel, R. Miles and W. Weber, Proceedings of "Actuator 94", June 15-17, 1994, Bremen, Germany, 56.
16. J. Fahrenberg, D. Maas, W. Menz, and W. K. Schomburg, Proceedings of Actuator'94, June 15-17, 1994, Bremen, Germany, 71.
17. M. Esashi, Sensors and Actuators, A21-A23 (1990) 161.
18. F. C. van de Pol, H. T. G. van Lintel, M. Elwenspoek and H. J. Fluitman, Sensors and Actuators A21-A23 (1990) 198.
19. J. G. Smits, Sensors and Actuators A21-A23 (1990) 203.

20. R. Zengerle, A. Richter, F. Brosinger, A. Richter and H. Sandmaier, Technical Digest of Transducer '93, Yokohama (1993) 106.
21. R. Rapp, W. K. Schomburg, D. Maas, J. Shultz and W. Stark, Sensors and Actuators A, 40 (1994) 57.
22. A. Richter, A. Plettner, K. A. Hofmann and H. Sandmaier, Proceedings of MEMS '91, January 30-February 2, 1991, Nara, Japan, 271.
23. I. Mizoguchi, M. Ando, T. Mizuno, T. Takagi, N. Nakajima, Proceedings of MEMS '92, Febrary 4-7, Travemünde, Germany, 1992, 31.
24. S. Hattori, T. Fukuda, S. Nagamori, H. Matsumura, Y. Katsurayama, H. Katayama, Proceedings of Second Int. Symp. on Micro Machine and Human Science, October 8-9, 1991, Nagoya, Japan, 113.
25. M. C. Carrozza, N. Croce, B. Magnani and P. Dario, J. Micromech. Microeng., 5 (1995) 177.
26. V. Gass, B.H. van der Schoot, S. Jeanneret and N.F. de Rooij, Sensors and Actuators A, 3 (1993) 214.
27. B. Büstgens, W. Bacher, W. Bier, R. Ehnes, D. Maas, R. Ruprecht and W. K. Schomburg, Proceedings of "Actuator 94", June 15-17, 1994, Bremen, Germany, 86.
28. R. Garabedian, C. Gonzales, J. Richards, A. Knoesen, R. Spencer. S.D. Collins and R.L. Smith, Sensors and Actuators A, 43 (1994) 202.
29. B. Liedberg, C. Nylander and I. Lundstrom, Sensors and Actuators, 4 (1983) 299.
30. Micro Parts GmbH, Dortmund, Germany.
31. R.R. Reston and E.S. Kolesar, J. Microelectromechanical Syst., 3, 4 (1994) 134.
32. I. Walther, B. H. van der Schoot, S. Jeanneret, P. Arquint, N. F. de Rooij, V. Gass, B. Bechler, G. Lorenzi, A. Cogoli, Journal of Biotechnology, 38 (1994) 21.
33. P. Dario and M.C. Carrozza, Proceedings of Sixth Int. Symp. on Micro Machine and Human Science, October 4-6, 1995, Nagoya, Japan (1995) 57.
34. W.F. Agnew and D.B. McCreery (eds), Neural Prostheses, Prentice Hall, Englewood Cliffs, NJ, U.S.A., 1990.
35. R.A. Normann, IEEE Engineering in Medicine and Biology Magazine (Jan/Feb 1995) 77.
36. K. Najafi, IEEE Engineering in Medicine and Biology Magazine (June/July 1994) 375.
37. K. E. Jones, P. K. Cambell, R. A. Norman, Annals of Biomedical Engineering, 20 (1992) 423.
38. G. Kovacs, Technical Report No. EO73-12, Integrated Circuits Laboratory, Stanford University, CA, 1990.
39. D. J. Edell, Doctoral Dissertation, U. C. Davis, 1980
40. D. J. Edell, IEEE Trans. on Biomedical Eng. BME-33, 2. (February 1986) 203.
41. G. T. A. Kovacs, C. W. Storment, M. Halks-Miller, IEEEE Trans. on Biomedical Eng. 41, (1994) 567.
42. T. Akin, K. Najafi. R.H. Smoke and R.M. Bradley, IEEE Trans. Biomedical Eng., 41 (1994) 305.
43. J.F. Hetke, J.L. Lund, K. Najafi, K.D. Wise and D.J. Anderson, IEEE Trans. on Biomedical Eng., 41, 4 (1994) 314.
44. J-U. Meyer, H. Beutel, E. Valderrama, E. Cabruja, P. Aebischer, G. Soldani, P. Dario, Proceedings of MEMS '95, January 29-February 2, 1995, Amsterdam, The Netherlands, 358.

45. P. Dario, M. Cocco, G. Soldani, E. Valderrama, G. Cabruja, J-U. Meyer, T. Giesler, H-J. Beutel, H. Scheithauer, M. Alavi, V. Burker, Proceedings of Micro System Technologies, '94, October 19-21, 1994, Berlin, Germany, 417.
46 P. Dario, E. Guglielmelli, V. Genovese, M. Toro, Robotics and Autonomous Systems (1996) 1.
47. K.S.J. Pister, M.W. Judy, S.R. Burgett and R.S. Fearing, Sensors and Actuators A, 33 (1992) 249.
48. K. Suzuki, I. Shimoyama and H. Miura, J. Microelectromechanical Syst., 3, 1 (1994) 4.
49. A. Teshigagara, M. Hisanaga, T. Hattori, Proceedings of the Third International Symposium on Micro Machine and Human Science, October 14-16, 1992, Nagoya, Japan, 137.
50. M. O. Schurr and G. Buess, MST News, 13, (July 1995) 1.
51. H. H. Narumiya, Proceedings of IARP Workshop on Micromachine Technologies and Systems, October 26-28, 1993, Tokyo, Japan, 50.
52. K. Ikuta, M. Tsukamoto and S. Hirose, Proceedings of 1988 IEEE Int. Conf. on Robotics and Automation, April 24-29, 1988, Philadelphia, USA, 427.
53. P. Dario, R. Valleggi, M. Pardini, A. Sabatini, Proceedings of MEMS '91, January 30-February 2, 1991, Nara, Japan, 171.
54. L. Lencioni, B. Magnani, M. C. Carrozza, P. Dario, B. Allotta, M. G. Trivella, A. Pietrabissa, Proceedings of the First AISEM Conference, February 19-20, 1996, Roma, Italy, World Scientific Publishing, Singapore, 1996.
55. T. Fukuda, S. Guo, K. Kosuge, F. Arai, M. Negoro and K. Nakabayashi, Proceedings of IEEE Int. Conf. on Robotics and Automation, May 8-13, 1994, San Diego, CA, 2290.
56. Micromachine, Micromachine Center (MMC), Tokyo, Japan, No. 10, March 20, 1995, 6.
57. Micromachine, Micromachine Center (MMC), Tokyo, Japan, No. 10, March 20, 1995, 16.
58. Micromachine, Micromachine Center (MMC), Tokyo, Japan, No. 7, May 30, 1994, 9.
59. Micromachine, Micromachine Center (MMC), Tokyo, Japan, No. 10, March 20, 1995, 15.
60. C. J. Kim, A. P. Pisano and R. S. Muller, J. of Microelectromechanical Syst., 1,1 (1992) 31.
61. Micromachine, Micromachine Center (MMC), Tokyo, Japan, No. 9, November 30, 1994, 15.
62. K. Ikuta, Proceedings of IROS '88, Oct. 31-Nov. 2, 1988, Tokyo, Japan, 9.
63. R. H. Sturges and S. Laowattana, Proceedings of IEEE Int. Conf. on Robotics and Automation, April 9-11, 1991, Sacramento, CA, 2582.
64. M. B. Cohn, L. S. Crawford, J. M. Wendlandt and S. S. Sastry, J. Robotic Systems, 12, 6 (1995) 401.
65. B. Slatkin, J. Burdick and W. Grundfest, Proceedings of IEEE '95 IROS Conf., August 5-9, 1995, Pittsburgh, PA, Vol. 2, 162.
66. Micromachine, Micromachine Center (MMC), Tokyo, Japan, No. 9, November 30, 1994, 4.
67. Micromachine, Micromachine Center (MMC), Tokyo, Japan, No. 10, March 20, 1995, 8.

68. Micromachine, Micromachine Center (MMC), Tokyo, Japan, No. 10, March 20, 1995, 10.
69. I. W. Hunter, S. Lafontaine, P. M. F. Nielsen, P. J. Hunter and J. M. Hollerbach, Proceedings of 1989 IEEE Int. Conf. on Robotics and Automation, May 14-19, 1989, Scottsdale (AZ), 1553.
70. T. Sato, T. Kameya, H. Miyazaki and Y. Hatamura Proceedings of 1995 IEEE Int. Conf. on Robotics and Automation, May 21-27, 1995, Nagoya, Japan, 59.
71. H. Morishita and Y. Hatamura, Proceedings of 1st IARP Workshop on Micro Robotics and Systems, June 15-16, 1993, Karlsruhe, Germany, 34.
72. Micromachine, Micromachine Center (MMC), Tokyo, Japan, No. 12, August 1995, 17.
73. G. Führ, R: Hagerdon, T. Müller, B. Wagner and W. Benecke, Proceedings of MEMS'91, January 30-February 2, 1991, Nara, Japan, 259.
74. M. Washizu, Integrated Micro-Motion Systems-Micromaching, Control and Application, F. Harashima (Ed.), Elsevier Science Publisher B. V. (1990) 417.
75. T. Kozuka, T. Tuziuti, H. Mitome and T. Fukuda, Proceedings of 5th Int. Symp. on Micro Machine and Human Science, October 2-4, 1994, Nagoya, Japan, 83.
76. S. Chiu, Science, 253 (August 23, 1991) 861
77. Micromachine, Micromachine Center (MMC), Tokyo, Japan, No. 12, August 1995, 11.

10. Future Problems

W. Menz

10.1 The complete Microsystem

When speaking about microsystems technology, one generally means micro-structure technology because a systems technology is not or not yet existing. In the early days of microsystems engineering the attention had been focused exclusively on silicon micromechanics so that, with reference to the microelectronic model, the opinion was that the microsystem would have to be a monolithically integrated solution with all elements - mechanical, optical, fluidic and electronic elements - integrated in one chip. Apart from the fact that on these assumptions microsystems engineering would have been a domain of only a few semiconductor producing firms of the world, such a technology is actually not realistic because the combination of so many differing processes would be associated with serious problems of yield. This initial state of euphoria of monolithic integration has therefore been substituted meanwhile with a more realistic way of looking at things. The hybrid solution will probably succeed as a technological middle course, with, as a matter of fact, the components obtained in different processes joined to a system on a substrate. This seems to constitute the compromise between a challenging performance profile and favorable costs of manufacture. Moreover, the hybrid solution allows, above all in small series production, higher flexibility to be achieved in case of alteration of the design.

So far, a complete microsystem similar to that discussed in Chapter 1 has been implemented and made known to the public in a few examples only. Analyzing e.g. the contributions to the international conferences on microsystems engineering, one finds that the majority of papers (approx. 70%) are dealing exclusively with sensor elements and sensor principles, in approx. 20% of the papers new actuator elements are presented, and only a negligibly small number of scientific contributions concentrate on the microsystems aspect.

However, to achieve an economic breakthrough on a large scale, it is of high importance to present the spectrum of the potential of microsystems engineering in full width. The development of microcomponents and the establishment of pilot manufacturing facilities for components create the prerequisites, but are not sufficient for a proliferation of microsystems engineering.

The individual subsystems of a microsystem will be discussed in the following paragraphs.

Sensors

The possibility of integrating sensors into a microsystem as arrays is probably the principal advantage of a microsystem. It allows, on the one hand, to improve by orders of magnitude the reliability of a system and, on the other hand, to refine considerably the evidencing quality of a measurement. If one looks e.g. at an array of similar sensor elements, information can be obtained, using statistical methods, about mean values, standard deviations, and many other items. Gradation of the sensor elements to cover different ranges of sensitivity allows the working range of the microsystem to be extended because for each measurement a sensor element with the optimum range of sensitivities is available. Above all chemical sensors with limited selectivity take advantage of the array technique, much to the benefit of a system. By suitable evaluation of sensors having different characteristic lines, the selectivity can be markedly improved and the evaluation software allows to assign dedicated tasks to the individual array.

In mathematics terms the advantages of a sensor array can be demonstrated by the following example. Let us assume that an unknown substance is to be measured which consists of the four components a, b, c, d. We now take a sensor responding in any way to all four components. Thus, the measured value of the sensor includes information about all four components a, b, c, d whereas a single component cannot be determined. Now, if one uses four sensors in the measurement whose characteristic lines respond independently of each other to the four components of the unknown medium, one obtains a system of equations with four equations and four unknowns which (hopefully) can be solved with the help of the microprocessor integrated. So, one is able in that case to determine separately the four components of the unknown substance. This is, as a matter of fact, a bit more difficult in practice because, generally, the parameter functions of the individual sensors are not exactly known and further unknown variables are involved in the measurement. But still, the value measured by a sensor array contains much more information than that of an individual sensor integrated in a conventional system.

Actuators

As microstructures can be integrated on a substrate at favorable costs and in a high packing density, it is possible to make actuator subsystems with redundancy feature. This possibility is of high interest in making safety related systems such as in medical engineering (implants, microendoscopes), process engineering, and traffic engineering (motor vehicles, aircraft). When individual actuators have become exhausted in service or destroyed by external impacts, redundant actuators can be activated and the systems returned into an operable state. Thus, microsystems engineering provides the means of making self-testing or even self-repairing systems.

Signal Processing

The tasks described before obviously impose more stringent requirements for signal processing in the microsystem. The problems faced here relate to the discrepancy between the exacting tasks, which have to be solved in real time during operation, and the technical means offered by a microprocessor which has to be integrated in a microsystem taking into account the performance profile, the dissipation, and the volume. New concepts of signal processing must be used such as processing of fuzzy information and application of neuronal nets.

Interfaces

As already outlined repeatedly, the interfaces play a crucial role in a microsystem, both those between the components within a microsystem and those coupling it to the macroscopic environment. Figure 10.1 is a survey in a graphic representation of the enormous number of possible interfaces in a microsystem. Whereas in microelectronics only the electric and thermal interfaces play a role, fluidic, mechanic, acoustic and optical interfaces are involved in addition in microsystems engineering. Besides the transmission of energy and information, also tasks related to the transport of substances (drugs, irrigation liquids, biological matter) must be solved in special applications. In most instances, a technology for developing such interfaces is still in its infancy. Great efforts in this respect must be made in the future in order to provide methods and optimize them with a view to manufacture of large series.

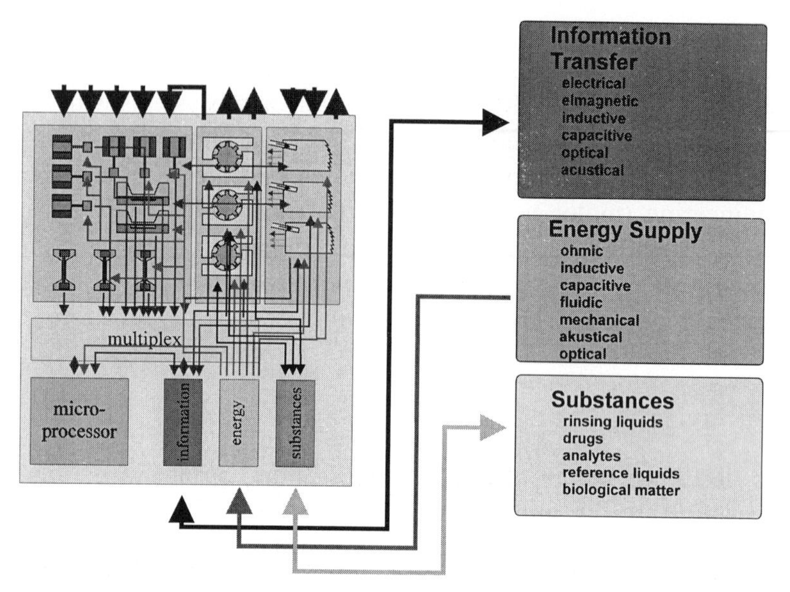

Fig.10.1 Interfaces to be delt with in a complete microsystem. The interfaces within a microsystem or between microsystem and macroscopic environment can be subdivided into: information interfaces, energy interfaces, and substances interfaces.

Packaging and Connection Technology

For the reasons indicated before, packaging and connection technology will play a key role in microsystems engineering commercialization on the industrial scale. The tasks to be solved are so multiple that one should seriously think of abandoning the term "packaging and connection technology" and replace it by a different term in order to avoid too close associations with the electric connection technique in microelectronics. Important future tasks of packaging and connection technology include:

- silicon-silicon bonding,
- anodic bonding of silicon on glass,
- electric connections (wire bonding) over wide gaps and with great difference in height;
- development of further techniques of joining components.

Adhesive bonding plays an important role in microsystems engineering. Meanwhile, techniques have been developed to improve, on the one hand, the

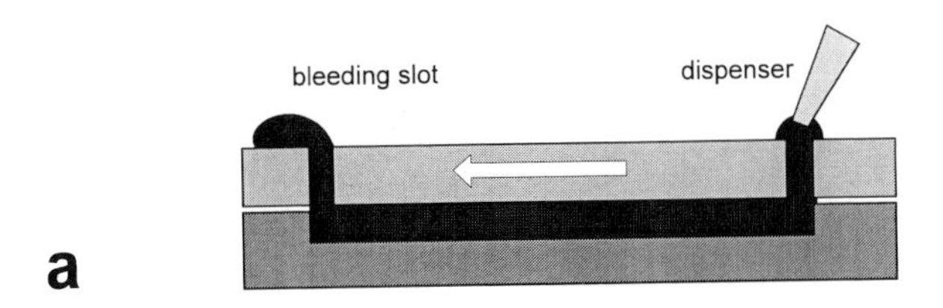

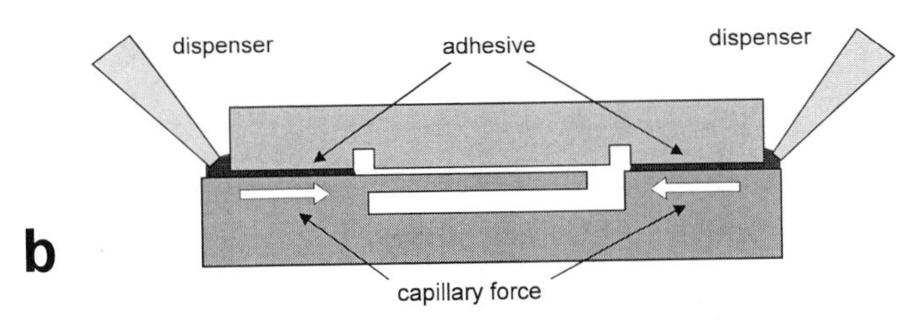

Fig. 10.2 Assembly of microcomponents or modules by different means of adhesive bonding:
a) channel technique with application of adhesive by dispenser a bleeding slot for the overflow
b) capillary technique with control of the capillary forces by release grooves.

chemistry of bonding and to design, on the other hand, microsystems in such a way that bonding of components or subsystems is feasible without sophisticated dosage of the adhesive. Bonding assisted by capillary forces is cited as an example. Some details are evident from Fig. 10.2.

A problem not yet solved satisfactorily to this day consists in dicing micro-components which, first, are manufactured batchwise on a large substrate. In many cases the dicing methods known from microelectronics cannot be applied here because many mechanically sensitive packages would not survive the rather rude scribe and break operation to which the silicon wafers are subjected.

10.2 The Industrial Potential of Microsystem Technology

The economic potential of microsystems engineering can be considered to lie in two different development strategies. Big industry, above all semi-conductor industry, will pursue the monolithical concept of an aggregate system on the silicon wafer. The products of the automobile industry and consumer goods industry are particularly suited for this purpose where large quantities of production can be anticipated, but also call for considerable expenditure before start of manufacture.

The second industrial group includes the small and medium sized enterprises (SMEs) which have either specialized on a certain product or tackle many niche solutions implying small quantities of production. For the latter group high flexibility in product development and quick development cycles with little preproduction expenditure and investments take top priority. For this group of users a modular concept for microsystems engineering is appropriate. Modules are made of individual components or subsystems, e.g. arrays of sensors, evaluation circuits, and other components. A module is capable of assuming a task comparable to that of a gate array which is known from microelectronics and used in many different ways, mainly by the SMEs. Also the module of microsystems engineering will mostly include more potentials than actually required in an individual application. Still, due to mass fabrication of such modules, the price will be lower than that for a dedicated application.

In conventional system packaging one is accustomed to employing expensive fluid components such as pumps, valves, switches in the most economical way feasible and making multiple use of them. Thus, fluidic networks are generated, determined by the costs of individual components. By contrast, such networks are conceived in microsystems technology for quite different reasons. A pump in microengineering is no longer an expensive single element of which multiple use must be made, but it is part of a module accommodating a multitude of pumps. The employment of pumps is determined here by the overall efficiency of a system and no longer by the cost of the individual element. Obviously, modules for valves or mixed fluid components can be similarly conceived.

Starting from these considerations, the modular concept has been elaborated with modules amenable to multiple applications and, nevertheless, suited for low cost manufacture. A system is then assembled by wiring different modules. Likewise, systems fulfilling different functions can be made from the same modules but a different type of wiring.

It is assumed in this module concept that some firms will specialize on the manufacture of dedicated modules and will thus be able to offer components with a high density of functions at favorable costs whereas other firms, in turn, will assemble these modules into individual microsystems. The prerequisites of making such a concept a success in industry will be listed below:

- System Simulation

  For an optimized aggregate system software tools with application oriented surfaces are needed which enable the person developing them to simulate on the computer the desired application before starting manufacture of microsystems. Such tools are not yet available on the market, but in the future they will be an important key to implementation on an industrial scale.

- Availability of Modules

  A sufficient number of different modules must be offered on the market before integration into a system on the industrial scale can be initiated. The respective components must be compatible with each other, although they are being offered by different suppliers.

- Technology of Assembly and Encasing

  It is decisive for the economic success, not only of the module concept but of microsystems engineering as a whole, to provide a high performance packaging and connection technique conforming to the demands of industry. Under the module concept described, this technology must be tailored to the SMEs, i.e. be flexible and capable of integration in the process of manufacturing without requiring much expenditure prior to manufacture.

- Standardization of the Interfaces

  A task of utmost importance to a more widespread use of microsystems engineering consists in standardization. As a matter of fact, standardization is a term frequently raising wrong associations and, by this, evoking prejudices not just to reality. One of the currently advanced arguments against too early a standardization is that laying down parameters in regulations will impede ongoing research and development in the respective field. In microsystems engineering and especially in case of a module concept it is, however, of extreme importance that all suppliers of components or modules reach an agreement concerning compatibility. Only then, components from various suppliers can be assembled into a microsystem.

Arrangements concerning interface standardization may relate first to a specification of external dimensions, location of the fluidic connections and the electric connections. Then, in order to standardize interfaces further, one can agree successively on specifying additional parameters.

10.3 The Importance of Standardization

Although standardization of the interfaces has some priority in the present state of development, the concept of global standardization must not be left out of consideration. It would be reasonable to standardize in steps, in parallel to industrial implementation. At the early stages, the partners in discussions will be the scientists from research and development sectors, whereas at the later stages rather the industrial partners are requested to reach an agreement. The following stepwise procedure is proposed

(1) Standardization of Terminology

This is the top level of standardization. In the first chapter of this book it has already been said that the term "microsystem" itself is understood and interpreted in completely different ways. In the international exchange of research results the semantic level plays a crucial role besides the technical-scientific definition. In languages so much remote from each other in their cultural backgrounds as Japanese and English are, the terms of microsystems engineering must be defined unambiguously in order to be able to establish a reasonable exchange.

(2) Standardization of Interfaces

It has already been mentioned that standardization of interfaces is of extraordinary importance to implementation on an industrial scale, but also to cooperation of the research laboratories at the international level. The prejudice that early standardization could stand in the way of free research and development activities can be disproved most conveniently by the example of the interfaces. It is actually the agreement preceding a mutual exchange of components which will reduce the development times for microsystems and initiate cooperation rather than delaying them by specification of interfaces.

(3) Standardization of Processes and Materials

It would be highly desirable with a view to making efficient research and development work if one could rely on certain standardized processes. Standardization of X-ray lithography would be of great help because then routine operations in manufacturing masks could be taken over by external suppliers, and laboratories as well as small firms could be freed from the burden of making considerable capital investments and employing staff specifically trained for these operations.

Likewise, specified standard substrates could be developed which would already be coated with the resist. Standardization in that case could also reduce the costs of such auxiliary substances and enhance their availability.

In electroplating - a further example - standard electrolytes could be developed which would contribute to stabilizing the processes. In the present situation the firms in the chemistry sector are obviously not interested in developing minimum amounts of a special electrolyte to a specified quality standard.

(4) Standardization of Components

Tender actions would by facilitated if reference could be made to standardized components. Mass products such as an accelerator sensor for the airbag or a pressure sensor for the intake manifold of a motor vehicle would have to conform to a quality standard because the respective orders are in most cases not carried out by one single subcontractor (second sourcing). But it is up to industry to make efforts to achieve standardization.

(5) Standardization of Systems

Within the framework of these activities the system performances of specified systems should be fixed in order to be capable of defining the interplay with other systems and with the marcroscopic environment. Above all in tackling safety relating tasks it will be necessary to know the behavior of the systems with respect to each other and to simulate it in order to avoid that coupled systems reach a state of chaos.

Besides, it is important that microsystems fit smoothly already existing standards, e.g. in communication engineering, medical engineering, etc.

(6) Standardization of Procedures

This means standardization of test routines and of methods of quality assurance. It is an area of high importance to industry and should be left to that sector.

It is shown once more in Fig. 10.3 how the responsibilities for the tasks of standardization should be distributed under the scheme outlined above. While, accordingly, the definition of terms, the interfaces and the standard processes of microsystems development are primarily the task to be fulfilled by the research community, the ensuing tasks should be assumed above all by industry.

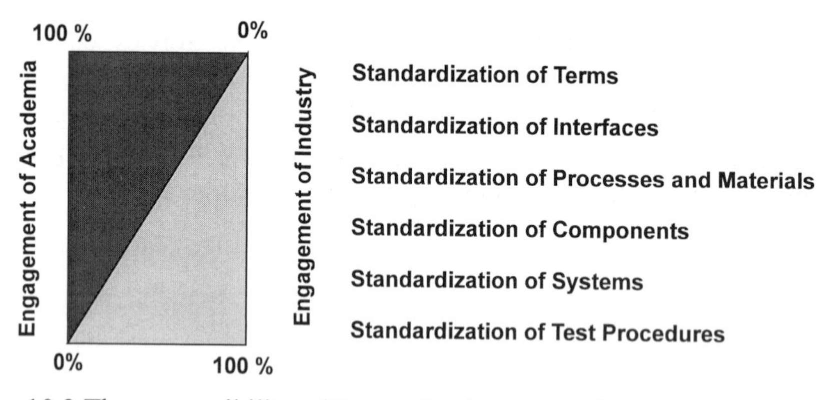

Fig. 10.3 The responsibility of the academic community versus industrial interests for standardization of microsystem technoilogy.

10.4 Outlook

Already now, the spectrum of potential applications of microsystems engineering is extremely vast. In a first estimate four areas of application can be recognized:

(1) General Instrumentation and Control

This includes process control engineering, i.e. the control of complex (chemical) processes with decentralized systems capable of making independent decisions in situ and initiating measures. Other areas in this category are traffic engineering, building installations, consumer goods and leisure equipment engineering, and the like.

(2) Telecommunications

Above all in optical information engineering, microsystems engineering is being applied to a growing extent. Besides the transmission and processing of information, navigation and position finding play a major role.

(3) Environmental Engineering

In environmental engineering the features of multifunctionality, adaptivity, and low energy consumption of microsystems take high priority. In air, water and soil monitoring battery operated autonomous systems capable of intercommunication could provide novel problem solutions.

(4) Medical Engineering

Also in medical engineering microsystems engineering will put emphasis on novel lines. Above all three fields of application get visible:

- implants,
- minimally invasive therapy,
- laboratory technique.

Regarding implants, experience has been accumulated for several decades with complex implantable microsystems: The first pacemaker was implanted as early as in 1960. Meanwhile, millions of these microsystems have been used. Other complex systems are in preparation. The small size, the reliability, the little power demand are properties characterizing the microsystem and playing a prominent role in implant technology. The role microsystems engineering will play in the future in this very field is undisputed.

Another important field of application of microsystems engineering is in minimally invasive therapy. Here small incisions or the natural ports of the body are used to approach the seat of a disease with intelligent remotely controlled microsystems in order to make complex interventions there. The advantage is that healthy tissue is only minimally injured and the intervention is quick causing little pain on the patient.

But it should not be left out of consideration that information available during an intervention of the conventional type must be made available to the surgeon by the microsystem in a minimally invasive operation. Above all the haptic and visual information must be processed accordingly and supplied to the surgeon in a high quality.

Besides in surgery, microsystems engineering plays an important role in early diagnostic as well. The simpler and less complicated a diagnosis is for the patient, the earlier seats of diseases can be detected routinely before they become a high risk to health. Thus, it should be possible with the help of microsystems engineering and appropriate diagnostic systems to detect tumors on inaccessible places in the body before metastases develop. Body regions such as the small intestine could be regularly monitored and, if necessary, therapy could start early.

In laboratory diagnostics microsystems could, first, free the patient from unnecessary pain and, second, make work easier for laboratory staff and provide quicker, more reliable results with a higher power of evidencing. The difficult questions concerning the long-term stability of microsystems play only a secondary role. In laboratory diagnostics also systems could be tested which, at a later date, could be incorporated into implants.

Besides these fields of application, which are known already now, completely new fields will certainly emerge in the future which, today, cannot even be imagined. Let us look again at the development of microelectronics as a parallel. It can be stated that 30 years ago nobody would have deemed the personal computer to be a potential product of microelectronics. The development will be similar in microsystems engineering as soon as a sufficiently great number of developers will focus their attention on the potentials inherent in microsystems engineering.

Already now some scientists consider nanotechnology to be microsystems engineering at a higher level. But it is often overlooked that nanotechnology is less a question of the geometric dimension than of technology. In nanotechnology use is made of effects which have their origins in single atoms, clusters of atoms or molecules. One certainly profits already now of partial aspects of nanotechnology, but the actual problems to be solved in microsystems engineering do not concentrate on further reducing the dimensions but on improving the technologies applied.